projective geometry with applications

LECTURE NOTES IN PURE AND APPLIED MATHEMATICS

1. *N. Jacobson,* Exceptional Lie Algebras
2. *L.-Å. Lindahl and F. Poulsen,* Thin Sets in Harmonic Analysis
3. *I. Satake,* Classification Theory of Semi-Simple Algebraic Groups
4. *F. Hirzebruch, W. D. Newmann, and S. S. Koh, Differentiable Manifolds and Quadratic Forms*
5. *I. Chavel,* Riemannian Symmetric Spaces of Rank One
6. *R. B. Burckel,* Characterization of C(X) Among Its Subalgebras
7. *B. R. McDonald, A. R. Magid, and K. C. Smith,* Ring Theory: Proceedings of the Oklahoma Conference
8. *Y.-T. Siu,* Techniques of Extension on Analytic Objects
9. *S. R. Caradus, W. E. Pfaffenberger, and B. Yood,* Calkin Algebras and Algebras of Operators on Banach Spaces
10. *E. O. Roxin, P.-T. Liu, and R. L. Sternberg,* Differential Games and Control Theory
11. *M. Orzech and C. Small,* The Brauer Group of Commutative Rings
12. *S. Thomier,* Topology and Its Applications
13. *J. M. Lopez and K. A. Ross,* Sidon Sets
14. *W. W. Comfort and S. Negrepontis,* Continuous Pseudometrics
15. *K. McKennon and J. M. Robertson,* Locally Convex Spaces
16. *M. Carmeli and S. Malin,* Representations of the Rotation and Lorentz Groups: An Introduction
17. *G. B. Seligman,* Rational Methods in Lie Algebras
18. *D. G. de Figueiredo,* Functional Analysis: Proceedings of the Brazilian Mathematical Society Symposium
19. *L. Cesari, R. Kannan, and J. D. Schuur,* Nonlinear Functional Analysis and Differential Equations: Proceedings of the Michigan State University Conference
20. *J. J. Schäffer,* Geometry of Spheres in Normed Spaces
21. *K. Yano and M. Kon,* Anti-Invariant Submanifolds
22. *W. V. Vasconcelos,* The Rings of Dimension Two
23. *R. E. Chandler,* Hausdorff Compactifications
24. *S. P. Franklin and B. V. S. Thomas,* Topology: Proceedings of the Memphis State University Conference
25. *S. K. Jain,* Ring Theory: Proceedings of the Ohio University Conference
26. *B. R. McDonald and R. A. Morris,* Ring Theory II: Proceedings of the Second Oklahoma Conference
27. *R. B. Mura and A. Rhemtulla,* Orderable Groups
28. *J. R. Graef,* Stability of Dynamical Systems: Theory and Applications
29. *H.-C. Wang,* Homogeneous Branch Algebras
30. *E. O. Roxin, P.-T. Liu, and R. L. Sternberg,* Differential Games and Control Theory II
31. *R. D. Porter,* Introduction to Fibre Bundles
32. *M. Altman,* Contractors and Contractor Directions Theory and Applications
33. *J. S. Golan,* Decomposition and Dimension in Module Categories
34. *G. Fairweather,* Finite Element Galerkin Methods for Differential Equations
35. *J. D. Sally,* Numbers of Generators of Ideals in Local Rings
36. *S. S. Miller,* Complex Analysis: Proceedings of the S.U.N.Y. Brockport Conference
37. *R. Gordon,* Representation Theory of Algebras: Proceedings of the Philadelphia Conference
38. *M. Goto and F. D. Grosshans,* Semisimple Lie Algebras
39. *A. I. Arruda, N. C. A. da Costa, and R. Chuaqui,* Mathematical Logic: Proceedings of the First Brazilian Conference
40. *F. Van Oystaeyen,* Ring Theory: Proceedings of the 1977 Antwerp Conference
41. *F. Van Oystaeyen and A. Verschoren,* Reflectors and Localization: Application to Sheaf Theory
42. *M. Satyanarayana,* Positively Ordered Semigroups
43. *D. L Russell,* Mathematics of Finite-Dimensional Control Systems
44. *P.-T. Liu and E. Roxin,* Differential Games and Control Theory III: Proceedings of the Third Kingston Conference, Part A
45. *A. Geramita and J. Seberry,* Orthogonal Designs: Quadratic Forms and Hadamard Matrices
46. *J. Cigler, V. Losert, and P. Michor,* Banach Modules and Functors on Categories of Banach Spaces

47. *P.-T. Liu and J. G. Sutinen,* Control Theory in Mathematical Economics: Proceedings of the Third Kingston Conference, Part B
48. *C. Byrnes,* Partial Differential Equations and Geometry
49. *G. Klambauer,* Problems and Propositions in Analysis
50. *J. Knopfmacher,* Analytic Arithmetic of Algebraic Function Fields
51. *F. Van Oystaeyen,* Ring Theory: Proceedings of the 1978 Antwerp Conference
52. *B. Kadem,* Binary Time Series
53. *J. Barros-Neto and R. A. Artino,* Hypoelliptic Boundary-Value Problems
54. *R. L. Sternberg, A. J. Kalinowski, and J. S. Papadakis,* Nonlinear Partial Differential Equations in Engineering and Applied Science
55. *B. R. McDonald,* Ring Theory and Algebra III: Proceedngs of the Third Oklahoma Conference
56. *J. S. Golan,* Structure Sheaves Over a Noncommutative Ring
57. *T. V. Narayana, J. G. Williams, and R. M. Mathsen,* Combinatorics, Representation Theory and Statistical Methods in Groups: YOUNG DAY Proceedings
58. *T. A. Burton,* Modeling and Differential Equations in Biology
59. *K. H. Kim and F. W. Roush,* Introduction to Mathematical Consensus Theory
60. *J. Banas and K. Goebel,* Measures of Noncompactness in Banach Spaces
61. *O. A. Nielson,* Direct Integral Theory
62. *J. E. Smith, G. O. Kenny, and R. N. Ball,* Ordered Groups: Proceedings of the Boise State Conference
63. *J. Cronin,* Mathematics of Cell Electrophysiology
64. *J. W. Brewer,* Power Series Over Commutative Rings
65. *P. K. Kamthan and M. Gupta,* Sequence Spaces and Series
66. *T. G. McLaughlin,* Regressive Sets and the Theory of Isols
67. *T. L. Herdman, S. M. Rankin III, and H. W. Stech,* Integral and Functional Differential Equations
68. *R. Draper,* Commutative Algebra: Analytic Methods
69. *W. G. McKay and J. Patera,* Tables of Dimensions, Indices, and Branching Rules for Representations of Simple Lie Algebras
70. *R. L. Devaney and Z. H. Nitecki,* Classical Mechanics and Dynamical Systems
71. *J. Van Geel,* Places and Valuations in Noncommutative Ring Theory
72. *C. Faith,* Injective Modules and Injective Quotient Rings
73. *A. Fiacco,* Mathematical Programming with Data Perturbations I
74. *P. Schultz, C. Praeger, and R. Sullivan,* Algebraic Structures and Applications: Proceedings of the First Western Australian Conference on Algebra
75. *L Bican, T. Kepka, and P. Nemec,* Rings, Modules, and Preradicals
76. *D. C. Kay and M. Breen,* Convexity and Related Combinatorial Geometry: Proceedings of the Second University of Oklahoma Conference
77. *P. Fletcher and W. F. Lindgren,* Quasi-Uniform Spaces
78. *C.-C. Yang,* Factorization Theory of Meromorphic Functions
79. *O. Taussky,* Ternary Quadratic Forms and Norms
80. *S. P. Singh and J. H. Burry,* Nonlinear Analysis and Applications
81. *K. B. Hannsgen, T. L. Herdman, H. W. Stech, and R. L. Wheeler,* Volterra and Functional Differential Equations
82. *N. L. Johnson, M. J. Kallaher, and C. T. Long,* Finite Geometries: Proceedings of a Conference in Honor of T. G. Ostrom
83. *G. I. Zapata,* Functional Analysis, Holomorphy, and Approximation Theory
84. *S. Greco and G. Valla,* Commutative Algebra: Proceedings of the Trento Conference
85. *A. V. Fiacco,* Mathematical Programming with Data Perturbations II
86. *J.-B. Hiriart-Urruty, W. Oettli, and J. Stoer,* Optimization: Theory and Algorithms
87. *A. Figa Talamanca and M. A. Picardello,* Harmonic Analysis on Free Groups
88. *M. Harada,* Factor Categories with Applications to Direct Decomposition of Modules
89. *V. I. Istrăţescu,* Strict Convexity and Complex Strict Convexity
90. *V. Lakshmikantham,* Trends in Theory and Practice of Nonlinear Differential Equations
91. *H. L. Manocha and J. B. Srivastava,* Algebra and Its Applications
92. *D. V. Chudnovsky and G. V. Chudnovsky,* Classical and Quantum Models and Arithmetic Problems
93. *J. W. Longley,* Least Squares Computations Using Orthogonalization Methods
94. *L. P. de Alcantara,* Mathematical Logic and Formal Systems
95. *C. E. Aull,* Rings of Continuous Functions
96. *R. Chuaqui,* Analysis, Geometry, and Probability
97. *L. Fuchs and L. Salce,* Modules Over Valuation Domains

98. *P. Fischer and W. R. Smith,* Chaos, Fractals, and Dynamics
99. *W. B. Powell and C. Tsinakis,* Ordered Algebraic Structures
100. *G. M. Rassias and T. M. Rassias,* Differential Geometry, Calculus of Variations, and Their Applications
101. *R.-E. Hoffmann and K. H. Hofmann,* Continuous Lattices and Their Applications
102. *J. H. Lightbourne III and S. M. Rankin III,* Physical Mathematics and Nonlinear Partial Differential Equations
103. *C. A. Baker and L, M. Batten,* Finite Geometrics
104. *J. W. Brewer, J. W. Bunce, and F. S. Van Vleck,* Linear Systems Over Commutative Rings
105. *C. McCrory and T. Shifrin,* Geometry and Topology: Manifolds, Varieties, and Knots
106. *D. W. Kueker, E. G. K. Lopez-Escobar, and C. H. Smith,* Mathematical Logic and Theoretical Computer Science
107. *B.-L. Lin and S. Simons,* Nonlinear and Convex Analysis: Proceedings in Honor of Ky Fan
108. *S. J. Lee,* Operator Methods for Optimal Control Problems
109. *V. Lakshmikantham,* Nonlinear Analysis and Applications
110. *S. F. McCormick,* Multigrid Methods: Theory, Applications, and Supercomputing
111. *M. C. Tangora,* Computers in Algebra
112. *D. V. Chudnovsky and G. V. Chudnovsky,* Search Theory: Some Recent Developments
113. *D. V. Chudnovsky and R. D. Jenks,* Computer Algebra
114. *M. C. Tangora,* Computers in Geometry and Topology
115. *P. Nelson, V. Faber, T. A. Manteuffel, D. L. Seth, and A. B. White, Jr.,* Transport Theory, Invariant Imbedding, and Integral Equations: Proceedings in Honor of G. M. Wing's 65th Birthday
116. *P. Clément, S. Invernizzi, E. Mitidieri, and I. I. Vrabie,* Semigroup Theory and Applications
117. *J. Vinuesa,* Orthogonal Polynomials and Their Applications: Proceedings of the International Congress
118. *C. M. Dafermos, G. Ladas, and G. Papanicolaou,* Differential Equations: Proceedings of the EQUADIFF Conference
119. *E. O. Roxin,* Modern Optimal Control: A Conference in Honor of Solomon Lefschetz and Joseph P. Lasalle
120. *J. C. Díaz,* Mathematics for Large Scale Computing
121. *P. S. Milojević,* Nonlinear Functional Analysis
122. *C. Sadosky,* Analysis and Partial Differential Equations: A Collection of Papers Dedicated to Mischa Cotlar
123. *R. M. Shortt,* General Topology and Applications: Proceedings of the 1988 Northeast Conference
124. *R. Wong,* Asymptotic and Computational Analysis: Conference in Honor of Frank W. J. Olver's 65th Birthday
125. *D. V. Chudnovsky and R. D. Jenks,* Computers in Mathematics
126. *W. D. Wallis, H. Shen, W. Wei, and L. Zhu,* Combinatorial Designs and Applications
127. *S. Elaydi,* Differential Equations: Stability and Control
128. *G. Chen, E. B. Lee, W. Littman, and L. Markus,* Distributed Parameter Control Systems: New Trends and Applications
129. *W. N. Everitt,* Inequalities: Fifty Years On from Hardy, Littlewood and Pólya
130. *H. G. Kaper and M. Garbey,* Asymptotic Analysis and the Numerical Solution of Partial Differential Equations
131. *O. Arino, D. E. Axelrod, and M. Kimmel,* Mathematical Population Dynamics: Proceedings of the Second International Conference
132. *S. Coen,* Geometry and Complex Variables
133. *J. A. Goldstein, F. Kappel, and W. Schappacher,* Differential Equations with Applications in Biology, Physics, and Engineering
134. *S. J. Andima, R. Kopperman, P. R. Misra, J. Z. Reichman, and A. R. Todd,* General Topology and Applications
135. *P Clément, E. Mitidieri, B. de Pagter,* Semigroup Theory and Evolution Equations: The Second International Conference
136. *K. Jarosz,* Function Spaces
137. *J. M. Bayod, N. De Grande-De Kimpe, and J. Martínez-Maurica,* *p*-adic Functional Analysis
138. *G. A. Anastassiou,* Approximation Theory: Proceedings of the Sixth Southeastern Approximation Theorists Annual Conference
139. *R. S. Rees,* Graphs, Matrices, and Designs: Festschrift in Honor of Norman J. Pullman
140. *G. Abrams, J. Haefner, and K. M. Rangaswamy,* Methods in Module Theory

141. *G. L. Mullen and P. J.-S. Shiue*, Finite Fields, Coding Theory, and Advances in Communications and Computing
142. *M. C. Joshi and A. V. Balakrishnan*, Mathematical Theory of Control: Proceedings of the International Conference
143. *G. Komatsu and Y. Sakane*, Complex Geometry: Proceedings of the Osaka International Conference
144. *I. J. Bakelman*, Geometric Analysis and Nonlinear Partial Differential Equations
145. *T. Mabuchi and S. Mukai*, Einstein Metrics and Yang–Mills Connections: Proceedings of the 27th Taniguchi International Symposium
146. *L. Fuchs and R. Göbel*, Abelian Groups: Proceedings of the 1991 Curaçao Conference
147. *A. D. Pollington and W. Moran*, Number Theory with an Emphasis on the Markoff Spectrum
148. *G. Dore, A. Favini, E. Obrecht, and A. Venni*, Differential Equations in Banach Spaces
149. *T. West*, Continuum Theory and Dynamical Systems
150. *K. D. Bierstedt, A. Pietsch, W. Ruess, and D. Vogt*, Functional Analysis
151. *K. G. Fischer, P. Loustaunau, J. Shapiro, E. L. Green, and D. Farkas*, Computational Algebra
152. *K. D. Elworthy, W. N. Everitt, and E. B. Lee*, Differential Equations, Dynamical Systems, and Control Science
153. *P.-J. Cahen, D. L. Costa, M. Fontana, and S.-E. Kabbaj*, Commutative Ring Theory
154. *S. C. Cooper and W. J. Thron*, Continued Fractions and Orthogonal Functions: Theory and Applications
155. *P. Clément and G. Lumer*, Evolution Equations, Control Theory, and Biomathematics
156. *M. Gyllenberg and L. Persson*, Analysis, Algebra, and Computers in Mathematical Research: Proceedings of the Twenty-First Nordic Congress of Mathematicians
157. *W. O. Bray, P. S. Milojević, and Č. V. Stanojević*, Fourier Analysis: Analytic and Geometric Aspects
158. *J. Bergen and S. Montgomery*, Advances in Hopf Algebras
159. *A. R. Magid*, Rings, Extensions, and Cohomology
160. *N. H. Pavel*, Optimal Control of Differential Equations
161. *M. Ikawa*, Spectral and Scattering Theory: Proceedings of the Taniguchi International Workshop
162. *X. Liu and D. Siegel*, Comparison Methods and Stability Theory
163. *J.-P. Zolésio*, Boundary Control and Variation
164. *M. Křížek, P. Neittaanmäki, and R. Stenberg*, Finite Element Methods: Fifty Years of the Courant Element
165. *G. Da Prato and L. Tubaro*, Control of Partial Differential Equations
166. *E. Ballico*, Projective Geometry with Applications
167. *M. Costabel, M. Dauge, and S. Nicaise*, Boundary Value Problems and Integral Equations in Nonsmooth Domains
168. *G. Ferreyra, G. R. Goldstein, and F. Neubrander*, Evolution Equations

Additional Volumes in Preparation

projective geometry with applications

edited by

Edoardo Ballico
University of Trento
Povo, Italy

Marcel Dekker, Inc. **New York • Basel • Hong Kong**

Library of Congress Cataloging-in-Publication Data

Projective geometry with applications / edited by Edoardo Ballico.
p. cm. — (Lecture notes in pure and applied mathematics; v. 166)
Includes bibliographical references.
ISBN 0-8247-9278-5
1. Geometry, Projective. I. Ballico, E. (Edoardo) II. Series.
QA471.P84 1994
516'.5—dc20 94-27247
CIP

The publisher offers discounts on this book when ordered in bulk quantities. For more information, write to Special Sales/Professional Marketing at the address below.

This book is printed on acid-free paper.

MARCEL DEKKER, INC.
270 Madison Avenue, New York, New York 10016

Current printing (last digit):
10 9 8 7 6 5 4 3 2 1

PRINTED IN THE UNITED STATES OF AMERICA

Preface

This book is a collection of 15 research papers and three survey articles concerning projective geometry as the focus of study, or as a tool, or as the subject to which other techniques, mainly from algebraic geometry, but also from complex analysis, commutative algebra, or number theory, are applied. The principal topics covered in these papers are adjunction theory, Fano manifolds, applications of Mori theory, subvarieties of Grassmannians, enumerative geometry of finite subsets of projective varieties, and Weierstrass points on Gorenstein curves and related ramification loci. For more details, see the Introduction. The Introduction also contains a very brief discussion of two topics related to projective geometry but not covered elsewhere in this book: complex hyperbolic geometry and connections with number theory, and an attempt to link classical concepts from algebraic geometry to theoretical vision.

Edoardo Ballico

Contents

Contributors

E. Ambrogio Università di Torino, Torino, Italy

E. Arrondo Facultad de Ciencias Matemáticas, Universidad Complutense de Madrid, Madrid, Spain

Lucian Badescu Institute of Mathematics of the Romanian Academy, Bucharest, Romania

Edoardo Ballico Università di Trento, Povo, Italy

Mauro C. Beltrametti Università degli Studi di Genova, Genova, Italy

M. Bertolini Università degli Studi di Milano, Milano, Italy

Luisella Caire Politecnico di Torino, Torino, Italy

Eduardo Esteves Massachusetts Institute of Technology, Cambridge, Massachusetts

Letterio Gatto Politecnico di Torino, Torino, Italy

F. Giovanetti Università di Trento, Povo, Italy

Lothar Göttsche Università di Trento, Povo, Italy

Masaaki Homma Yamaguchi University, Yamaguchi, Japan

Sheldon Katz Oklahoma State University, Stillwater, Oklahoma

Fabrizio Ponzo Politecnico di Torino, Torino, Italy

Rosa M. Miró-Roig University of Trento, Povo, Italy

L. Picco Botta Università di Torino, Torino, Italy

B. Russo University of Trento, Povo, Italy

Andrew J. Sommese University of Notre Dame, Notre Dame, Indiana

Robert Speiser Brigham Young University, Provo, Utah

C. Turrini Università degli Studi di Milano, Milano, Italy

Jaroslaw A. Wiśniewski Institute of Mathematics, Warsaw University, Warsaw, Poland

Introduction

E. Ballico
Dept. of Mathematics, Università di Trento
38050 Povo (TN), Italy
e-mail: ballico@itncisca.bitnet or ballico@itnvax.science.unitn.it
fax: italy + 461881624

I believe that it is pointless to define (and hence restrict) the specific subject of Projective Geometry. Projective Geometry certainly contains the study of the subvarieties of $\mathbf{P}^n$ as subvarieties of $\mathbf{P}^n$. It certainly contains (or overlaps with) also papers on twin subjects which use "projective" tools and "projective" concepts. And there are very strong projective tools (both old and new); in many "applied" areas elementary "projective" concepts are very useful. Viceversa, of course the projective study of subvarieties of $\mathbf{P}^n$ uses as much as possible mathematics coming from other aereas. I believe that some of the best recent work which can be called "Projective Geometry" arised from interactions with other aereas. One of the aim of this volume is to show the spread of mathematics which can be classified (as primary or as secondary 1991 Mathematical Subject Classification) as "Projective Geometry": 14N05, 14N010, 14N99. The papers printed in this volume are grouped according to their subtopic. The second part of this introduction (subsections (C), (D), (E), (F), (G), (H), (I), (J) and (K)) contains a very brief discussion of each of these subtopics and of each paper. The first part of this introduction (subsections (A) and (B)) contains a very brief discussion of some "projective" subtopics which are not represented in this volume.

A bit of propaganda. Europroj is a scientific organization which links around 200 algebraic geometers in Universities belonging to around 16 European States plus "isolated" researchers, mainly in US. Inside Europroj there is a very active group of specialists in projective geometry which organizes workshops and advanced schools, spread problems and results, and support young researchers. Europroj publishes a newsletter, Euronews.

All the papers in this volume were refereed. I want to thank here all the referees. All the papers in this volume are in final form and no part of them will be submitted or published elsewhere.

(A) In recent years some complex analytic properties of quasi-projective varieties acttracted the attention of mathematicians with strong background in Analysis. One of the topic is "Hyperbolic Geometry". Here we consider very briefly this topic when the playground of the research (now very active) is a pair ($\mathbf{P}^n$,X) with X hypersurface. For the basic definitions (and much more) on hyperbolic complex spaces, see for instance [La1] or [La2]. For a survey of research up to 1992, see [No1] and [Za]; for some papers vaguely related see [No2].A weaker notion of hyperbolicity was introduced in [Za]; an integral algebraic variety X is said to be algebraically hyperbolic if there is no non constant morphism from an Abelian variety to X ; this notion can be studied using only algebraic tools. Many papers are concerned with the hyperbolicity of X and of $\mathbf{P}^n\backslash X$. Usually for $\mathbf{P}^n\backslash X$ the important assumptions are on deg(X), the degree of the irreducible components of X and on the singularities of X (see e.g. [DSW] for a feeling of the subject); for n = 2 there is very promising work in progress by G. Dethloff and P.-M. Wong. Recently the picture was clarified. In [B], it was proved that a "generic" (i.e. outside countably many proper subvarieties) hypersurface of high degree is algebraic hyperbolic; simultaneously and independently K. Masuda and J. Noguchi ([MN]) proved the hyperbolicity of such manifolds (i.e. a much stronger result).

There is an arithmetic analogue of the work outlined in the the first part of this subsection (A). And again, projective geometry is the playground. Recall that a set of q hyperplanes of $\mathbf{P}^n$ are said to be in linear general position (or in general position in the terminology of [RW] and [Wo]) if for every integer t with $0<t\leq\min(n+1,q)$ the intersection of any t of these hyperplanes has codimension t. We need another definition. Let D be a very ample effective divisor on a projective variety V defined over a number field $\mathbf{K}$; fix a finite set S of valuations of $\mathbf{K}$ and let A_S be the set of S-integers of $\mathbf{K}$; take a basis $x_0,\dots,x_n$ of $H^0(V, D)$ with $D = \{x_0\}$; the rational function $x_1/x_0,\dots,x_n/x_0$ induces an embedding $\mathbf{i}$ of $V(\mathbf{K})\backslash D$ into $\mathbf{K}^n$; a point of $V(\mathbf{K})\backslash D$ is said to be *D-integral* if if $\mathbf{i}(P)\in A_S{}^n$.The following striking theorem is the main result of [RW] and shows the goals of the subject.

Theorem *Let* $\mathbf{K}$ *be a number field, S a finite set of valuations of* $\mathbf{K}$ *and D the union of q hyperplanes of* $\mathbf{P}^n(\mathbf{K})$ *in linear general position. Fix an integer k with* $1\leq k\leq n$ *and* $q>2n-k+1$*.Then the set of D-integral points of* $\mathbf{P}^n\backslash D$ *is contained in a finite union of linear subspaces of dimension k-1. In particular if* $q\geq 2n+1$ *the set of D-integral points of* $\mathbf{P}^n(\mathbf{K})\backslash D$ *is finite.*

(B) There was an attempt to link "projective geometry" and "theoretical vision". It was organized by O. Faugeras and D. Laksov a workshop "Algebraic Geometry and

Vision School" at Domaine des Courmettes, Nice, France, June 1 - 4 1993, as one of the workshops and schools of Europroj. Since the attempt was not a failure, the effort to link much more researchers in the two fields is going on. I will quote almost verbatim a report by D. Laksov about that workshop (Euronews n. 10, July 1993, page 8). "The geometers learned that much of the old literature, that is out of fashion but still interesting from the point of view of algebraic geometry, is very useful in vision." "Some of the problems in calibration in vision can be formulated as classical enumerative geometric problems. Some of these were already discussed by Charles, Hesse and Sturm. This problems are special cases of Schubert's theory of correlations." Here the Key Word is "Complete correlations".

(C) The papers by Wisniewski and by Beltrametti - Sommese are concerned with the use of Mori theory in adjunction theory and the use of both of them to give several strong classification results. There is a shift of emphasis (thanks mainly to several fundamental results on Mori theory) from the study of very ample line bundles (e.g. for adjunction type properties) to the study of ample line bundles. Hence projective techniques are now used less heavily. However, these techniques are not superseded, since often the assumption of very ampleness leads to much more refined classificationa (with very nice examples and very nice geometric constructions). Furthermore, for certain ranges the assumption of very ampleness is still essential. Much effort is devoted to the search of the "right" category in which to work between ampleness and very ampleness and a second "right" category in which even the ampleness condition is relaxed. The paper by Beltrametti and Sommese is devoted to the search of this second "right" category.

Recall that a smooth projective variety X of dimension n is called a Fano manifold if the anti-canonical divisor $-K_X$ is ample; the index of the Fano manifold X is the largest integer r:= r(X) dividing $-K_X$ in $Pic(X) \cong \mathbf{Z}^\rho$. The paper of Wisniewski is essentially a survey of very recent work (with announcements of new results by the author) on the classification of complex Fano manifolds of index r and dimension 2r with second Betti number $b_2(X) \geq 2$. In their classification a very relevant role is played by certain types of vector bundles and "nice" sheaves (the sheaves, E, with $\mathbf{P}(E)$ smooth), which are called Banica sheaves. Here the reader will find also the announcements (and the relevant definitions) of several joint results on the theory of Banica sheaves.

(D) The paper of E. Esteves and M. Homma concerns the geometry of a birational morphism f: $C \to \mathbf{P}^n$ with C smooth curve (or equivalently of the integral curve $Y := f(C) \subset \mathbf{P}^n$). A very important geometric information is the number and weights of the Weierstrass points of the pair (C,f). In this paper in particular there is a complete classification (in arbitrary characteristic, but the result is new even in characteristic 0) of pairs (C,f) with at most 2 Weierstrass points. Furthermore in characteristic 0 there is a lower bound on the number of Weierstrass points (as set, not counting multiplicity) in terms of n, the genus of C and the degree of f(C). Then several features relating to "characteristic p geometry/pathology" are considered. Most results are sharp, with complete description of borderline examples.

Let G(1,n) be the Grassmannian of lines in $\mathbf{P}^n$. An integral (n-1)-dimensional subvariety of G(1,n) is called a *congruence* (or a congruence of lines). The study of congruences of lines was very active in the last century; since a few years it is very active, again (with different methods, problems and motivations). The introduction of the paper by Arrondo, Bertolini and Turrini contains (very briefly) key references (papers by G. Fano) for the old works and for the state of the art on the subject. The paper by Arrondo, Bertolini and Turrini gives a complete classification and construction of all smooth congrueces , V, such that all the lines parametrized by V meet a fixed curve. As a byproduct they obtain a complete classification of all scrolls of dimension n-1 over a curve which can be embedded in G(1,n).

One of the very interesting, classical and difficult problems of the projective geometry is the classification of all possible extensions of a given subvariety Y of $\mathbf{P}^n$, i.e., seeing $\mathbf{P}^n$ as a hyperplane, H, of $\mathbf{P}^{n+1}$ all possible $Z \subset \mathbf{P}^{n+1}$ with $Y = Z \cap H$ (transversal intersection). The paper by Badescu gives a criterion on Y which forces every such Z to be a cone with base Y. This paper is a generalization (with other tools) of previous remarkable results of Zak-L'vovsky.

The paper by Ballico, Giovanetti and Russo extends in positive characteristic a very nice result ([DK1], Th. 7.2) on a class of rank n vector bundles on $\mathbf{P}^n$ ("the logarithmic bundles"). This class was introduced and studied in detail in [DK1]. We want to stress here how much classical projective geometry is used in [DK1] and [DK2] and how much new projective geometry is introduced in these beautyful papers. The reader interested in other far reaching extensions of classical projective geometry may find a huge number of papers and books by Dolgachev, Kapranov, Sturmfels, Gelfand (and his many coworkers) and many others listed in their papers.

(E) The papers by Göttsche and by Speiser have a common background: the search and use of a workable nice compactification for the set of all "d points of the variety V". We stress the word WORKABLE. The main motivation comes from enumerative geometry (computation of characteristic numbers, and so on) and for this it is extremely important to be able to compute intersection numbers in the Chow ring of the compactification. There is not a unique solution (or even a "best possible solution") to this important problem; there must be general solutions, but for a particular (but important) problem/computation a very particular compactifications may be necessary. For the case of smooth surfaces the best approach till now is the 1991 thesis of Göttsche (quoted in his paper in this volume); an english version of this thesis will appear soon. In any dimension a general "inductive" approach was proposed by Arrondo, Sols and Speiser (see references in the paper by Speiser in this volume); the paper by Speiser in this volume is a continuation of their joint work. Let $X \subset \mathbf{P}^N$ be a smooth projective variety defined over an algebraically closed field of characteristic $\neq 2$ and 3; the paper by Göttsche in this volume gives general formulas for the trisecant lines to X fulfilling Schubert conditions (i.e. intersecting a given general (a+1)-codimensional linear subspace of $\mathbf{P}^N$). These formulas are very good to make explicit calculations; Göttsche in this paper gives two applications of these formulas. The first application is to projections of X into $\mathbf{P}^{N-1}$ from a point $P \in X$. The second application is to questions on 2-spannedness of the linear system of hyperplanes on X; this kind of problem is very interesting from the point of view of adjunction theory and the results in the present paper much more general and strong than previously known ones.

(F) The following four papers by Gatto, Ponza, Gatto and Ballico are concerned with Weierstrass points and ramification loci. The first paper by Gatto and the paper by Ponza consider the Weierstrass points of a Gorenstein curve C. This paper by Gatto contains several key example on how the singular points of C influences the Weierstrass points of C. The paper by Ponza computes some formulas for extraweights at n-branched singular points, using as key tool the notion of weight sequence previously introduced by Gatto. The introduction of the paper by Ponza contains not only the needed definitions and references but also a detailed, clear and very useful explanation of the method, how to use it, its connection with other authors' approaches and its technical . The other two papers are surveys; they are completely skew (their intersection is empty) and their union covers both the classical case and several extensions of it to other situations. The survey "Weierstrass loci and generalizations, I" by L. Gatto is on the Weierstrass points with respect to a linar system on a curve. In the first part of this survey the author consider the classical case of a smooth curve. In the second part it consider the case of a singular curve (both the case of Gorenstein curves and the case of arbitrary singular curves). In the last few years several papers by several mathematicians appeared on Weierstrass points on singular curves (almost all only for Gorenstein ones). This very up to date part of the survey will bring the reader to the frontier of the research on this topic. The style of both parts of this survey is very plain and detailed, with the aim to give a some suggestions to the non-specialists in the field. The second survey (by myself) is written in a more condensed style. It contains a discussion of generalizations of the notion of Weierstrass point to the case of a vector bundle on a curve, on higher dimensional varieties and a related notion ("hyperplane versality") linked to Singularity Theory. Most of this survey is related to very recent researsch, still (when these lines are written) in preprint form or in the form of Ph. D. thesis.

(G) The paper by Ambrogio and Picco Botta is on singular Fano 3-folds. Hence the subject is related to the paper by Wisniewski, but the aim and the techniques are completely different. Let $W \subset \mathbf{P}^n$ be a dimension 3 complex variety whose general hyperplane section is a smooth Enriques surface, W is projectively normal, W is not a cone, the general curve section has genus $g>5$ (hence $n = g$ and $\deg(W) = 2g-2$). Under suitable assumptions on Sing(W) in [CM] to W it was associated a variety $V \subset \mathbf{P}^{g-1}$ with $\deg(V) = g$ and Sing(V) finite. If $g = 6$ (under the same assumptions on Sing(W)) the situation was completely analized in [CM] (V is the intersection of a quadric and a cubic hypersurface); in particular if $g = 6$, V is trigonal. In the paper by Ambrogio and Picco Botta it is proved (under the same assumptions on Sing(W)), that if $g>6$ the general such V is not trigonal.

(H) The paper "Families of singular meromorphic foliations" by E. Ballico contains bounds for the number of the irreducible components of the "set of meromorphic foliations with given invariants" on a projective manifolds.

(I) The papers E. Ballico "On stable sheaves on algebraic surfaces" and E. Ballico - R. Mirò-Roig "Cohomology of generic rank 2 stable bundles on Fano 3-folds of index 2" study essentially vector bundles on projective varieties. The first one consider stable reflexive sheaves on a singular surface. For related but not intersecting results and other techniques useful for the study of the same objects, see [Is]. In the paper "Cohomology of generic rank 2 stable bundles on Fano 3-folds of index 2" the authors consider rank 2 stavle vector bundle on a Fano 3-fold of index 2 X. They prove that for most Chern classes c_1 and c_2 the "generic" rank 2 stable bundle on X with Chern classes c_1, c_2 has the best possible cohomological properties.

(J) In her paper "Linear series of low degree on plane curves and the Castelnuovo function" L. Caire gives conditions for the existence and the uniqueness of some linear series on a plane curve C containing a fixed 0-dimensional subscheme X of $\mathbf{P}^2$. The proof uses a careful study of the Castelnuovo function of X. More informations are obtained if C is assumed to have high geometric genus and only nodes and ordinary cusps as singularities or having only one multiple ordinary point.

(K) The paper of S. Katz is on a very hot topic: Calabi-Yau manifolds. In the last few decades there were several cases of new and fundamentals ideas created by a close interaction and cooperation among mathematicians and physicist. This had and will have even more in the future deep consequences for the devolpement of mathematics. The case of Calabi - Yau manifolds is an excellent and succesful example of the potentiality of these interactions. Mathematically, a Calabi - Yau 3-fold is a complex compact Kähler manifold V (or a complex orbifold) with K_V trivial and $H^1(V,\mathcal{O}_V) = 0$. The consistency of string theory requires an additional six-dimensional space: exactly a Calabi - Yau manifold. Two such manifolds are called a mirror pair if they induce the same conformal theory. Intuitions coming from physics and physicist's mathematically non rigorous path integral calculations lead to general conjectures. It is possible to verify these conjectures in a few specific examples. The computations of the number of rational curves on a quintic hypersurface of $\mathbf{P}^4$ was the first example of the power of this approach. The paper by S. Katz in this volume contains the calculation of the number of rational curves on several Calabi-Yau manifolds. For several papers on this subject, see [Y]. For a general introduction aimed to mathematicians see [Mo]. Very interesting results in this field are appearing almost every day. There was a workshop/school in Dirkolbotn, Norway (10-14 August 1993) organized for Europroj by Strømme with 3 series of lectures on "Calabi - Yau manifolds and their appearence in Enumerative Geometry" (by D. Morrison, S. Katz and P.M.H. Wilson) and several seminars. A long report on that school will appear (or appeared) on an issue of Euronews; the report contains abstracts of the seminars; both the report and long abstracts of the 3 series of lectures are available through Strømme (Stein.Stromme@mi.uib.no); these reports contain very long lists of very useful references. Another advanced school and workshop (with D. Morrison and Y. Manin among the speakers) will be held in Trento (Italy) from 15 to 24 June 1994.

All the papers in this volume had a referee, except (for lack of time) my survey paper "Weierstrass loci and generalizations, II", which however was distributed and read by a few friends before sending it to Marcel Dekker. All the papers in this volume are in final form and no part of them will be submitted elsewhere.

References

[B] E. Ballico, *Algebraic hyperbolicity of generic high degree hypersurfaces,* preprint.

[CM] A. Conte, J. P. Murre, *Three - dimensional algebraic varieties whose hyperplane sections are Enriques surfaces,* Annali Scuola Norm. Sup. Pisa, (4) **12** (1985), 43-80.

[DSW] G. Dethloff, G. Schumaker, P.-M. Wong, *Hyperbolicity of the complements of plane algebraic curves,* Amer. J. Math. (to appear).

[DK1] I. Dolgachev, M. Kapranov, *Arrangements of hyperplanes and vector bundles on $\boldsymbol{P}^n$,* Duke Math. J. **71** (1993), 633-664.

[DK2] I. Dolgachev, M. Kapranov, *Schur quadrics and their generalization in the theory of vector bundles on $\boldsymbol{P}^2$,* preprint.

[Is] A. Ishi, *On the moduli of reflexive sheaves on a surface with rational double points,* Math. Ann. **294** (1992), 125-150.

[La1] S. Lang, *Hyperbolic and Diophantine analysis,* Bull. Amer. Math. Soc. **14** (1986), 159-205.

[La2] S. Lang, *Introduction to complex hyperbolic spaces,* Springer-Verlag, 1987.

[MN] K. Masuda, J. Noguchi, *A construction of hyperbolic hypersurface in $\boldsymbol{P}^n(\boldsymbol{C})$*, preprint.

[Mo] D. Morrison, *Mirror symmetry and rational curves on quintic threefolds: A guide for mathematicians,* J. Amer. Math. Soc. **6** (1993), 223-247.

[No1] J. Noguchi, *Some problems in value distributions and hyperbolic geometry,* in: Proceedings volume of the International Symposium Holomorphic Mappings, Diophantine Geometry and Related Topics (in Honor of S. Kobayashi), R.I.M.S., Kyoto Univ. J. Noguchi Ed., pp. 66-79.

[No2] J. Noguchi (Ed.), Proceedings volume of the International Symposium Holomorphic Mappings, Diophantine Geometry and Related Topics (in Honor of S. Kobayashi), R.I.M.S., Kyoto Univ., (26-30 October 1992).

[Ru1] M. Ru, *Geometric and arithmetic aspects on $\boldsymbol{P}^n$ minus hyperplanes*, preprint.

[Ru2] M. Ru, *Integral points and the hyperbolicity of the complement of hypersurfaces,* J. reine angew. Math. **442** (1993), 163-176.

[RW] M. Ru, P.-M. Wong, *Integral points of $\boldsymbol{P}^n \backslash \{2n+1$ hyperplanes in general position$\}$*, Invent. Math. **106** (1991), 195-216.

[Y] S.-T. Yau (Ed.), *Essays on Mirror Manifolds,* International Press, Hong Kong, 1992.

[Wo] P.-M. Wong, *Recent results in hyperbolic geometry and Diophantine geometry,* in: Proceedings volume of the International Symposium Holomorphic Mappings, Diophantine Geometry and Related Topics (in Honor of S. Kobayashi), R.I.M.S., Kyoto Univ. , J. Noguchi Ed., pp. 120-135.

[Za] M. Zaidenberg, *Hyperbolicity in projective spaces*, in: Proceedings volume of the International Symposium Holomorphic Mappings, Diophantine Geometry and Related Topics (in Honor of S. Kobayashi), R.I.M.S., Kyoto Univ. , J. Noguchi Ed., pp. 136-156, or Prépublication 226 (1992) de l'Institute Fourier, Grenoble.

Remarks on Numerically Positive and Big Line Bundles

Mauro C. Beltrametti
Dipartimento di Matematica
Università degli Studi di Genova
Via L.B. Alberti 4
I-16132 Genova, Italy
beltrame@dima.unige.it
fax: Italy + 10–3538769

Andrew J. Sommese
Department of Mathematics
University of Notre Dame
Notre Dame, Indiana 46556, U.S.A.
sommese.1@nd.edu
fax: U.S.A. + 219–631–6579

Introduction. Let X be an n-dimensional smooth connected projective variety over the complex field $\mathbb{C}$. A line bundle L is said to be *numerically effective* (*nef*, for short) if $L \cdot C \geq 0$ for any irreducible curve C in X. A line bundle is said to *big* if $\kappa(L) = n$, where $\kappa(L)$ denotes the Kodaira dimension of L. If L is nef, then L is big if and only if $L^n > 0$. According to [4], when L is nef and big, the pair (X, L) will be called a *quasi-polarized* variety. A line bundle L is said to be *numerically positive* if $L \cdot C > 0$ for any irreducible curve C in X. Following [11] we call *nup* a numerically positive line bundle L.

A nice example due to C.P. Ramanujam and Mumford (see [6], p. 56, 57) shows that there exists a smooth 3-fold X and an effective numerically positive but not ample line bundle L on X. It turns out that this line bundle, L, is also big (see §1). Thus the natural question to study quasi-polarized varieties (X, L) with L nup and big but not ample arises. It should be noted that, in view of Kleiman's ampleness criterion [10], any such a pair (X, L) gives rise to an example where the cone $NE(X)$, in $N_1(X) := (\{1\text{-cycles}\}/\sim) \otimes \mathbb{R}$, generated by the effective 1-cycles is not closed with respect to the real topology. Here "$\sim$" denotes the numerical equivalence of cycles.

Note also that if the canonical bundle K_X is nup and big then K_X is ample. This is an immediate consequence of the Kawamata-Shokurov basepoint free theorem. As a consequence of the results of [3] and [4] the following partial answer to the question above can be given.

Theorem. *Let (X, L) be a smooth quasi-polarized variety of dimension $n \geq 3$. Assume that L is nup. Then L is ample unless (possibly) either:*

1. *There exists an ample line bundle, $\mathcal{L}$, on X such that $(X, \mathcal{L})$ is a $\mathbb{P}^{n-1}$-bundle, indeed a scroll, $\varphi : X \to C$, over a smooth curve, C, and $K_X + nL \approx \varphi^*H$ for some (not necessarily ample) line bundle H on C;*
2. *(X, L) is a scroll over a normal surface;*
3. *(X, L) is a quadric fibration over a smooth curve, or*
4. *$K_X + (n-1)L$ is nef and $\Gamma(m(K_X + (n-1)L))$, $m \gg 0$, gives a birational morphism, $\phi : X \to X'$. If ϕ is not an isomorphism, then ϕ is the simultaneous contraction*

to distinct smooth points of divisors $E_i \cong \mathbb{P}^{n-1}$ *such that* $\mathcal{O}_{E_i}(E_i) \cong \mathcal{O}_{\mathbb{P}^{n-1}}(-1)$ *and* $L_{E_i} \cong \mathcal{O}_{\mathbb{P}^{n-1}}(1)$ *for* $i, \ldots, t$. *Furthermore* $L' := (\phi_* L)^{**}$ *is nup and big,* $L \approx \phi^* L' - \sum_{i=1}^{t} E_i$, $K_{X'} + (n-1)L'$ *is ample and* $K_X + (n-1)L \approx \phi^*(K_{X'} + (n-1)L')$.

Note that the Mumford-Ramanujam example belongs to class 2). In Examples (3.13) and (3.14) we construct new examples of nup and big but not ample line bundles as in 1) and 3) respectively.

Note also that in the 2-dimensional case any numerically positive and big line bundle is ample in view of Nakai's ampleness criterion. We refer to [11] for results on nup line bundles on surfaces, and for a new example of a nup non-ample line bundle on a rational surface ([11, (3.3)]).

We shall denote by "$\approx$" the linear equivalence of line bundles and by "$\cdot$" the intersection of cycles. For any quasi-polarized variety (X, L), with $n = \dim X$, the *sectional genus*, $g(X, L)$, is defined by the equality

$$(K_X + (n-1)L) \cdot L^{n-1} = 2g(X, L) - 2,$$

where, as above, K_X denotes a canonical line bundle on X. Note that $g(X, L)$ is an integer and $g(X, L) \geq 0$ for $n \leq 3$. This is shown in [4], where it is also conjectured that $g(X, L) \geq 0$ for any quasi-polarized variety (X, L). For any coherent sheaf $\mathcal{F}$ on X let $h^i(\mathcal{F})$ be the complex dimension of $H^i(X, \mathcal{F})$, $i \geq 0$. The Δ-*genus*, $\Delta(X, L)$, of (X, L) is defined as $\Delta(X, L) = n + L^n - h^0(L)$.

The quasi-polarized variety (X, L) is said to be a *quasi-*$\mathbb{P}^r$*-bundle* over a variety Y if there is a surjective morphism $\varphi : X \to Y$ such that $(F, L_F) \cong (\mathbb{P}^r, \mathcal{O}_{\mathbb{P}^r}(1))$, $r = n - \dim Y$, for every fiber F of φ. We say that (X, L) is a *quasi-scroll* (respectively a *quasi-quadric fibration*) over a variety Y if there exists a surjective morphism with connected fibers $\varphi : X \to Y$ such that $K_X + (n - \dim Y + 1)L \approx \varphi^* \mathcal{L}$ (respectively $K_X + (n - \dim Y)L \approx \varphi^* \mathcal{L}$), for some ample line bundle $\mathcal{L}$ on Y.

Through this article, besides the definitions and notation above, we employ the standard terminology. As usual we don't distinguish between line bundles and Cartier divisors. We refer to [9] for the definitions of terminal, canonical, $\mathbb{Q}$-factorial singularities. We also refer to [13] and [9] for some Mori's theory we use, as definition of extremal ray, Cone Theorem, Contraction Theorem and Kawamata-Shokurov basepoint free theorem. For any further background material and notation we refer to [3] and [4].

The first author would like to thank Takao Fujita for helpful discussions during his visit at the University of Genova in June 1988. In particular he suggested the construction in Example (3.13), and the conjecture at the end of §3 come out from these discussions. This paper is a revised version of preliminary notes written years ago. Moreover we both greatly thank Takao Fujita for pointing out some gaps in the version of the paper before this one, and suggesting improvements of a number of results in that paper. We both would like to thank Antonio Lanteri for sending us the preprint [11], which stimulated our interest to come back to this topic. Both authors would also like to thank the Max-Planck-Institut in Bonn for its support during the final stages of writing up this paper. The second author also thanks the National Science Foundation (NSF Grant DMS93-02121) for their support.

1. The Mumford-Ramanujam example. First consider the case of surfaces. The following fact shows that the Mumford example of a nup non-ample line bundle (see [6], p. 56 and also [10], p. 326) is essentially unique for $\mathbb{P}^1$-bundles and corresponds to the case

$a = 1$ (from the Proposition). Note also that in the Proposition below the condition $e < 0$ implies, since by Nagata $e \geq -q$, that $q \geq 1$ where q denotes the irregularity of the surface S (see also [11, (1.2),(3.1)]).

PROPOSITION **1.1** (Mumford example.) *Let S be a smooth surface which is a $\mathbb{P}^1$-bundle $p : S \to B$, of invariant e, over a smooth curve B. Let $L \sim aE + bf$ be a line bundle on S, where a, b are integers and E, f denote a section of self-intersection $E^2 = -e$ and a fiber of p respectively. Then L is nup non-ample if and only if $L \sim a(E + \frac{e}{2}f)$ with $a > 0$, $e < 0$ and there are no effective curves on S numerically equivalent to L.*

Proof. Assume L is nup and not ample. Then $L \cdot f > 0$, $L \cdot E > 0$ give $a > 0$, $b > ae$. Therefore $e < 0$, since otherwise L would be ample by the ampleness criterion [7], p. 380. From $0 = L \cdot L = -a^2e + 2ab$ we get $b = ae/2$. Clearly there are no effective curves, $C \subset S$, such that $C \sim L$, since otherwise we would have the contradiction $0 = L \cdot L = L \cdot C > 0$.

To show the converse, let $L \sim a(E + \frac{e}{2}f)$, $e < 0$, $a > 0$ and let C be any irreducible curve on S. Note that L is not ample since $L \cdot L = 0$. From [7, Prop. 2.21, p. 382] we know that either $C \sim E + \beta f$ with $\beta \geq 0$, or $C \sim \alpha E + \beta f$ with $\alpha \geq 2$, $\beta \geq \alpha e/2$, α, β integers.

In the first case,

$$L \cdot C = (aE + \frac{ae}{2}f) \cdot (E + \beta f) = -ae + \frac{ae}{2} + \beta a > 0$$

if and only if $\beta > e/2$. Since $\beta \geq 0$ and $e < 0$ the condition $\beta > e/2$ is satisfied, so that $L \cdot C > 0$.

In the second case,

$$L \cdot C = (aE + \frac{ae}{2}f) \cdot (\alpha E + \beta f) = a(-\alpha e + \frac{\alpha e}{2} + \beta) > 0$$

if and only if $\beta > \frac{\alpha e}{2}$. Then either $L \cdot C > 0$ or $\beta = \alpha e/2$. But in this case $C \sim L$, contradicting our assumption. Q.E.D.

Example 1.2 (C.P. Ramanujam, [6], p. 57.) Let S be a nonsingular surface and let L be a nup non-ample divisor on S. Let H be an effective ample divisor on S. Then we define $X := \mathbb{P}(L \otimes \mathcal{O}_S(-H) \oplus \mathcal{O}_S)$ and let $\pi : X \to S$ be the projection. Let X_0 be the zero section of $\pi : X \to S$, corresponding to the quotient

$$\mathcal{E} := L \otimes \mathcal{O}_S(-H) \oplus \mathcal{O}_S \to L \otimes \mathcal{O}_S(-H) \to 0.$$

Then $X_0 = \xi$, ξ the tautological line bundle on X. Since $c_1(\mathcal{E}) = L - H$, $c_2(\mathcal{E}) = 0$ we have

$$X_0^2 = \xi^2 = c_1(\mathcal{E}) \cdot \xi - c_2(\mathcal{E}) = (L \otimes \mathcal{O}_S(-H))_{X_0}.$$

We define $D := X_0 + \pi^*H$. Note that D is effective by construction and a direct computation shows that D is nup (see [6], p. 57). On the other hand D is not ample because $D_{X_0} = (X_0 + \pi^*H)_{X_0} = L_{X_0}$ and therefore

$$D^2 \cdot X_0 = D_{X_0} \cdot D_{X_0} = L^2_{X_0} = 0.$$

Note that D is big since

$$D^3 = D^2 \cdot (X_0 + \pi^*H) = D^2 \cdot \pi^*H = D \cdot (X_0 + \pi^*H) \cdot \pi^*H$$

and the last term is positive since $\pi^*H \cdot \pi^*H$, $X_0 \cdot \pi^*H$ are effective curves and D is nup.

2. Some general facts on quasi-polarized varieties. We need to restate for a quasi-polarized variety (X, L) of dimension $n \geq 3$ some results which are well known if L is ample. They are corollaries of the results and the methods of [3] and [4].

LEMMA **2.3** *Let (X, L) be a smooth quasi-polarized variety of dimension $n \geq 3$ with L nup. If $K_X + tL$ is not nef for some $t > 0$ then there exists an extremal ray R and a morphism $\varphi = \mathrm{cont}_R : X \to Y$ with connected fibers and normal image, such that*

1. $(K_X + tL) \cdot R < 0$;
2. *$M \in \mathrm{Pic}(X)$ is relatively ample with respect to φ if and only if $M \cdot R > 0$;*
3. *φ is not birational if $t \geq n - 1$.*

Proof. Let C be a curve in X such that $(K_X + tL) \cdot C < 0$. By Mori's Cone Theorem we can write, in the closure $\overline{NE}(X)$ of the cone $NE(X)$ of the effective 1-cycles,

$$C = u + \sum_i R_i,$$

where $K_X \cdot u \geq 0$, the sum is finite and the R_i's are extremal rays. Therefore, since $L \cdot u \geq 0$, we have $(K_X + tL) \cdot R_i < 0$ for some index i. Let $R := R_i$ and let $\varphi : X \to Y$ be the contraction of R. Then we have (2.3.1) and Mori's Contraction Theorem applies to give (2.3.2).

Let $t \geq n - 1$ and assume φ birational. Let $\ell \in R_i$ be a rational curve with $-K_X \cdot \ell$ minimal. Since L is nup we see that $-K_X \cdot \ell > tL \cdot \ell \geq t \geq n - 1$. Thus $-K_X \cdot \ell \geq n$ and the contraction associated to R has an image of dimension at most 1 by [15, (2.4)], which contradicts the birationality of φ. Q.E.D.

PROPOSITION **2.4** *Let (X, L) be a smooth quasi-polarized variety of dimension $n \geq 3$ with L nup. Then $K_X + nL$ is nef unless $(X, L) \cong (\mathbb{P}^n, \mathcal{O}_{\mathbb{P}^n}(1)$. In particular if L is nup and not ample, then $K_X + nL$ is nef.*

Proof. If $K_X + nL$ is not nef, let R be an extremal ray such that $(K_X + nL) \cdot R < 0$ and let $\varphi : X \to Y$ be the contraction of R. Let $\ell \in R_i$ be a rational curve with $-K_X \cdot \ell$ minimal. Since L is nup we see that $-K_X \cdot \ell > nL \cdot \ell \geq t \geq n$. Thus $-K_X \cdot \ell \geq n + 1$, and since $-K_X \cdot \ell \leq n + 1$ we see that $(K_X + (n + 1)L) \cdot \ell = 0$. By [15, (2.4)], $\mathrm{Pic}(X) = \mathbb{Z}$. Thus $-K_X \cong (n + 1)L$ and L is ample. The theorem now follows from the Kobayashi-Ochiai theorem. Q.E.D.

Let us recall the following general standard fact we need.

LEMMA **2.5** *Let M, V be normal projective connected varieties of dimension $n \geq 3$. Assume that M is smooth and V has at worst terminal $\mathbb{Q}$-factorial singularities. Let $\varphi : M \to V$ be a birational morphism. Then if φ is not an isomorphism there exists an effective curve C in M such that $\dim \varphi(C) = 0$ and $K_M \cdot C < 0$.*

Proof. Let $E = \{x \in M, \varphi \text{ is not an isomorphism at } x\}$ be the exceptional locus of φ. From [8, 5.8] we know that each irreducible component of E is a prime divisor, E_i. Since V has terminal singularities we have

$$K_M \approx \varphi^* K_V + \sum_i a_i E_i,$$

where a_i are positive integers for each index i.

We use induction on $\dim \varphi(E)$. Assume $\dim \varphi(E) = 0$. Let $H_1, \ldots, H_{n-2}$ be general hyperplane sections in M. Then $S := H_1 \cap \ldots \cap H_{n-2}$ is a smooth surface by Bertini's theorem.

Consider the restriction, $\varphi_S : S \to \varphi(S)$, of φ to S. Then φ_S is a birational morphism whose exceptional locus is the cycle $\mathcal{Z} = \sum_i E_{i|S}$. Thus any irreducible component, C, of $\mathcal{Z}$ is the requested curve. In fact $\sum_i a_i E_i \cdot C = \sum_i a_i E_{i|S} \cdot C < 0$ by a well known result due to Mumford [14].

Assume $\dim \varphi(E) = k$ and let $A_1, \ldots, A_k$ be general hyperplane sections of V. Let $V_k := A_1 \cap \ldots \cap A_k$. Then $\dim(V_k \cap \varphi(E)) = 0$ and $\varphi_k : \varphi^{-1}(V_k) \to V_k$ is a birational morphism whose exceptional locus is $E_k := E \cap \varphi^{-1}(V_k)$ and

$$\dim \varphi(E_k) = \dim(V_k \cap \varphi(E)) = 0.$$

Note that $\varphi^{-1}(A_i)$, $i = 1, \ldots, k$, and $\varphi^{-1}(V_k)$ are smooth by Bertini's theorem.

Take the surface section $S_k := H'_1 \cap \ldots \cap H'_{n-k-2}$ obtained as transversal intersection of $n-k-2$ general hyperplane sections $H'_1, \ldots, H'_{n-k-2}$ of $\varphi^{-1}(V_k)$. Since $\varphi^{-1}(V_k)$ is smooth we conclude that the surface S_k is smooth. Thus we fall in the previous induction step. Q.E.D.

COROLLARY **2.6** (Kobayashi-Ochiai Theorem.) *Let (X, L) be a smooth quasi-polarized variety of dimension $n \geq 3$. Assume that L is nup and $K_X \sim -tL$ for some $t > 0$. Then L is ample. Thus $t \leq n+1$. if $t = n+1$, one has $(X, L) \cong (\mathbb{P}^n, \mathcal{O}_{\mathbb{P}^n}(1))$, and if $t = n$, (X, L) is isomorphic to a hyperquadric Q in $\mathbb{P}^{n+1}$ such that $\mathcal{O}_Q(1) \approx L$.*

Proof. By the Kawamata-Shokurov basepoint free theorem, $K_X + NL$ is semi-ample for any integer $N > t$. Since $K_X + NL \sim (N-t)L$ is nup it is easily seen that $(N-t)L$ and hence L is ample. The rest of the result is just the usual Kobayashi-Ochiai theorem. Q.E.D.

Let us give an explicit example of an hyperquadric with terminal not $\mathbb{Q}$-factorial singularities.

Example 2.7 Let Q be the smooth quadric in $\mathbb{P}^3$ and let V be the cone projecting Q from a point $v \in \mathbb{P}^4 \setminus \mathbb{P}^3$. V has terminal singularities since the vertex, v, is an ordinary double point and it is not $\mathbb{Q}$-factorial. To see this let D be the cone projecting a line ℓ on Q from v. Choose a second line, ℓ', on Q with $\ell \cap \ell' = \emptyset$ and let D' be the cone projecting ℓ' from v. Note that $D \cap D' = \{v\}$. Thus neither D nor D' can be $\mathbb{Q}$-Cartier divisors. In particular v is not a $\mathbb{Q}$-factorial point of V. Let $\varphi : M \to V$ be a minimal desingularization of V. The exceptional locus of φ is a curve, C. Then $K_M \approx \varphi^* K_V$, $K_V \approx -3L$, $L = \mathcal{O}_V(1)$. In particular $K_M \cdot C = 0$, so that C is not an extremal ray. Note that $K_M \approx -3\mathcal{L}$ with $\mathcal{L} = \varphi^* L$ nef and big but not ample.

3. Nup and big non-ample line bundles. In this section we deal with smooth quasi-polarized varieties (X, L) of dimension $n \geq 3$ with L nup and big line bundle on X.

First we need the following preliminary Lemma.

LEMMA **3.8** *Let (X, L) be as above. Further assume that $-K_X$ is nef. Then L is ample.*

Proof. Since $L - K_X$ is nef and big, the Kawamata- Shokurov basepoint free theorem applies to say that $\mathrm{Bs}|mL| = \emptyset$ for $m \gg 0$. Since $L \cdot C > 0$ for every effective curve, there can be no positive dimensional fibers of the map associated to $|mL|$. Q.E.D.

We can now prove the main result of the paper. Note that the pair (X', L'), from 4) of the theorem below, plays the role of "first reduction" in the adjunction thoretic sense (see e.g., [2]).

THEOREM **3.9** *Let (X, L) be a smooth quasi-polarized variety of dimension $n \geq 3$. Assume that L is nup. Then L is ample unless (possibly) either:*

1. *There exists an ample line bundle, $\mathcal{L}$, on X such that $(X,\mathcal{L})$ is a $\mathbb{P}^{n-1}$-bundle, indeed a scroll, $\varphi : X \to C$, over a smooth curve, C, and $K_X + nL \approx \varphi^* H$ for some (not necessarily ample) line bundle H on C;*
2. *(X,L) is a quasi-scroll over a normal surface;*
3. *(X,L) is a quasi-quadric fibration over a smooth curve, or*
4. *$K_X+(n-1)L$ is nef and $\Gamma(m(K_X+(n-1)L))$, $m \gg 0$, gives a birational morphism, $\phi : X \to X'$. If ϕ is not an isomorphism, then ϕ is the simoultaneous contraction to distinct smooth points of divisors $E_i \cong \mathbb{P}^{n-1}$ such that $\mathcal{O}_{E_i}(E_i) \cong \mathcal{O}_{\mathbb{P}^{n-1}}(-1)$ and $L_{E_i} \cong \mathcal{O}_{\mathbb{P}^{n-1}}(1)$ for $i,\dots,t$. Furthermore $L' := (\phi_* L)^{**}$ is nup and big, $L \approx \phi^* L' - \sum_{i=1}^{t} E_i$, $K_{X'} + (n-1)L'$ is ample and $K_X + (n-1)L \approx \phi^*(K_{X'} + (n-1)L')$.*

Proof. First, assume that $K_X + (n-1)L$ is not nef. Then from Lemma (2.3) we know that there exists an extremal ray, R, such that $(K_X + (n-1)L) \cdot R < 0$ and the contraction $\varphi = \mathrm{cont}_R : X \to Y$ is not a birational morphism. If $\dim Y = 0$, Mori's theory says that X is a Fano n-fold with Picard number $\rho(X) = 1$. Hence in particular L is ample. Assume $\dim Y > 0$ and let F be a general fiber of φ. Then L_F is nup and big on F (see e.g., [4, (1.4)]) and $K_F+(n-1)L_F$ is not nef since $(K_X+(n-1)L)\cdot R < 0$. Let ℓ be a curve which generates R. Since $L \cdot \ell > 0$ we conclude by Lemma (2.3) that L is relatively ample with respect to φ. Let $\mathcal{L} := L + \varphi^*\mathcal{H}$ for some ample line bundle, $\mathcal{H}$, on Y. Then $\mathcal{L}$ is ample. Furthermore, by the above, $K_F+(n-1)\mathcal{L}_F \sim K_F+(n-1)L_F$ is not nef and hence Proposition (2.4) applies to give $\dim F = n-1$ and $(F, \mathcal{L}_F) \cong (\mathbb{P}^{n-1}, \mathcal{O}_{\mathbb{P}^{n-1}}(1))$. Then in particular φ is equidimensional, so that by [3, (2.12)] we conclude that $(X,\mathcal{L})$ is a $\mathbb{P}^{n-1}$-bundle. Moreover $(X,\mathcal{L})$ is a scroll in the adjunction theoretic sense by [1, (2.2)]. Therefore $K_X + n\mathcal{L} \approx \varphi^* A$ for some ample line bundle A on Y. By taking $H := A - n\mathcal{H}$ we have $K_X + nL \approx \varphi^* H$ and we are in case (3.9.1).

Now, we assume that $K_X + (n-1)L$ is nef. Then $\mathrm{Bs}|m(K_X + (n-1)L)| = \emptyset$ for every $m \gg 0$ by the Kawamata-Shokurov basepoint free theorem. Let $\phi : X \to Y$ be the morphism associated to $\Gamma(m(K_X+(n-1)L))$, $m \gg 0$. Then ϕ has connected fibers and normal image. If $K_X + (n-1)L$ is ample, ϕ is an isomorphism and we are in a special case of (3.9.4).

Thus we can assume that $K_X + (n-1)L$ is not ample. Let F be a general fiber of ϕ. Therefore $K_{X|F} \cong K_F$ and $K_F \sim -(n-1)L_F$. Hence, by Corollary (2.6), $\dim Y = 0, 1, 2$ or n. If $\dim Y = 0$, then $-K_X \sim (n-1)L$ is nef and big so that Lemma (3.8) applies to say that L is ample. Assume $\dim Y = 1$. Hence by Corollary (2.6), $(F, L_F) \cong (Q, \mathcal{O}_Q(1))$, Q smooth hyperquadric in $\mathbb{P}^n$, and we are in class (3.9.3). If $\dim Y = 2$, again Corollary (2.6) yields $(F, L_F) \cong (\mathbb{P}^{n-2}, \mathcal{O}_{\mathbb{P}^{n-2}}(1))$ and we are in class (3.9.2).

Let $\dim Y = n$. Then the morphism ϕ is birational and there exists an irreducible reduced curve, $C \subset X$, such that $(K_X + (n-1)L) \cdot C = 0$. The usual argument as in the proof of Lemma (2.3), by using Mori's Cone Theorem, shows that there exists an extremal ray, R, such that $(K_X + (n-1)L) \cdot R = 0$ and the contraction $\rho := \mathrm{cont}_R : X \to Z$ is a birational morphism, since ϕ factors through ρ. Let E be the locus of R. Let ℓ be a curve which generated R. Since $\ell^2 > 0$ we conclude by Lemma (2.3) that L is relatively ample with respect to ρ. Then [2, Proposition (1.5)] applies to say that $(E, L_E) \cong (\mathbb{P}^{n-1}, \mathcal{O}_{\mathbb{P}^{n-1}}(1))$. Now, exactly the same argument as in the proof of [2, (3.1.4)] applies to give the result. Note that [2, Corollary (0.6.1)] holds true under our present assumption that L is nup and big. Q.E.D.

Remark 3.10 Let (X,L) be as in Theorem (3.9) with $n = 3$ and assume L to be not ample.

Then Nakai's ampleness criterion says that there exists some effective divisor D such that $L^2 \cdot D = 0$. Hence from [4, §2] we know that D has to verify the following conditions: $h^0(L) = h^0(L + D)$, $h^0(D) = 1$, D is not nef. ■

Theorem (3.9) above has two consequences.

COROLLARY **3.11** *Let L be a nup and big, not ample, line bundle on a smooth connected variety X of dimension $n \geq 3$. Then $K_X + (n-1)L$ is nef unless (X, L) is as in class 1) of Theorem (3.9).*

Proof. It follows immediately from Theorem (3.9). Q.E.D.

T. Fujita pointed the following out to us.

COROLLARY **3.12** *Let L be a nup and big line bundle on a projective variety X with terminal singularities. Assume that K_X is nef. Then $K_X + tL$ is ample for any rational $t > 0$.*

Proof. By the Kawamata-Shokurov basepoint free theorem, $K_X + tL$ is semiample. Since $K_X + tL$ is nup it is easy to see that there can be no effective curves in the fibers of the map associated to any $N(K_X + tL)$ where N is chosen so that $N(K_X + tL)$ is Cartier and spanned. Q.E.D.

Let us propose the following conjecture.

Conjecture. Let L be a nup line bundle on a smooth connected variety, X. Assume that $\kappa(L) \geq 0$ and $L - K_X$ is nef. Then L is ample. ■

To give some evidence to the conjecture above note that the result is true if X is a 3-fold and $L = K_X$ [12]. Moreover it is also true in the special case when $\kappa(L) = n$ and $L - K_X$ is nef and big. This is a consequence of the proof of Lemma (3.8).

The Mumford-Ramanujam construction recalled in §1 gives an explicit example as in case (3.9.2) of the theorem above.

Following Mumford's construction we can easily produce examples as in case (3.9.1) too.

Example 3.13 Let C be a nonsingular curve of genus ≥ 2. Then there exists a stable vector bundle $\mathcal{E}$ of rank 2 and degree zero on C such that the line bundle $\mathcal{O}_{\mathbb{P}(\mathcal{E})}(1)$ is nup and not big (see [6, Ex. 10. 6, p. 56]). Let A be an ample line bundle on C and let $X := \mathbb{P}(\mathcal{E} \oplus \mathbb{A})$. Let $L := \mathcal{O}_X(1)$ be the tautological bundle of X. Note that $\mathcal{E} \oplus A$ is not ample since the direct summand $\mathcal{E}$ has degree zero (see e.g., [6, Chap. III, §1]). Then L is not ample. Note also that L is clearly nef. We claim that L is nup and big. Note that

$$L^3 = c_1(\pi^*(\mathcal{E} \oplus A)) \cdot L^2 = (\deg A) f \cdot L^2 = \deg A > 0,$$

where f denotes a fiber of the bundle projection $\pi : X \to C$. Hence L is big. To see that L is nup assume otherwise. Then $L \cdot \gamma = 0$ for some curve, γ, on X. Since $L = \mathcal{O}_X(1)$, L is ample on the fibers of π and hence the restriction, $p := \pi_\gamma : \gamma \to C$ is onto. Consider the quotient $p^*(\mathcal{E} \oplus A) \to L_\gamma \to 0$. Since p^*A is ample and $\deg L_\gamma = 0$ it follows that the image of p^*A in L_γ is the zero section and hence $p^*\mathcal{E}$ has a degree zero quotient $p^*\mathcal{E} \to L_\gamma \to 0$. To finish we follow an argument suggested to us by T. Fujita. The surjection $p^*\mathcal{E} \to L_\gamma$ yields a section Z of $\mathbb{P}(p^*\mathcal{E}) \to \gamma$ such that $\xi_{p^*\mathcal{E}} \cdot Z = 0$. Let $q : \mathbb{P}(p^*\mathcal{E}) \to \mathbb{P}(\mathcal{E})$ be the map induced by p, which is finite to one. Then $\xi_{p^*\mathcal{E}} \sim q^*\xi_\mathcal{E}$, hence $\xi_\mathcal{E} \cdot q_*Z = 0$. This contradicts the nupness of $\xi_\mathcal{E}$. Thus (X, L) is a 3-dimensional quasi-polarized pair as in Theorem (3.9.1).

Note that the same construction above, by taking an ample vector bundle A of rank r on C, gives an example of a $(r+2)$-dimensional quasi-polarized variety (X, L) as in Theorem (3.9.1). ■

By using the construction as in the example above, we can give a new example as in (3.9.3).

Example 3.14 Let C, $\mathcal{E}$ be as in Example (3.13). Let L be an ample line bundle on C and let $\mathcal{F} := \mathcal{O}_C \oplus L$. Let X be the fiber product of $\mathbb{P}(\mathcal{E})$, $\mathbb{P}(\mathcal{F})$ over C and consider the commutative diagram

$$\begin{array}{ccccc} & & X := \mathbb{P}(\mathcal{E}) \times_C \mathbb{P}(\mathcal{F}) & & \\ & \overset{q_1}{\swarrow} & \cap & \overset{q_2}{\searrow} & \\ \mathbb{P}(\mathcal{E}) & \overset{\pi_1}{\longleftarrow} & \mathbb{P}(\mathcal{E}) \times \mathbb{P}(\mathcal{F}) & \overset{\pi_2}{\longrightarrow} & \mathbb{P}(\mathcal{F}) \\ & \underset{p_1}{\searrow} & \downarrow & \underset{p_2}{\swarrow} & \\ & & C & & \end{array}$$

Let $\xi_{\mathcal{E}}$ and $\xi_{\mathcal{F}}$ be the tautological bundles of $\mathbb{P}(\mathcal{E})$, and $\mathbb{P}(\mathcal{F})$ respectively. Denote $\mathcal{L} := q_1^*\xi_{\mathcal{E}} \otimes q_2^*\xi_{\mathcal{F}}$. Note that we have a commutative diagram

$$\begin{array}{ccc} X & \hookrightarrow & \mathbb{P}(\mathcal{E}) \times \mathbb{P}(\mathcal{F}) \\ \sigma \downarrow & & \downarrow p \\ \Delta & \hookrightarrow & C \times C \end{array}$$

where Δ is the diagonal and $X = p^{-1}(\Delta)$. Let $V := \mathbb{P}(\mathcal{E}) \times \mathbb{P}(\mathcal{F})$. Let $i_1 : C \times C \to C$, $i_2 : C \times C \to C$ be the projections on the first and the second factor respectively. Note that $p_1 \circ \pi_1 = i_1 \circ p$, $p_2 \circ \pi_2 = i_2 \circ p$. Compute

$$\begin{aligned} K_V &\cong \pi_1^* K_{\mathbb{P}(\mathcal{E})} + \pi_2^* K_{\mathbb{P}(\mathcal{F})} \\ &\cong \pi_1^*(p_1^*(K_C + \det \mathcal{E}) - 2\xi_{\mathcal{E}}) + \pi_2^*(p_2^*(K_C + \det \mathcal{F}) - 2\xi_{\mathcal{F}}) \\ &\cong p^* K_{C\times C} + p^*(i_1^* \det \mathcal{E} + i_2^* \det \mathcal{F}) - 2(\pi_1^*\xi_{\mathcal{E}} + \pi_2^*\xi_{\mathcal{F}}). \end{aligned}$$

Therefore, by the adjunction formula and since $X = p^{-1}(\Delta)$,

$$\begin{aligned} K_X &\cong (K_V + p^{-1}(\Delta))_{p^{-1}(\Delta)} \cong (K_V + p^*(\Delta))_{p^{-1}(\Delta)} \\ &\cong \sigma^*(K_{C\times C|\Delta} + i_1^* \det \mathcal{E}_{|\Delta} + i_2^* \det \mathcal{F}_{|\Delta} + \mathcal{N}_{\Delta}^{C\times C}) - 2(\pi_1^*\xi_{\mathcal{E}} + \pi_2^*\xi_{\mathcal{F}})_{p^{-1}(\Delta)}. \end{aligned}$$

By using the identification $\Delta \cong C$, one has $K_{C\times C} \cong 2K_C$ and $\mathcal{N}_{\Delta}^{C\times C} \cong -K_C$. Thus we find

$$K_X \cong \sigma^*(K_C + \det \mathcal{E} + \det \mathcal{F}) - 2\mathcal{L}.$$

Note that $K_C + \det \mathcal{E} + \det \mathcal{F}$ is ample since C has genus ≥ 2, $\det \mathcal{E} \cong \mathcal{O}_C$ since $\mathcal{E}$ is of degree zero and $\xi_{\mathcal{E}}$ is nup and not big by assumption, and $\det \mathcal{F} = L$ is ample. Therefore we conclude that $(X, \mathcal{L})$ is a quadric fibration under σ. We claim that $\mathcal{L}$ is nup and big but not ample.

To show that $\mathcal{L}$ is nup note first that $\mathcal{L}$ is nef since, by definition, $\mathcal{L}$ is the restriction to X of a nef line bundle. Assume that there exists a curve, γ, on X such that

$$\mathcal{L} \cdot \gamma = (\pi_1^*\xi_{\mathcal{E}} + \pi_2^*\xi_{\mathcal{F}}) \cdot \gamma = 0.$$

Then $\pi_1^*\xi_{\mathcal{E}} \cdot \gamma = \pi_2^*\xi_{\mathcal{F}} \cdot \gamma = 0$. Hence $\xi_{\mathcal{E}} \cdot \pi_1(\gamma) = 0$. Since $\xi_{\mathcal{E}}$ is nup this implies that γ is contained in a fiber of π_1 and therefore γ is contained in a fiber, $\beta^{-1}(c)$, of $\beta := p_1 \circ q_1$:

$X \to C$, for some $c \in C$. Note that $\beta^{-1}(c) \cong \mathbb{P}^1 \times \mathbb{P}^1$ and $\mathcal{L}_{\beta^{-1}(c)} \cong \mathcal{O}_{\mathbb{P}^1\times\mathbb{P}^1}(1,1)$ is ample on $\beta^{-1}(c)$. This contradiction gives nupness.

To show that $\mathcal{L}$ is not ample, take a section, γ, of $\mathbb{P}(\mathcal{F}) \to C$ corresponding to the quotient $\mathcal{F} \to \mathcal{O}_C \to 0$. Then $\xi_{\mathcal{F}|\gamma} \cong \mathcal{O}_\gamma$. Note that $\mathbb{P}(\mathcal{E}) \cong q_2^{-1}(\gamma)$ under q_1 and hence $q_1^*\xi_{\mathcal{E}|q_2^{-1}(\gamma)} \cong \xi_{\mathcal{E}}$. Compute

$$\mathcal{L} \cdot q_2^{-1}(\gamma) = (q_1^*\xi_{\mathcal{E}} + q_2^*\xi_{\mathcal{F}})_{q_2^{-1}(\gamma)} \cong \xi_{\mathcal{E}} + q_2^*(\xi_{\mathcal{F}|\gamma}) \cong \xi_{\mathcal{E}}.$$

Therefore $\mathcal{L}$ is not ample on $q_2^{-1}(\gamma)$ since $\xi_{\mathcal{E}}$ is not ample by assumption.

To show that $\mathcal{L}$ is big, compute

$$\begin{aligned}
\mathcal{L}^3 &= (\pi_1^*\xi_{\mathcal{E}} + \pi_2^*\xi_{\mathcal{F}})^3 \cdot p^{-1}(\Delta) \\
&= (3\pi_1^*\xi_{\mathcal{E}}^2 \cdot \pi_2^*\xi_{\mathcal{F}} + 3\pi_1^*\xi_{\mathcal{E}} \cdot \pi_2^*\xi_{\mathcal{F}}^2) \cdot p^{-1}(\Delta) \\
&= (3\pi_1^*\xi_{\mathcal{E}} \cdot \pi_2^*\xi_{\mathcal{F}}^2) \cdot p^{-1}(\Delta),
\end{aligned}$$

where the last equality follows from the assumption $\xi_{\mathcal{E}}^2 = 0$. Note that $\xi_{\mathcal{F}} = p_2^*c_1(\mathcal{F})$ and therefore

$$\pi_2^*\xi_{\mathcal{F}}^2 = (p_2 \circ \pi_2)^* c_1(\mathcal{F}) \cdot \pi_2^*\xi_{\mathcal{F}}.$$

Thus we conclude that

$$\mathcal{L}^3 = 3\pi_1^*\xi_{\mathcal{E}} \cdot \pi_2^*\xi_{\mathcal{F}} \cdot (p_2 \circ \pi_2)^* c_1(\mathcal{F}) \cdot p^{-1}(\Delta).$$

Since $(p_2 \circ \pi_2)^* c_1(\mathcal{F}) \cdot p^{-1}(\Delta) = \deg c_1(\mathcal{F}) \cdot F$, where F is a fiber of $p_2 \circ \pi_2$, we get

$$\mathcal{L}^3 = 3 \deg c_1(\mathcal{F}) \cdot \pi_1^*\xi_{\mathcal{E}} \cdot \pi_2^*\xi_{\mathcal{F}} \cdot F.$$

By construction, we have $F \cong \mathbb{P}^1 \times \mathbb{P}^1$, $\pi_1^*\xi_{\mathcal{E}|F} \cong \mathcal{O}_F(1,0)$, $\pi_2^*\xi_{\mathcal{F}|F} \cong \mathcal{O}_F(0,1)$. Therefore we find $\mathcal{L}^3 = 3 \deg c_1(\mathcal{F}) = 3 \deg L > 0$, and we are done. ∎

The above construction generalizes to give nup and big line bundles, L, on n-folds X with n odd such that $K_X + \frac{n+1}{2}L \approx p^*H$, where H is an ample line bundle over a curve, C, and X is a $\mathbb{P}^{(n-1)/2} \times \mathbb{P}^{(n-1)/2}$ fiber bundle, $p : X \to C$, over C.

Question. Do there exist n-folds, (X, L), with $n \geq 4$, L nup and big line bundle on X, such that (X, L) is a quasi-quadric fibration over a curve?

REFERENCES

[1] M.C. Beltrametti, A.J. Sommese, "Comparing the classical and the adjunction theoretic definition of scrolls," Geometry of Complex Projective Varieties, Proceedings Cetraro, June 1990, Seminars and Conferences, n. 9, Mediterranean Press, (1993), 56–74.

[2] M.C. Beltrametti, A.J. Sommese, "On the adjunction theoretic classification of polarized varieties," J. reine angew. Math., 427 (1993), 157–192.

[3] T. Fujita, "On polarized manifolds whose adjoint bundles are not semipositive," Algebraic Geometry, Sendai 1985, Advanced Studies in Pure Mathematics, 10 (1987), 167–178.

[4] T. Fujita, "Remarks on quasi-polarized varieties," Nagoya Math. J., 115 (1989), 105–123.

[5] T. Fujita, *Classification Theories of Polarized Varieties*, London Math. Soc. Lecture Note Series, 155, Cambridge University Press, (1990).

[6] R. Hartshorne, *Ample subvarieties of algebraic varieties*, Lecture Notes in Math., 156 (1970), Springer-Verlag.

[7] R. Hartshorne, *Algebraic Geometry*, G.T.M., 52 (1978), Springer-Verlag.
[8] Y. Kawamata, "The cone of curves of algebraic varieties," Ann. of Math., 119 (1984), 603–633.
[9] Y. Kawamata, K. Matsuda, K. Matsuki, "Introduction to the minimal model problem," Algebraic Geometry, Sendai 1985, Advanced Studies in Pure Mathematics, 10 (1987), 283–360.
[10] S. Kleiman, "Towards a numerical theory of ampleness," Ann. of Math., 84 (1966), 293–344.
[11] A. Lanteri, B. Rondena, "Numerically positive divisors on algebraic surfaces," preprint.
[12] K. Matsuki, "A criterion for the canonical bundle of a threefold to be ample," Math. Ann., 276 (1987), 557–564.
[13] S. Mori, "Threefolds whose canonical bundles are not numerically effective," Ann. of Math., 116 (1982), 133–176.
[14] D. Mumford, "The topology of normal singularities of an algebraic surface and a criterion for simplicity," Publ. Math. I.H.E.S., 9 (1961), 5–22.
[15] J.A. Wiśniewski, "Length of extremal rays and generalized adjunction," Math. Z., 200 (1989), 409–427.

A Report on Fano Manifolds of Middle Index and $b_2 \geq 2$

Jarosław A. Wiśniewski: Institute of Mathematics, Warsaw University
Banacha 2, 02-097 Warszawa, Poland; e-mail: `jarekw@plearn.bitnet`

Let X be a smooth projective variety of dimension n defined over complex numbers. We say that X is Fano if its anti-canonical divisor $-K_X$ (or, equivalently, its first Chern class c_1X) is ample. The index $r = r(X)$ of the Fano manifold X is defined as the largest integer dividing $-K_X$ in the Picard group $PicX \cong \mathbf{Z}^\rho$. The task of this report is to present a classification of Fano manifolds of index r and dimension $n = 2r$ whose second Betti number $b_2(X) = \rho$ is at least 2. *

(1.1). It is known that the index r of a Fano manifold X is *at most* $dimX + 1$. Moreover if either $r = dimX + 1$ or $r = dimX$ then X is a either the projective space $\mathbf{P}^n$ or the smooth quadric $\mathbf{Q}^n$, respectively (see [KO]). If X is a Fano surface of index 1 then it is called del Pezzo surface and is obtained by blowing-up $b_2(X) - 1$ general points on $\mathbf{P}^2$ (where $2 \leq b_2(X) \leq 8$). The degree $d(X) = (-K_X)^2$ of the del Pezzo surface X is then equal to $9 - b_2(X)$. The complete linear system $|-K_X|$ is base point free for $d = d(X) \geq 2$ and gives a map of X into $\mathbf{P}^d$; for $d \geq 3$ this is an embedding.

(1.2). Fano manifolds of of dimension $n \geq 3$ and index $n - 1$ are also called del Pezzo manifolds and were classified by Fujita, see for example [F1, I.8.11]. From Fujita's classification it follows that, if X is a del Pezzo manifold of dimension $n \geq 4$ then it is one of the following:

$V_1 =$ a weighted hypersurface of degree 6 in the weighted projective space $\mathbf{P}(1, \ldots, 1, 2, 3)$,
$V_2 =$ a weighted hypersurface of degree 4 in the weighted projective space $\mathbf{P}(1, \ldots, 1, 2, 2)$ (equivalently: a double covering of $\mathbf{P}^n$ ramified along a quartic),
$V_3 =$ a hypercubic in $\mathbf{P}^{n+1}$,

$V_4 =$ a complete intersection of type (2,2) in $\mathbf{P}^{n+2}$,
$V_5 =$ a linear section of Grasmanian $G(2, W) \subset \mathbf{P}(\Lambda^2 W)$ of linear spaces of dimension 2 in a 5-dimensional linear space W, $dim V_5 \leq 6$,
$V_6 = \mathbf{P}^2 \times \mathbf{P}^2$.

In dimension 3 we have also $\mathbf{P}^1 \times \mathbf{P}^1 \times \mathbf{P}^1$ and projective bundles: $\mathbf{P}(T\mathbf{P}^2)$, where $T\mathbf{P}^2$ denotes the tangent bundle to $\mathbf{P}^2$, and $\mathbf{P}(\mathcal{O}_{\mathbf{P}^2}(2) \oplus \mathcal{O}_{\mathbf{P}^2}(1))$.

The subscript d of the del Pezzo manifold V_d denotes its degree, that is the self-intersection of the first Chern class of the line bundle $\mathcal{O}_V(1)$ such that $\mathcal{O}_V(1)^{\otimes(n-1)} = \mathcal{O}_V(-K_X)$.

Fano n-folds of index $n-2$ were classified by Mukai [Mu] under an assumption on existence of smooth descending sequence of submanifolds, so-called *smooth ladder*.

(2.1). Based on the above examples and on the classification of Fano n-folds of index $n-2$ Mukai conjectured that if $r(X) > n/2 + 1$ then $b_2(X) = 1$ and if $r(X) = n/2 + 1$ and $b_2(X) \geq 2$ then

$$X \cong \mathbf{P}^{r-1} \times \mathbf{P}^{r-1}.$$

The conjecture was proved in [W1] and, subsequently, in [W2] Fano n-folds with $r(X) = (n+1)/2$ and $b_2(X) \geq 2$ were studied. The result was the following

Theorem 2.2. *Let X be a Fano n-fold (where n is odd ≥ 3) of index $r = (n+1)/2$. If $b_2(X) \geq 2$ then X has a $\mathbf{P}^{r-1}$-bundle structure over either projective space $\mathbf{P}^r$ or a smooth quadric $\mathbf{Q}^r$. The structure of X is described as follows:*

No.	*Base*	*projective bundle*	*another description*
1.	$\mathbf{P}^r$	$\mathbf{P}(\mathcal{O}(2) \oplus \mathcal{O}(1)^{\oplus(r-1)})$	*blow-up of* $\mathbf{P}^{r-1} \subset \mathbf{P}^{2r-1}$
2.	$\mathbf{P}^r$	$\mathbf{P}(T\mathbf{P}^r)$	*divisor* $(1,1) \subset \mathbf{P}^r \times \mathbf{P}^r$
3.	$\mathbf{Q}^r$	$\mathbf{P}(\mathcal{O}(1)^{\oplus r})$	$\mathbf{P}^{r-1} \times \mathbf{Q}^r$

The proof of the above result is a three-step argument. Firstly, using the deformations of rational curves contracted by elementary contractions of X one was able to establish that there exists an elementary contraction of X (in terms of Mori theory) whose fibers are of dimension $\leq r-1$. Secondly, the following result of Fujita was used to establish the geometric structure of the contraction.

Proposition 2.3. [F2, Lemma 2.12], [I], *Let $p : X \to Y$ be a map of a smooth projective variety X onto a normal variety Y such that $p_* \mathcal{O}_X = \mathcal{O}_Y$ (a contraction morphism with connected fibers). Assume that on X there exists an ample line bundle L such that $K_X + rL$ is trivial on fibers of p and all fibers of p are of* *dimension* $\leq r-1$. *Then* Y *is smooth and* $p : X \to Y$ *is a projective bundle over* Y, *that is, there exists a vector bundle* $\mathcal{E}$ *over* Y *such that* $X = \mathbf{P}(\mathcal{E})$ *(one may assume* $\mathcal{E} \cong p_*L$*).*

The proposition was used for $L = \mathcal{O}_X(-K_X/r)$ so that, by adjunction, the resulting bundle $\mathcal{E} := p_*L$ was ample of rank $r = dimY$ and $c_1\mathcal{E} = c_1Y$. Studying pairs $(Y, \mathcal{E})$ satisfying these conditions was the ultimate step of the proof of (2.2), this problem was treated by Fujita, Peternell, Ye and Zhang, see [F3], [P1], [P2] and [YZ].

(3.1). Fano manifolds of index r and dimension $n = 2r$ we will call for short Fano n-folds of middle index. By H we will denote the ample divisor such that $rH = -K_X$; the self-intersection of H we call the degree of such X and will be denoted by $d(X)$.

The above mentioned results may suggest to extend the method of classification to Fano manifolds of middle index which have $b_2 \geq 2$. Indeed, similar arguments can be applied. To make the exposition more transparent let us assume that $r \geq 3$ so that X is of dimension ≥ 6, though it should be noted that the arguments can be extended to the case $r \leq 2$ as well, and the result coincides with the classification of Mukai [Mu], see also [W3] and [W4]. Also, as it was proved in [W5] and in [K], for $n \geq 6$ we may assume that $b_2(X) = 2$ with the only exception in dimension 6 when $X \cong \mathbf{P}^2 \times \mathbf{P}^2 \times \mathbf{P}^2$.

Let us recall that an elementary contraction $\varphi : X \to Y$ is a morphism with connected fibers onto a normal variety Y (so that $\varphi_*\mathcal{O}_X = \mathcal{O}_Y$) and all curves contracted by φ are proportional in $H_2(X, \mathbf{Z})$ and they have negative intersection with the canonical divisor K_X. Elementary contractions of smooth 3-folds are known due to Mori [M2]. The classification of Fano 3-folds with $b_2 \geq 2$ done by Mori and Mukai [MM] was based on studying the elementary contractions. We follow this idea.

A Fano n-fold with $b_2 = 2$ has two elementary contractions. Studying the interplay between the deformations of rational curves contracted by these maps (similarly as in [M1]) allows to derive the following description :

Proposition 3.2. [W5, Theorem I], *Let* X *be a Fano manifold of middle index with* $b_2 \geq 2$. *There exists an elementary contraction* $p : X \to Y$ *such that*
(i) $dimY < dimX$ *and all fibers of* p *are of dimension* $\leq r$.
"The other" contraction $\varphi : X \to Z$ *is either of type (i), or*
(ii) φ *is birational and its exceptional set is an irreducible divisor* E *and* $\varphi_{|E}$ *has all fibers of dimension* r, *or*
(iii) there exists a fiber of φ *of dimension* $= r+1$, *but then all fibers of* p *are of dimension* $r-1$.

The next ingredient of the proof comes from the study of contraction morphisms:

Proposition 3.3. [AW Thm 4.1], [ABW], *Let* $p : X \to Y$ *be an elementary contraction of a smooth variety* X *onto a normal variety* Y *of smaller dimension.*

Assume that there exists an ample line bundle L on X such that $K_X + rL$ is trivial on fibers of p and all fibers of p are of dimension $\leq r$. Then Y is smooth and one of the following occurs:

(i) $dimY = r+1$ and $\mathcal{E} := p_*L$ is a locally free sheaf of rank r and $p : X \to Y$ is a projective bundle, $X \cong \mathbf{P}(\mathcal{E})$, (this is the case of (2.3));

(ii) $dimY = r$ and $\mathcal{E} := p_*L$ is a locally free sheaf of rank $r+2$ and $p : X \to Y$ is a quadric bundle so that X can be embedded as a divisor of relative degree 2 into $\mathbf{P}(\mathcal{E})$;

(iii) $dimY = r+1$ and $\mathcal{E} := p_*L$ is a reflexive non-locally free sheaf of rank r with isolated singularities and $X \cong \mathbf{P}(\mathcal{E})$.

(3.4). The singularities of the non-locally free sheaf $\mathcal{E}$ occuring in the last case of the above proposition are very special. Namely, locally, at a point $y \in Y$ we have an exact sequence of germs:

$$0 \longrightarrow \mathcal{O}_{Y,y} \xrightarrow{s} \mathcal{O}_{Y,y}^{\oplus(r+1)} \longrightarrow \mathcal{E}_y \longrightarrow 0$$

with $s(1) = (t_1, \ldots, t_{r+1})$ where $(t_1, \ldots, t_{r+1})$ are regular generators of the maximal ideal of the local ring $\mathcal{O}_{Y,y}$. Bănică called such sheaves *smooth* and he studied them in [B], see also [BW].

(3.5). According to (3.4) the study of Fano manifolds of middle index and $b_2 \geq 2$ can be split into three cases. We will do it below. The notation is consistent with (3.1)—(3.4) and $\mathcal{E}$ is always $p_*\mathcal{O}_X(H) = p_*\mathcal{O}_X(-K_X/r)$.

(4.1). First we present the case of projective bundles. The classification was done in a joint paper with Thomas Peternell and Michał Szurek [PSW1]. The vector bundle $\mathcal{E}$ is then ample of rank $r = dimY - 1$ and $c_1\mathcal{E} = -K_Y$. The result is as follows.

No.	Base of p	projective bundle	"the other contraction"
1.	V_d	$\mathbf{P}(\mathcal{O}_V(1)^{\oplus r})$	$V \times \mathbf{P}^{r-1} \to \mathbf{P}^{r-1}$
2.	$\mathbf{P}^{r+1}$	$\mathbf{P}(\mathcal{O}(2)^{\oplus 2} \oplus \mathcal{O}(1)^{\oplus(r-2)})$	small
3.	$\mathbf{P}^{r+1}$	$\mathbf{P}(\mathcal{O}(3) \oplus \mathcal{O}(1)^{\oplus(r-1)})$	divisorial
4.	$\mathbf{Q}^{r+1}$	$\mathbf{P}(\mathcal{O}(2) \oplus \mathcal{O}(1)^{\oplus(r-1)})$	divisorial
5.	$\mathbf{Q}^4$	$\mathbf{P}(\mathbf{E}(1) \oplus \mathcal{O}(1))$	scroll over $\mathbf{P}^4$

(4.2). Remarks. The variety V_d in the case 1 is a del Pezzo variety od degree d and $1 \leq d \leq 5$. The contraction φ is birational in cases 2—4, its exceptional set is a set of codimension 2 in case 2 or a divisor in cases 3 and 4, and the target is in these cases a cone. In case 5 the vector bundle $\mathbf{E}$ is spinor bundle over $\mathbf{Q}^4$ (see [O]), the contraction φ makes X a scroll with a 4-dimensional fiber, as described in [BSW, Example 3.2.4].

(4.3). It should be noted that in [PSW1] apart from the above varieties one more possibility was not excluded: X may have two $\mathbf{P}^2$-bundle structures over smooth Fano 4-folds Y_1 and Y_2 which are of index 1 and for any rational curve $C \subset Y_i$ there is $-K_{Y_i}.C \geq 4$. (The case $-K_Y.C = 3$ can be actually ruled out by arguments similar to those from [PSW1].) This is because the proof of the classification of the above projective bundles is based on a "comparison lemma" which compares on $\mathbf{P}(\mathcal{E})$ three families of rational curves: these in the fibers of p and φ, and minimal sections over minimal rational curves in Y. The argument, based on comparing deformations of these, fails if these families are too small, which is the case if $dimY = 4$.

However, using some recent results of Cho and Miyaoka [CM] one can rule out this remaining possibility as well. Indeed, from their result it follows that any 4-fold Y satisfying the condition:

$$\text{for any rational curve } C \subset Y \text{ it holds } -K_Y.C \geq 4$$

is isomorphic to either $\mathbf{P}^4$ or $\mathbf{Q}^4$. Therefore one may use the classification of bundles from [PSW2] to exclude this possibility.

(5.1). A classification of Fano manifolds of middle index which have quadric bundle structure is done in [W5]. The Fano n-fold is then a divisor in $\mathbf{P}(\mathcal{E})$ of relative degree 2 over Y. More precisely, by adjunction

$$X \in |\mathcal{O}_{\mathbf{P}(\mathcal{E})}(2) \otimes \bar{p}^*\mathcal{O}_Y(-K_Y - det\mathcal{E})|$$

where $\bar{p} : \mathbf{P}(\mathcal{E}) \to Y$ is the projection from the projective bundle. The result is as follows

No.	Base of p	$\mathcal{E}$	other description of X
1.	$\mathbf{Q}^r$	$\mathcal{O}(1)^{r+2}$	$\mathbf{Q}^r \times \mathbf{Q}^r$
2.	$\mathbf{P}^r$	$\mathcal{O}(1)^{r+2}$	divisor $(1,2) \subset \mathbf{P}^r \times \mathbf{P}^{r+1}$
3.	$\mathbf{P}^r$	$\mathcal{O}(2) \oplus \mathcal{O}(1)^{r+1}$	blow-up of $\mathbf{Q}^{r-1} \subset \mathbf{Q}^{2r}$
4.	$\mathbf{P}^r$	$T\mathbf{P}^r \oplus \mathcal{O}(1)^2$	divisor $(1,1) \subset \mathbf{P}^r \times \mathbf{Q}^{r+1}$
5.	$\mathbf{P}^r$	$\mathcal{O} \oplus \mathcal{O}(1)^{r+1}$	double cover of $\mathbf{P}^r \times \mathbf{P}^r$

(5.2). Remarks. Above, in the third case, the quadric $\mathbf{Q}^{2r}$ is blown-up along its linear section $\mathbf{Q}^{r-1} = \mathbf{Q}^{2r} \cap \mathbf{P}^r$. The branch divisor in case 5 is of bidegree $(2,2)$; moreover, it is worthwile to note that the bundle $\mathcal{E}$ is not ample in this case.

(6.1). In a joint paper with Edoardo Ballico [BW] we study non-locally free sheaves whose projectivization gives Fano manifolds of middle index. It is proved that the local extension enjoyed by the sheaves from (3.4) can be then extended to a global locally free sheaf $\mathcal{F}$ of rank $r+1$, that is, we have a sequence

$$0 \longrightarrow \mathcal{O} \xrightarrow{s} \mathcal{F} \longrightarrow \mathcal{E} \otimes \mathcal{L} \longrightarrow 0 \tag{6.2}$$

where $\mathcal{L}$ is a line bundle over Y. The above extension is not unique, nor is the choice of the line bundle $\mathcal{L}$. The section s of the locally free sheaf $\mathcal{F}$ vanishes exactly at $c_{r+1}(\mathcal{F})$ different points which correspond to singular fibers of the map p.

(6.3). The study of sheaves occuring in the case *(iii)* of (3.4) is similar to the one concerning locally free sheaves. Namely, we use a comparison lemma to study "the other" contraction of X and subsequently to prove that Y is either $\mathbf{P}^{r+1}$ or $\mathbf{Q}^{r+1}$. Then we extend, as in (6.2), $\mathcal{E}(-1)$ and $\mathcal{E}$ to a locally free sheaf which we call, respectively, $\mathcal{F}_1$ and $\mathcal{F}_0$. Since $\mathcal{E}(-1)$ is numerically effective, it follows that $\mathcal{F}_1$ is numerically efective as well and thus we may use results from [PSW2]. Consequenly we can realise X as a divisor in a projective bundle which is a Fano manifold. We obtain the following examples:

No.	Base of p	$\mathcal{F}_0$	$\mathcal{F}_1$	other description of X
1.	$\mathbf{Q}^{r+1}$	$\mathcal{O}(1)^{r+1}$	$\mathcal{O}^{r+2}/\mathcal{O}(-1)$	quadric bundle #4
2.	$\mathbf{P}^{r+1}$	$\mathcal{G}$	$\mathcal{O}^{r+2}/\mathcal{O}(-2)$	quadric bundle #2
3.	$\mathbf{P}^{r+1}$	$\mathcal{O}(2) \oplus \mathcal{O}(1)^r$	$(T\mathbf{P}(-1) \oplus \mathcal{O}(1))/\mathcal{O}$	blow-up of $\mathbf{P}^{r-1} \subset \mathbf{Q}^{2r}$
4.	$\mathbf{P}^{r+1}$	$T\mathbf{P}^{r+1}$	$\mathcal{O}^{r+3}/\mathcal{O}(-1)^2$	$(1,1) \cap (1,1) \subset \mathbf{P}^{r+1} \times \mathbf{P}^{r+1}$

(6.4). Remarks. Note that 1, 3 and 4 are divisors in varieties from Theorem (2.2). All projective bundles $\mathbf{P}(\mathcal{F}_i)$ occurring in the table, except $\mathbf{P}(\mathcal{G})$, are Fano manifolds. $-K_{\mathbf{P}(\mathcal{G})} = \mathcal{O}_{\mathbf{P}(\mathcal{G})}(r)$ is not ample but it is nef; actually, $\mathcal{G}$ is spanned and the map associated to $\mathcal{O}_{\mathbf{P}(\mathcal{G})}(1)$ contracts to point a section of the projective bundle over a hyperplane in $\mathbf{P}^{r+1}$. In case 3 the map $p : X \to \mathbf{P}^{r+1}$ comes from a rational map of $\mathbf{Q}^{2r}$ which is defined by a linear subsystem of $\mathcal{O}(1)$ whose base point locus is a given $B = \mathbf{P}^{r-1} \subset \mathbf{Q}^{2r}$. Then any $r-1$ dimensional fiber of p comes from a $\mathbf{P}^{r-1} \subset \mathbf{Q}^{2r}$ such that $\mathbf{P}^{r-1} \cup B = \mathbf{Q}^{2r} \cap \mathbf{P}^r$ and two r dimensional fibers of p come from two $\mathbf{P}^r \subset \mathbf{Q}^{2r}$ containing B.

(7.1). The above classification shows that the structure of Fano manifolds of the middle index is richer in lower dimensions. For example, the variety $V_5 \times \mathbf{P}^{r-1}$ occurs only for $n \leq 10$ and there are two special 6-folds: the case 5 in (4.1) and $\mathbf{P}^2 \times \mathbf{P}^2 \times \mathbf{P}^2$.

In dimension 4 the second example in the list of projective bundles does not occur. On the other hand we have the following list of Fano manifolds of index 2 which do not occur in higher dimensions (see [Mu] or [W4]):

$b_2 = 2$, $\mathbf{P}(\mathbf{N}(1)) = \mathbf{P}(\mathbf{E}(1))$, two $\mathbf{P}^1$ structures, where $\mathbf{N} = \Omega\mathbf{P}^3(2)/\mathcal{O}$ is a null-correlation bundle on $\mathbf{P}^3$ and $\mathbf{E}$ is a spinor bundle on $\mathbf{Q}^3$,
$b_2 = 2$, $\mathbf{P}^1 \times \mathbf{P}^3$,
$b_2 = 3$, $\mathbf{P}^1 \times \mathbf{P}(T\mathbf{P}^2)$, three $\mathbf{P}^1$-bundle structures,
$b_2 = 3$, $\mathbf{P}^1 \times \mathbf{P}(\mathcal{O}_{\mathbf{P}^2}(2) \oplus \mathcal{O}_{\mathbf{P}^2}(1))$,
$b_2 = 4$, $\mathbf{P}^1 \times \mathbf{P}^1 \times \mathbf{P}^1 \times \mathbf{P}^1$.

(7.2). Note that we obtain the following

Corollary. *Let X be a Fano manifold of middle index and $b_2 = 2$. Then $H = -K_X/r$ is a sum of two numerically effective divisors which define elementary contractions of X.*

Acknowledgements. The present report as well as some related results from [BW] were prepared while I was a guest of SFB 170 *"Geometrie und Analysis"* in Göttingen. During this visit I benefited a lot from discussions with other participants of SFB and I appreciate the time spent in Göttingen very much. I would like also to acknowledge a support from a Polish grant KBN 2/1093/91/01 (GR 54).

References.

[ABW] Andreatta, Ballico, Wiśniewski, Two theorems on elementary contractions, to appear in Math. Ann.
[AW] Andreatta, Wiśniewski, A note on non-vanishing and applications, to appear in Duke Math. J.
[B] Bănică, Smooth reflexive sheaves, Revue Romaine Math. Pures Appl. **63** (1991) 571—573.
[BW] Ballico, Wiśniewski, Bănică sheaves and Fano manifolds, in preparation.
[BSW] Beltrametti, Sommese, Wiśniewski, Results on varieties with many lines and their applications to adjunction theory, in *Complex Algebraic Manifolds, Bayreuth 1990*, Lecture Notes in Math. **1507** Springer-Verlag.
[CM] Cho, Miyaoka, Characterizations of projective spaces and hyperquadrics in terms of the minimal degrees of rational curves, preprint (1993).
[F1] Fujita, Classification theories of polarized varieties, London Lect. Notes **115**, Cambridge Press 1990
[F2] —,On polarized manifolds whose adjoint bundles are not semipositive, in *Algebraic Geometry, Sendai 1985*, Adv. Studies in Math. **10**, pp. 167–178, Kinokuniya 1987.
[F3] —, On adjoint bundle of ample vector bundles, in *Complex Algebraic Manifolds, Bayreuth 1990*, Lecture Notes in Math. **1507** Springer-Verlag.
[I] Ionescu, Generalized adjunction and applications, Math. Proc. Camb. Phil. Soc. **99** (1988), 457—472.

[KO] Kobayashi, Ochiai, A characterisation of complex projective spaces and quadrics, J. Math. Kyoto Univ. **13-1** (1973), 31—47.

[K] Kontani, A characterization of $\mathbf{P}^2 \times \mathbf{P}^2 \times \mathbf{P}^2$, preprint.

[M1] Mori, Manifolds with ample tangent bundles, Ann. Math. **110** (1979), 593—606,

[M2] —, Threefolds whose canonical bundles are not numerically effective, Ann. Math. **116** (1982), 133—176.

[MM] Mori, Mukai, Classification of Fano 3-folds with $b_2 \geq 2$, Manuscr. Math. **36** (1981), 147—162.

[Mu] Mukai, Fano manifolds of coindex 3, Proc. Natl. Acad. USA **86** (1989), 3000—3002.

[O] Ottaviani, Spinor bundles on quadrics, Trans. AMS **307** (1988), 301—316.

[P1] Peternell, A characterisation of $\mathbf{P}^n$ by vector bundles, Math. Zeit. **205** (1990), 487—490.

[P2] —, Ample vector bundles on Fano manifolds, Int. J. Math. **2** (1991), 311—322.

[PSW1] Peternell, Szurek, Wiśniewski, Fano manifolds and vector bundles, Math. Ann. **294** (1992), 151—165.

[PSW2] —, Numerically effective vector bundles with small Chern classes, in *Complex Algebraic Manifolds, Bayreuth 1990*, Lecture Notes in Math. **1507** Springer-Verlag.

[W1] Wiśniewski, On a conjecture of Mukai, Manuscr. Math. **68** (1990), 135–141.

[W2] —, On Fano manifolds of large index, Manuscr. Math. **70** (1991), 145–152.

[W3] —, Ruled Fano 4-folds of index 2, Proc. AMS **105** (1989), 55—61.

[W4] —, On Fano 4-folds of index 2. A contribution to Mukai classification, Bull. PAN **38** (1990), 173—183.

[W5] —, Fano manifolds and quadric bundles, to appear in Math. Zeit.

[YZ] Ye, Zhang, On ample vector bundles whose adjoint bundles are not numerically effective, Duke Math. J. **60** (1960), 671—688.

Order Sequences and Rational Curves

EDUARDO ESTEVES[1] Department of Mathematics, Massachusetts Institute of Technology, Cambridge MA02139, USA

MASAAKI HOMMA[2] Department of Mathematics, Yamaguchi University, Yamaguchi 753, Japan

1 Introduction

The theory of Weierstrass points on a smooth curve in arbitrary characteristic was initiated by F. K. Schmidt [Sc1], [Sc2] about a half century ago. Strangely enough, although he has been renowned as a specialist in the theory of algebraic functions, it seems that these works had been forgotten until the papers of Matzat [M] and Komiya [Km] appeared in the '70s. In the '80s, this characteristic-free approach to the theory of Weierstrass points (of arbitrary linear systems) was completely restored to an active area in several contexts ([Lk1], [Lk2], [StV]). Nowadays, the theory in positive characteristic has even been extended to linear systems on curves with Gorenstein singularities by Laksov and Thorup [LkT1], [LkT2] and Garcia and Lax [GLx]. However,these generalizations are beyond this paper; we consider linear systems on complete *smooth* curves over an algebraically closed field.

In the positive characteristic theory of Weierstrass points, the *order-sequence* plays an important role as an invariant of a linear system. Relations between the geometry of projective curves and their order-sequences have been studied by several authors ([BR], [GV], [HfKk], [HfV], [Hm1], [Hm2], [HmKj], [Kj2], [Kj3]), and the nature of order-sequences themselves has been also investigated ([Sc1], [G], [E], [Hm3], [Hm4]).

A fundamental and combinatorial property of order-sequences in characteristic $p > 0$ is to satisfy the p-adic criterion, which is stated as follows. Let $b_0 < b_1 < \cdots < b_N$ be a

[1]Supported in part by CNPq, Proc. No. 202151/90.5.
[2]Supported in part by JAMS.

sequence of nonnegative integers.

p-adic criterion. If a nonnegative integer m satisfies $\binom{b_j}{m} \not\equiv 0 \bmod p$ for some j, then $m = b_i$ for some i.

Conversely, any sequence of nonnegative integers $b_0 < b_1 < \cdots < b_N$ satisfying the p-adic criterion is the order-sequence of a linear system on a curve. In fact, the sequence is the order-sequence of the linear system on $\mathbf{P}^1$ corresponding to the morphism

$$\varphi = \varphi^{<b_0 \dots b_N>} : \mathbf{P}^1 \to \mathbf{P}^N$$

defined by $\varphi(t) = (t^{b_0}, \dots, t^{b_N})$.

In this paper, we direct our attention to nondegenerate rational curves in projective spaces, or linear systems on $\mathbf{P}^1$ and study them in a context of order-sequences.

A linear system $\mathcal{G}$ of degree d on $\mathbf{P}^1$ is said to be *diagonalizable*, if there is a basis S, T of $H^0(\mathbf{P}^1, \mathcal{O}(1))$ such that the vector space $V \subset H^0(\mathbf{P}^1, \mathcal{O}(d))$ corresponding to $\mathcal{G}$ is of the form

$$V = \bigoplus_{j=0}^{N} k \cdot S^{d-\lambda_j} T^{\lambda_j}$$

for some integers $\lambda_0 < \cdots < \lambda_N$. Obviously, the linear system corresponding to $\varphi^{<b_0 \dots b_N>}$ is diagonalizable.

In Section 2, we give characterizations of diagonalizable linear systems on $\mathbf{P}^1$ and the curves $\varphi^{<b_0 \dots b_N>}$ in terms of Weierstrass points: (1) *A linear system $\mathcal{G}$ on a curve is a diagonalizable linear system on $\mathbf{P}^1$ if and only if $\mathcal{G}$ is tame, that is, the coefficient of its Wronskian divisor $W(\mathcal{G})$ at each point is equal to the weight at the point, and $\mathcal{G}$ has at most two Weierstrass points* (Theorem 2); (2) *The morphism associated to a base-point-free linear system $\mathcal{G}$ with order-sequence $b_0, \dots, b_N$ coincides with $\varphi^{<b_0 \dots b_N>}$ up to projective transformations of $\mathbf{P}^1$ and $\mathbf{P}^N$ if and only if $\mathcal{G}$ is tame and has at most one Weierstrass point* (Corollary 1).

To establish these characterizations, we need a simple lemma on Hermite invariants at two distinct points (Lemma 1), which seems useful even in characteristic 0. For example, it implies the following theorem.

Theorem 1 *Let $\mathcal{G}$ be a base-point-free linear system of degree d and of dimension $N > 0$ on a curve of genus g over an algebraically closed field of characteristic 0. Denote by ν the (set-theoretic) number of $\mathcal{G}$-Weierstrass points. Then*

$$(\nu - 2)(\frac{d}{N} - 1) \geq 2g.$$

When $\mathcal{G}$ is the canonical linear system, the inequality means $\nu \geq 2g + 2$, which is sharp. When $\mathcal{G} = g_d^1$, it says $\nu \geq \frac{2}{d-1} g + 2$, which is almost sharp by Hurwitz formula.

Section 3 concerns the intersection $Z^{(m)}$ of general osculating m-planes of a curve, which was introduced by Kaji [Kj2] in a context of higher order strangeness. Denote by $\phi_{\mathcal{G}} : C \to \mathbf{P}^N$ the morphism defined by a base-point-free linear system $\mathcal{G}$ on a smooth curve C. We give an upper bound for dim $Z^{(m)}$ for an arbitrary nondegenerate curve $\phi_{\mathcal{G}}(C)$ in terms of its order-sequence (Theorem 3) and show this bound is sharp by describing explicitly $Z^{(m)}$ for $\varphi^{<b_0 \dots b_N>}(\mathbf{P}^1)$, which is an extension of a result by Ballico and Russo [BR, Proposition 2.2].

In Section 4, we establish a characteristic-free version of Theorem 1: *If a base-point-free linear system g_d^N with order-sequence $b_0, \ldots, b_N$ is tame and $Z^{(N-1)}$ for g_d^N is empty, then the inequality*

$$(\nu - 2)(\frac{d}{b_N} - 1) \geq 2g.$$

holds except for linear systems corresponding to $\varphi^{<b_0 \ldots b_N>}$'s (Corollary 3).

Finally, we mention another rare behavior of tangent lines of a certain space curve. Let C be a nondegenerate curve in $\mathbf{P}^N$ with $N \geq 3$. The curve is said to be *tangentially degenerate* if for a general point $P \in C$, the tangent line $T_P^{(1)}$ meets C at another point. A strange curve with center is on is, by definition, tangentially degenerate. So there are a lot of tangentially degenerate curves in positive chracteristic. However, the existence of a tangentially degenerate curve in characteristic 0 is a classical problem [C]. Many tangential pathologies of a projective curve C in characteristic $p > 0$ come from the inseparability of some Gauss map of C and/or the nonreflexivity of C or some tangent variety of C, which are equivalent to saying that $b_j \equiv 0 \bmod p$ for some order b_j of the curve ([HfKk], [HmKj], [Hm1]). So it seems reasonable to guess that first the existence of tangentially degenerate space curves is one of the pathologies in positive characteristic, and secondly some order of such a curve is congruent to 0 modulo the characteristic. The first part of this conjecture was proved by Kaji [Kj1] for immersed curves, but is still open in general. The second part of it is the topic of the final section; surprisingly, it is false!

2 Weierstrass points

2.1 A quick review

To begin with, we go briefly over the theory of Weierstrass points in arbitrary characteristic. For more details, see [StV].

Let C be a smooth curve of genus g over an algebraically closed field k of characteristic $p \geq 0$, and let $\mathcal{G}$ be the linear system of (projective) dimension N corresponding to a k-subspace V of $H^0(C, L)$, where L is an invertible sheaf of degree d. The language we use here is a little bit classical. For example, we often denote by g_d^N such a linear system $\mathcal{G}$, and we employ a module of functions instead of V itself; for a fixed $D \in \mathcal{G}$, there is a canonical isomorphism of k-vector spaces between V and the module of rational functions

$$\mathcal{L}(\mathcal{G}; D) := \{f \in k(C)^* \mid \text{div} f + D \in \mathcal{G}\} \cup \{0\},$$

where $k(C)^*$ is the set of nonzero rational functions on C.

A nonnegative integer μ is a *$\mathcal{G}$-Hermite invariant* at P (or *$(\mathcal{G}, P)$-order*) if there is a divisor $D \in \mathcal{G}$ such that its coefficient $v_P(D)$ at P is μ. Note that for any $P \in C$, there are $N + 1$ orders. We denote them by the sequence

$$\mu_0(P) < \cdots < \mu_N(P)$$

and call it the *$\mathcal{G}$-Hermite invariant-sequence* at P (or *$(\mathcal{G}, P)$-order-sequence*). For each i with $0 \leq i \leq N$, the function $\mu_i : C \to \mathbf{Z}$ is upper semicontinuous. The Hermite invariant-sequence at a general point is called the *order-sequence* of $\mathcal{G}$ and is denoted

$$b_0 < b_1 < \cdots < b_N.$$

An order-sequence $b_0 < b_1 < \cdots < b_N$ is said to be *classical* if $b_j = j$ for any $j = 0, 1, \ldots, N$. It should note that every order sequence is classical in characteristic 0. The $\mathcal{G}$-*weight* $w(P)$ at $P \in C$ is defined by

$$w(P) := \sum_{i=0}^{N} (\mu_i(P) - b_i).$$

For each point $P \in C$, choose a divisor $D_P \in \mathcal{G}$ such that $v_P(D_P) = \mu_0(P)$, a basis $f_{P,0}, \ldots, f_{P,N}$ of $\mathcal{L}(\mathcal{G}, D_P)$, and a local parameter t_P at P. Then the functions

$$t_P^{\mu_0(P)} f_{P,0}, \ldots, t_P^{\mu_0(P)} f_{P,N}$$

are the local equations of a basis of V in a certain neighborhood of P. Thus the collection of local regular functions

$$\left\{ \det \left(D_{t_P}^{(b_j)} \left(t_P^{\mu_0(P)} f_{P,i} \right) \right) \mid P \in C \right\}$$

determines an effective Cartier divisor, where $D_{t_P}^{(\alpha)}$ means the α-th Hasse-Schmidt derivation with respect to t_P (cf. [HsSc]). This Cartier divisor is denoted by $W(\mathcal{G})$, and is called the *Wronskian divisor* (or *ramification divisor*) of $\mathcal{G}$. It is obvious by definition that $v_P(W(\mathcal{G})) \geq w(P)$, and the equality holds if and only if

$$\det \left(\binom{\mu_i(P)}{b_j} \right) \not\equiv 0 \bmod p\ .$$

In particular, $v_P(W(\mathcal{G})) = w(P)$ in characteristic 0, and $v_P(W(\mathcal{G})) > 0$ if and only if $w(P) > 0$ in arbitrary characteristic.

Definition A linear system $\mathcal{G}$ is said to be *tame* if

$$v_P(W(\mathcal{G})) = w(P) \quad \text{for any} \quad P \in C.$$

A point of the support of $W(\mathcal{G})$ is called a $\mathcal{G}$-*Weierstrass point*; in other words, a point $P \in C$ is a Weierstrass point if and only if $w(P) > 0$. A global representation of $W(\mathcal{G})$ as a Weil divisor is given by

$$(N+1)D + \operatorname{div}\left(\det(D_t^{(b_j)} f_i)\right) + (b_0 + \cdots + b_N)\operatorname{div} dt$$

for an arbitrary choice of a divisor $D \in \mathcal{G}$, a basis $f_0, \ldots, f_N$ of $\mathcal{L}(\mathcal{G}; D)$ and a separating variable t; in particular,

$$\deg W(\mathcal{G}) = (N+1)d + (b_0 + \cdots + b_N)(2g-2).$$

Remark. When

$$B := \sum_{P \in C} \mu_0(P) P$$

is not zero divisor, it is the base locus of $\mathcal{G}$. Let

$$\mathcal{G}(-B) := \{D - B \mid D \in \mathcal{G}\}.$$

Then $\mathcal{G}(-B)$ is free from base points and the $(\mathcal{G}(-B), P)$-order-sequence is

$$0 = \mu_0(P) - \mu_0(P), \mu_1(P) - \mu_0(P), \ldots, \mu_N(P) - \mu_0(P).$$

In particular, the order-sequence of $\mathcal{G}(-B)$ coincides with that of $\mathcal{G}$.

On the other hand, since

$$\mathcal{L}(\mathcal{G}; D) = \mathcal{L}(\mathcal{G}(-B); D - B),$$

we have

$$W(\mathcal{G}) = W(\mathcal{G}(-B)) + (N+1)B.$$

In particular, $\mathcal{G}$ is tame if and only if so is $\mathcal{G}(-B)$.

2.2 The characterizations

We start with a simple but useful lemma.

Lemma 1 *For distinct points $P, Q \in C$, the inequality*

$$\mu_j(P) + \mu_{N-j}(Q) \le d$$

holds for any $j = 0, 1, \ldots, N$.

Proof. Set

$$\begin{aligned} \Gamma: &= \{D \in \mathcal{G} \mid D \succ \mu_j(P)P\}; \\ \Delta: &= \{D \in \mathcal{G} \mid D \succ \mu_{N-j}(Q)Q\}. \end{aligned}$$

Since Γ and Δ are linear subspaces of $\mathcal{G}$ of dimensions $N-j$ and j, respectively, $\Gamma \cap \Delta \neq \emptyset$; that is, there exists a divisor $D \in \mathcal{G}$ such that

$$D \succ \mu_j(P)P + \mu_{N-j}(Q)Q.$$

Therefore we have $d = \deg D \ge \mu_j(P) + \mu_{N-j}(Q)$. ||

The first application of Lemma 1 is the following theorem, which gives a characterization of diagonalizable linear systems on $\mathbf{P}^1$.

Theorem 2 *For a linear system $\mathcal{G} = g_d^N$ on a smooth curve C, the following conditions are equivalent:*

(1) *$\mathcal{G}$ is tame and has at most two Weierstrass points, that is there are two distinct points $P, Q \in C$ such that*

$$W(\mathcal{G}) = w(P)P + w(Q)Q$$

(*where $w(P)$ and/or $w(Q)$ can be* 0);

(2) *$C = \mathbf{P}^1$, and for a suitable choice of a basis S, T of $H^0(\mathbf{P}^1, \mathcal{O}(1))$ the linear system $\mathcal{G}$ corresponds to the subspace*

$$\bigoplus_{j=0}^{N} k \cdot S^{d-\lambda_j} T^{\lambda_j}$$

of $H^0(\mathbf{P}^1, \mathcal{O}(d))$ for some integers $\lambda_0 < \cdots < \lambda_N$.

In this case, the sequence $\lambda_0, \ldots, \lambda_N$ coincides with the Hermite invariant-sequence P or Q, and each of them can occur.

Proof. (2) $\Rightarrow$ (1) : Let $b_0 < b_1 < \cdots < b_N$ be the order-sequence of $\mathcal{G}$. Set $[0] := (1,0) \in \mathbf{P}^1$ and $[\infty] := (0,1) \in \mathbf{P}^1$. Then $\lambda_0[0] + (d - \lambda_0)[\infty] \in \mathcal{G}$ and

$$\mathcal{L}(\mathcal{G}; \lambda_0[0] + (d-\lambda_0)[\infty]) = \bigoplus_{j=0}^{N} k \cdot t^{\lambda_j - \lambda_0},$$

where $t := T/S$. Since

$$\det\left(D_t^{(b_i)} t^{\lambda_j - \lambda_0}\right) \neq 0$$

by definition, we have

$$\begin{aligned} W(\mathcal{G}) &= (N+1)\{\lambda_0[0] + (d-\lambda_0)[\infty]\} \\ &\quad + \operatorname{div}\left(\det\left(\binom{\lambda_j - \lambda_0}{b_i}\right) t^{\sum_k (\lambda_k - \lambda_0 - b_k)}\right) \\ &\quad + (b_0 + \cdots + b_N)\operatorname{div} dt \\ &= (N+1)\{\lambda_0[0] + (d-\lambda_0)[\infty]\} \\ &\quad + \sum_k (\lambda_k - \lambda_0 - b_k)[0] - \sum_k (\lambda_k - \lambda_0 - b_k)[\infty] \\ &\quad - 2\sum_k b_k[\infty] \\ &= \sum_k (\lambda_k - b_k)[0] + \sum_k \left((d - \lambda_{N-k}) - b_k\right)[\infty] \\ &= w([0])[0] + w([\infty])[\infty]. \end{aligned}$$

(1) $\Rightarrow$ (2) : Since $\deg W(\mathcal{G}) = w(P) + w(Q)$,

$$\sum_{i=0}^{N} (\mu_i(P) + \mu_{N-i}(Q)) = (N+1)d + 2g \sum_{i=0}^{N} b_i.$$

On the other hand, since

$$(\mu_i(P) + \mu_{N-i}(Q)) \leq d$$

by Lemma 1, we have $g = 0$ and

$$d = \mu_i(P) + \mu_{N-i}(Q) \quad (i = 0, \ldots, N).$$

Therefore, thanks to [Hm4, Lemma] (or [Hm3, Proof of Lemma]), we get the conclusion (2) and the additional statement. ||

As a corollary of Theorem 2, we can give a characterization of the curve $\varphi^{<b_0 \ldots b_N>}$.

Corollary 1 *Let $\mathcal{G}$ be a base-point-free linear system on C with the order-sequence $b_0, \ldots, b_N$. Then the following conditions are equivalent:*

(1) *$\mathcal{G}$ is tame and has at most one Weierstrass point;*

(2) *$C = \mathbf{P}^1$, and for suitable choices of coordinates of $\mathbf{P}^1$ and $\mathbf{P}^N$ the morphism $\phi_{\mathcal{G}} : \mathbf{P}^1 \to \mathbf{P}^N$ is given by $\phi_{\mathcal{G}}(t) = (t^{b_0}, \ldots, t^{b_N})$.*

Proof. (1) $\Rightarrow$ (2) : By the assumption, $W(\mathcal{G}) = w(Q)Q$, where $w(Q) = 0$ if $\mathcal{G}$ has no Weierstrass points. Choose a point $P \in C$ distinct from Q. Then $W(\mathcal{G}) = w(P)P + w(Q)Q$ with $P \neq Q$ and $\mu_i(P) = b_i$ $(i = 0, \ldots, N)$; so we can apply Theorem 2 to $\mathcal{G}$. Note that since $\mathcal{G}$ is free from base points, $d - b_N = \mu_0(Q) = 0$.
(2) $\Rightarrow$ (1) : From the proof of (2) $\Rightarrow$ (1) of Theorem 2, we have $W(\mathcal{G}) = w([\infty])[\infty]$, since $w([0]) = 0$ in our case. ||

Corollary 2 *Let g_d^N be a linear system on a curve C. Then the followings are equivalent:*

(1) *g_d^N is tame and at most one Weierstrass point;*

(2) *$C = \mathbf{P}^1$, and for a suitable choice of a basis S, T of $H^0(\mathbf{P}^1, \mathcal{O}(1))$ the linear system g_d^N corresponds to the subspace*

$$\bigoplus_{j=0}^{N} k \cdot S^{d - b_j} T^{b_j}$$

of $H^0(\mathbf{P}^1, \mathcal{O}(d))$, where $b_0 < b_1 < \cdots < b_N$ is the order-sequence of g_d^N.

Proof. It is similar to the proof of Corollary 1.

Remark. As for the linear system without Weierstrass points, see [Hm3, Theorem II]

Corollary 3 *Let $\mathcal{G}$ be a linear system of degree d with order sequence $b_0, \ldots, b_N$ on a curve of genus g. Denote by ν the number of $\mathcal{G}$-Weierstrass points. If $\mathcal{G}$ is tame, then*

$$(\nu - 2)\left(\frac{(N+1)d}{2\sum_{i=0}^{N} b_i} - 1\right) \geq 2g$$

unless $\mathcal{G}$ is one of the linear systems described in Corollary 2.

Proof. Let $P_1, \ldots, P_\nu$ be the $\mathcal{G}$-Weierstrass points, and put $P_{\nu+1} := P_1$. Note that $\nu \geq 2$ by our assumption. Since $v_P(W(\mathcal{G})) = w(P)$ for each $P \in C$, we have

$$\begin{aligned}
(N+1)d \;+\; \sum_{i=0}^{N} b_i(2g-2) &= \sum_{\alpha=1}^{\nu}\sum_{i=0}^{N}(\mu_i(P_\alpha) - b_i) \\
&= \frac{1}{2}\sum_{\alpha=1}^{\nu}\sum_{i=0}^{N}(\mu_i(P_\alpha) + \mu_{N-i}(P_{\alpha+1})) - \nu\sum_{i=0}^{N} b_i \\
&\leq \frac{1}{2}\nu(N+1)d - \nu\sum_{i=0}^{N} b_i \quad \text{(by Lemma 1).}
\end{aligned}$$

This completes the proof.

Proof of Theorem 1. Since every linear system is tame and every order-sequence is classical in characteristic 0, we have

$$(\nu - 2)(\frac{d}{N} - 1) \geq 2g$$

for any linear systems g_d^N with $\nu \geq 2$, by Corollary 3. From Corollary 1, the only base-point-free linear system g_d^N with at most one Weierstrass point in characteristic 0 is $g_d^N = | \mathcal{O}_{\mathbf{P}^1}(N) |$, for which the above inequality holds because $N = d$.

3 The intersection of general osculating m-planes

3.1 Dimension theorem

Let C be a complete smooth curve and

$$\phi_{\mathcal{G}} : C \to \mathbf{P}^N$$

a morphism corresponding to a base-point-free linear system $\mathcal{G}$ on C. For a point $P \in C$, by the osculating m-plane $T_P^{(m)}$ to C at P, we understand the intersection of all hyperplanes H with $v_P(\phi_{\mathcal{G}}{}^*H) > \mu_m(P)$. Set

$$\text{Reg}^{(m)}C := \{P \in C \mid \mu_i(P) = b_i \text{ for any } i \text{ with } 0 \leq i \leq m\},$$

and

$$Z^{(m)}(\phi_{\mathcal{G}}(C)) := \bigcap_{P \in \text{Reg}^{(m)}C} T_P^{(m)}.$$

We add a few words about the definition of $Z^{(m)}$. The map $\gamma^{(m)}$ from $\text{Reg}^{(m)}C$ to the Grassmannian $\mathbf{G}(\mathbf{P}^N, m)$ of m-planes in $\mathbf{P}^N$ defined by $\gamma^{(m)}(P) := [T_P^{(m)}]$ is a morphism because the composite of $\gamma^{(m)}$ and the Plücker embedding of $\mathbf{G}(\mathbf{P}^N, m)$ can be locally represented by the $(m+1)$-minors of the matrix

$$\begin{pmatrix} D_t^{(b_0)} f_0 & \cdots & D_t^{(b_0)} f_N \\ \vdots & \cdots & \vdots \\ D_t^{(b_m)} f_0 & \cdots & D_t^{(b_m)} f_N \end{pmatrix}$$

where $(f_0, \ldots, f_N)$ is a local representation of $\phi_{\mathcal{G}}$ around P, and t is a separating variable with $v_P(\text{div } dt) = 0$.

Since C is smooth, there is a unique extension of $\gamma^{(m)}$ to a morphism

$$\tilde{\gamma}^{(m)} : C \to \mathbf{G}(\mathbf{P}^N, m) ,$$

but $\tilde{\gamma}^{(m)}(P)$ does not always coincide with $[T_P^{(m)}]$ if $P \notin \text{Reg}^{(m)}C$, which means that $Z^{(m)}(\phi_{\mathcal{G}}(C))$ does not always coincide with

$$\bigcap_{P \in C} T_P^{(m)} .$$

However, the following holds.

Lemma 2 *If U is a nonempty open subset of* $\mathrm{Reg}^{(m)}C$, *then*

$$\bigcap_{P\in U} T_P^{(m)} = Z^{(m)}(\phi_{\mathcal{G}}(C))$$

Proof. To prove our assertion, it suffices to show that

$$\bigcap_{P\in U} T_P^{(m)} \subset T_Q^{(m)}$$

for each $Q \in \mathrm{Reg}^{(m)}C$. Set

$$\Sigma := \{[L] \in \mathbf{G}(\mathbf{P}^N, m) \mid L \supset \bigcap_{P\in U} T_P^{(m)}\}.$$

Since Σ is closed in $\mathbf{G}(\mathbf{P}^N, m)$, the inverse image ${\gamma^{(m)}}^{-1}(\Sigma)$ is also closed in $\mathrm{Reg}^{(m)}C$. On the other hand, ${\gamma^{(m)}}^{-1}(\Sigma) \supset U$ by definition; hence ${\gamma^{(m)}}^{-1}(\Sigma) = \mathrm{Reg}^{(m)}C$ because $\mathrm{Reg}^{(m)}C$ is irreducible. ||

By definition, $\dim Z^{(m)}(\phi_{\mathcal{G}}(C)) \leq m-1$ for any $(C, \mathcal{G})$. However, we can get a better upper bound, if the order-sequence of $\mathcal{G}$ is assigned.

For a fixed integer m with $1 \leq m \leq N-1$, set

$$\Lambda_p^{(m)}(b_0, \ldots, b_N) := \{0 \leq i \leq m \mid \binom{b_{m+1}}{b_i} \equiv \cdots \equiv \binom{b_N}{b_i} \equiv 0 \bmod p\}.$$

Theorem 3 *Let $\mathcal{G}$ be a base-point-free linear syatem on a curve C over a field of characteristic $p > 0$, and let $b_0 < b_1 < \cdots < b_N$ be the order-sequence of $\mathcal{G}$. Then*

$$\dim Z^{(m)}(\phi_{\mathcal{G}}(C)) \leq {}^{\#}\Lambda_p^{(m)}(b_0, \ldots, b_N) - 1.$$

Moreover, this inequality is sharp.

Proof. Let $k := \dim Z^{(m)}(\phi_{\mathcal{G}}(C)) + 1$. Then there is a subset B consisting of k elements of $\{b_0, \ldots, b_m\}$ such that $\{b_0, \ldots, b_N\} \setminus B$ satisfies the p-adic criterion [E, Proposition 8]. Since $b_{m+1}, \ldots, b_N \in \{b_0, \ldots, b_N\} \setminus B$, we have

$$\binom{b_{m+1}}{b_i} \equiv \cdots \equiv \binom{b_N}{b_i} \equiv 0 \bmod p$$

for any $b_i \in B$ by the p-adic criterion. Hence $B \subset \Lambda_p^{(m)}(b_0, \ldots, b_N)$, and therefore $k \leq {}^{\#}\Lambda_p^{(m)}(b_0, \ldots, b_N)$. The last statement will be proved in the next subsection.

3.2 An extension of Ballico-Russo's result

Let us consider the curve

$$\begin{aligned} \varphi = \varphi^{<b_0 \ldots b_N>} : \mathbf{P}^1 &\to \mathbf{P}^N \\ t &\mapsto (t^{b_0}, \ldots, t^{b_N}) \end{aligned}$$

over an algebraically closed field of characteristic $p > 0$, and set

$$Z^{(m)} := Z^{(m)}(\varphi^{<b_0 \ldots b_N>}(\mathbf{P}^1)),$$
$$\Lambda^{(m)} := \Lambda_p^{(m)}(b_0, \ldots, b_N).$$

Proposition 1 *Under the above notation,*

$$Z^{(m)} = \{(x_0, \dots, x_N) \mid x_i = 0 \text{ for any } i \in \{0, 1, \dots, N\} \setminus \Lambda^{(m)}\}.$$

In particular,

$$\dim Z^{(m)} = {}^{\#}\Lambda^{(m)} - 1.$$

Remark. As was mentioned in Introduction, Ballico and Russo proved this for $Z^{(N-1)}(\varphi^{<0,1,\dots,N>}(\mathbf{P}^1))$.

To prove Proposition 1, we need the following lemma.

Lemma 3 *Let* $0 \le a_0 \le a_1 \le \dots \le a_n$ *be integers. Then*

$$\binom{a_n}{a_{n-1}}\binom{a_{n-1}}{a_{n-2}} \cdots \binom{a_1}{a_0} = \binom{a_n}{a_0}\binom{a_n - a_0}{a_{n-1} - a_0} \cdots \binom{a_2 - a_0}{a_1 - a_0}.$$

Proof. This is a consequence of the identity

$$\binom{\alpha}{\beta}\binom{\beta}{\gamma} = \binom{\alpha}{\gamma}\binom{\alpha - \gamma}{\beta - \gamma}$$

for $\alpha \ge \beta \ge \gamma \ge 0$. ||

Proof of Proposition 1. We use the following notation. Set

$$\begin{array}{llllllll}
\mathbf{a}_0(t) := & (1, & t^{b_1}, & t^{b_2}, & \dots, & t^{b_m}, & \dots, & t^{b_N} \quad) \\
\mathbf{a}_1(t) := & (0, & 1, & \binom{b_2}{b_1} t^{b_2 - b_1}, & \dots, & \binom{b_m}{b_1} t^{b_m - b_1}, & \dots, & \binom{b_N}{b_1} t^{b_N - b_1} \quad) \\
\vdots & & & \ddots & & & & \\
\vdots & & & & \ddots & & & \\
\mathbf{a}_m(t) := & (0, & 0, & 0, & \dots, & 1, & \dots, & \binom{b_N}{b_m} t^{b_N - b_m} \quad)
\end{array}$$

and

$$\begin{aligned}
\mathbf{e}_m(t) &:= \mathbf{a}_m(t) \\
\mathbf{e}_{m-1}(t) &:= \mathbf{a}_{m-1}(t) - \binom{b_m}{b_{m-1}} t^{b_m - b_{m-1}} \mathbf{e}_m(t) \\
&\vdots \\
\mathbf{e}_i(t) &:= \mathbf{a}_i(t) - \sum_{j=i+1}^{m} \binom{b_j}{b_i} t^{b_j - b_i} \mathbf{e}_j(t) \\
&\vdots \\
\mathbf{e}_0(t) &:= \mathbf{a}_0(t) - \sum_{j=1}^{m} \binom{b_j}{b_0} t^{b_j - b_0} \mathbf{e}_j(t).
\end{aligned}$$

Then

$$T_{\varphi(t)}^{(m)} = < \mathbf{a}_0(t), \dots, \mathbf{a}_m(t) > = < \mathbf{e}_0(t), \dots, \mathbf{e}_m(t) >$$

and

$$\mathbf{e}_i(t) = (\overbrace{0, \ldots, 0}^{i}, 1, \overbrace{0, \ldots, 0}^{m-i}, e_{i,m+1}t^{b_{m+1}-b_i}, \ldots, e_{i,N}t^{b_N-b_i}),$$

where the $e_{i,j}$'s are independent of t.

For a moment, we assume that

$$(*) \qquad e_{i,m+1} = \cdots = e_{i,N} = 0 \text{ for any } i \in \Lambda^{(m)}.$$

Then the point $(x_0, \ldots, x_m, 0, \ldots, 0)$ with $x_\alpha = 0$ if $\alpha \notin \Lambda^{(m)}$ can be represented as $\sum_{\beta \in \Lambda^{(m)}} x_\beta \mathbf{e}_\beta(t)$. Hence

$$\{(x_0, \ldots, x_N) \mid x_i = 0 \text{ for any } i \in \{0, 1, \ldots, N\} \setminus \Lambda^{(m)}\} \subset Z^{(m)}$$

by Lemma 2. Since $\dim Z^{(m)} \leq \#\Lambda^{(m)} - 1$ by Theorem 3, the assertion follows.

Now we prove $(*)$. It is easy to see that for integers i, k with $0 \leq i \leq m < k \leq N$,

$$\begin{aligned} e_{i,k} = \binom{b_k}{b_i} \; &- \sum_{i<j\leq m} \binom{b_k}{b_j}\binom{b_j}{b_i} \\ &+ \sum_{i<j_1<j_2\leq m} \binom{b_k}{b_{j_2}}\binom{b_{j_2}}{b_{j_1}}\binom{b_{j_1}}{b_i} - \cdots \\ &+ (-1)^s \sum_{i<j_1<\cdots<j_s\leq m} \binom{b_k}{b_{j_s}}\binom{b_{j_s}}{b_{j_{s-1}}}\cdots\binom{b_{j_1}}{b_i} + \cdots . \end{aligned}$$

Since $\binom{b_k}{b_i}$ divides $e_{i,k}$ by Lemma 3, the property $(*)$ is proved.

4 Number of Weierstrass points, again

We return to the question of estimating a lower bound for the number of Weierstrass points. Our aim in this section is to prove a characteristic-free version of Theorem 1.

We fix a base-point-free linear system $\mathcal{G}$ of degree d and dimension N on a smooth curve C of genus g. We set $Z^{(m)} = Z^{(m)}(\phi_{\mathcal{G}}(C))$.

Before stating Theorem 4, we remark on definition of $T_P^{(m)}$; for a point $P \in C$, $T_P^{(0)}$ means $\{\phi_{\mathcal{G}}(P)\}$ and $T_P^{(N)}$ means $\mathbf{P}^N$, by definition.

Theorem 4 *Let $\mathcal{G}$ be a base-point-free linear system with order-sequence $b_0 < b_1 < \cdots < b_N$ on a curve C of genus g, and let ν be the number of $\mathcal{G}$-Weierstrass points. Assume that $\mathcal{G}$ is tame and $\nu \geq 2$. If there are two distinct Weierstrass points $P_1, P_2 \in C$ and an integer k with $0 \leq k \leq N$ such that*

$$\left(T_{P_1}^{(N-k)} \cap T_{P_2}^{(k)}\right) \setminus Z^{(m-1)} \neq \emptyset$$

for some m with $1 \leq m \leq N$, then we have

$$(\nu - 2)\left(\frac{d}{b_m} - 1\right) \geq 2g.$$

Proof. Choose a point

$$R \in \left(T_{P_1}^{(N-k)} \cap T_{P_2}^{(k)}\right) \setminus Z^{(m-1)},$$

and consider the projection $\mathbf{P}^N \cdots \overset{\pi}{\to} \mathbf{P}^{N-1}$ with center R. Let $\mathcal{G}^{(R)}$ be the sublinear system of $\mathcal{G}$ corresponding to the rational map $\pi \circ \phi_{\mathcal{G}} : C \cdots \to \mathbf{P}^{N-1}$.

Let $\{P_1, P_2, \ldots, P_\nu\}$ be the set of $\mathcal{G}$-Weierstrass points, and let

$$\mu_0(P_j) < \cdots < \mu_N(P_j)$$

be the $\mathcal{G}$-Hermite invariant-sequence at P_j $(1 \le j \le \nu)$.

Then, since $\mathcal{G}^{(R)}$ is a sublinear system of $\mathcal{G}$, the $\mathcal{G}^{(R)}$-Hermite invariant-sequence at P_j consists of N elements of the $\mathcal{G}$-Hermite invariants at P_j, say

$$\mu_0(P_j) < \cdots \widehat{\mu_{\alpha_j}(P_j)} \cdots < \mu_N(P_j).$$

Note that since $R \in T_{P_1}^{(N-k)} \cap T_{P_2}^{(k)}$, the inequalities

$$\alpha_1 \le N-k \quad \text{and} \quad \alpha_2 \le k$$

hold [Kj2, Lemma 1]. Moreover, since $R \notin Z^{(m)}$, the $\mathcal{G}^{(R)}$-order-sequence coincides with

$$b_0 < \cdots < b_{m-1} < \cdots \widehat{b_\beta} \cdots < b_N$$

for some $\beta \ge m$ [Kj2, Proposition 2].

By the formula for the degree of the Wronskian divisor (cf. Section 2.1) and using the tameness of $\mathcal{G}$, we have

$$\begin{aligned}
\deg W(\mathcal{G}^{(R)}) &\ge \sum_{j=1}^{\nu} w_{\mathcal{G}^{(R)}}(P_j) \\
&= \deg W(\mathcal{G}) - \sum_{j=1}^{\nu} (\mu_{\alpha_j}(P_j) - b_\beta) && (1) \\
\deg W(\mathcal{G}^{(R)}) &= (b_0 + \cdots \widehat{b_\beta} \cdots + b_N)(2g-2) + Nd \\
&= \deg W(\mathcal{G}) - b_\beta(2g-2) - d. && (2)
\end{aligned}$$

From (1) and (2),

$$\sum_{j=1}^{\nu} (\mu_{\alpha_j}(P_j) - b_\beta) \ge b_\beta(2g-2) + d.$$

On the other hand, since $\alpha_1 \le N-k$ and $\alpha_2 \le k$ and $\mu_{\alpha_j}(P_j) \le d$, we have

$$\begin{aligned}
\sum_{j=1}^{\nu} (\mu_{\alpha_j}(P_j) - b_\beta) &\le \mu_{N-k}(P_1) + \mu_k(P_2) + (\nu-2)d - \nu b_\beta \\
&\le d + (\nu-2)d - \nu b_\beta \quad \text{(by Lemma 1)}.
\end{aligned}$$

Hence

$$(\nu-2)(d-b_\beta) \ge 2g b_\beta.$$

Since $b_m \le b_\beta$ and $\nu \ge 2$, we have

$$(\nu-2)(d-b_m) \ge 2g b_m.$$

This completes the proof.

Corollary 4 *Let g_d^N be a base-point-free linear system with order-sequence $b_0, \ldots, b_N$ on a curve of genus g, and let ν be the number of g_d^N-Weierstrass points. If g_d^N is tame and $Z^{(m)} = \emptyset$, then*

$$(\nu - 2)\left(\frac{d}{b_m} - 1\right) \geq 2g$$

except for linear systems corresponding to $\varphi^{<b_0 \ldots b_N>}$'s.

Proof. This follows from Theorem 4 and Corollary 1.

Remark. For a small m, the inequality in Corollary 4 is not better than that of Corollary 3. But at least for $m = N$, it is better.

5 Tangentially degenerate curves

Let C be a nondegenerate curve in $\mathbf{P}^N$ with $N \geq 3$.

Definition The curve C is *tangentially degenerate*, if $\left(T_P^{(1)} \setminus \{P\}\right) \cap C \neq \emptyset$ for a general point P in C.

As was mentioned in Introduction, it is still open whether or not there exists a tangentially degenarate curve in characteristic 0. On the other hand, in characteristic $p > 0$, some examples [Kj1, Example 4.1 4.2], [GV, Example 3], [Lv, Example after 1.4] of tangentially degenerate curves have been known. One should note that those curves are nonreflexive, i.e., $b_2 \equiv 0 \bmod p$. Here we give a new example of a smooth rational tangentially degenerate curve, which has the property that no orders of the linear system corresponding to the embedding are congruent to 0 modulo p.

Let us consider the curve

$$\begin{array}{cccc} \varphi : & \mathbf{P}^1 & \rightarrow & \mathbf{P}^3 \\ & t & \mapsto & (1 : t : t^2 - t^p : t^3 + 2t^p - 3t^{p+1}) \end{array}$$

over an algebraically closed field of characteristic $p > 3$, and let $\mathcal{G}_\varphi$ be the linear system on $\mathbf{P}^1$ corresponding to φ.

Proposition 2 *Under the above notation, we have*

(1) *φ is an embedding;*

(2) *The $\mathcal{G}_\varphi$-order-sequence is $\{0, 1, 2, 3\}$;*

(3) *$\varphi(\mathbf{P}^1)$ is tangentially degenerate.*

Furthermore, let $\check{\varphi} : \mathbf{P}^1 \rightarrow \check{\mathbf{P}}^3$ be the strict dual of the curve $\varphi : \mathbf{P}^1 \rightarrow \mathbf{P}^3$. Then there are an automorphism θ of $\mathbf{P}^1$ and an isomorphism Θ from $\mathbf{P}^3$ to $\check{\mathbf{P}}^3$ such that the diagram

$$\begin{array}{ccc} \mathbf{P}^1 & \xrightarrow{\varphi} & \mathbf{P}^3 \\ \theta \downarrow & & \downarrow \Theta \\ \mathbf{P}^1 & \xrightarrow{\check{\varphi}} & \check{\mathbf{P}}^3 \end{array}$$

is commutative.

Proof. (1): Easy.
(2): Obviously, the $\mathcal{G}_\varphi$-Hermite invariant-sequence at $0 \in \mathbf{P}^1$ coincides with $0, 1, 2, 3$. Since $j \le b_j \le \mu_j(0) = j$, the $\mathcal{G}_\varphi$- order-sequence is $0, 1, 2, 3$.
(3): Let $a \in \mathbf{P}^1 \setminus \{\infty\}$. Then the tangent line $T_{\varphi(a)}$ to $\varphi(\mathbf{P}^1)$ at $\varphi(a)$ is spanned by the vectors

$$\begin{aligned} \varphi(a) &= (1, a, a^2 - a^p, a^3 + 2a^p - 3a^{p+1}) \\ D_t^{(1)}\varphi(a) &= (0, 1, \quad 2a, \quad 3a^2 - 3a^p). \end{aligned}$$

Since $\varphi(a+1) = \varphi(a) + D_t^{(1)}\varphi(a)$ by direct computation, $\varphi(\mathbf{P}^1)$ is tangentially degenerate.
The strict dual $\check{\varphi} : \mathbf{P}^1 \to \check{\mathbf{P}}^3$ is given by the 3×3 minors of the matrix

$$\begin{pmatrix} 1 & t & t^2 - t^p & t^3 + 2t^p - 3t^{p+1} \\ 0 & 1 & 2t & 3t^2 - 3t^p \\ 0 & 0 & 1 & 3t \end{pmatrix},$$

that is,

$$\check{\varphi}(t) = (t^3 + 2t^p + 3t^{p+1}, -3t^2 - 3t^p, 3t, -1).$$

Choose the automorphism θ of $\mathbf{P}^1$ as

$$\theta(t) := -t\ ,$$

and the isomorphism $\Theta : \mathbf{P}^3 \to \check{\mathbf{P}}^3$ as

$$\Theta(X_0, \ldots, X_3) := (X_0, \ldots, X_3) \begin{pmatrix} & & & 1 \\ & & \frac{1}{3} & \\ & \frac{1}{3} & & \\ 1 & & & \end{pmatrix}.$$

Then we have the commutative diagram. ||

Remark. It is easy to see that the only $\mathcal{G}_\varphi$-Weierstrass point is $\infty \in \mathbf{P}^1$. Since the Hermite invariant-sequence at ∞ is $0, 1, p, p+1$, the weight at ∞ is $2(p-2)$. On the otherhand, since $\deg W(\mathcal{G}_\varphi) = 4(p-2)$, the linear system $\mathcal{G}_\varphi$ is not tame.

References

[BR] E. Ballico and B. Russo, *On the general osculating flag to a projective curve in characteristic p,* Comm. Algebra **20**(1992), 3729–3740

[C] C. Ciliberto, *Review of Kaji's paper* [Kj1], Math. Review 1987 i:14027

[E] E. Esteves, *A geometric proof of an inequality of order sequences,* Comm. Algebra **21**(1993), 231–238

[G] A. Garcia, *Some arithmetic properties of order-sequences of algebraic curves,* J. Pure Appl. Algebra **85**(1993), 259–269

[GLx] A. Garcia and R. F. Lax, *Weierstrass weight of Gorenstein singularities with one or two branches,* Preprint 1992

[GV] A. Garcia and J. F. Voloch, *Duality for projective curves,* Bol. Soc. Bras. Mat. **21**(1991), 159–175

[HsSc] H. Hasse und F. K. Schmidt, *Noch eine Begründung der Theorie der höheren Differentialquotienten in einen algebraischen Funktionenkörper einer Unbestimmten,* J. Reine Angew Math. **177**(1937), 215–237

[HfKk] A. Hefez and N. Kakuta, *On the geometry of non-classical curves,* Bol. Soc. Bras. Mat. **23**(1992), 79–91

[HfV] A. Hefez and J. F. Voloch, *Frobenius non-classical curves,* Arch. Math. **54**(1990), 263–273; correction **57**(1991), p. 416

[Hm1] M. Homma, *Reflexivity of tangent varieties associated with a curve,* Ann. Mat. Pura Appl. **156**(1990), 195–210

[Hm2] M. Homma, *Duality of space curves and their tangent surfaces in characteristic* $p > 0$, Ark. Mat. **29**(1991), 221–235

[Hm3] M. Homma, *Linear systems on curves with no Weierstrass points,* Bol. Soc. Bras. Mat. **23**(1992), 93–108

[Hm4] M. Homma, *On Esteves' inequality of order sequences of curves,* Comm. Algebra **21**(1993), 3685–3689

[HmKj] M. Homma and H. Kaji, *On the inseparable degree of the Gauss map of higher order for space curves,* Proc. Japan Acad. **68**, Ser. A (1992), 11–14

[Kj1] H.Kaji, *On the tangentially degenerate curves,* J. London Math. Soc. (2) **33**(1986), 430–440

[Kj2] H. Kaji, *Strangeness of higher order for space curves,* Comm. Algebra **20**(1992), 1535–1548

[Kj3] H. Kaji, *On the inseparable degree of the Gauss map and the projection of the conormal variety to the dual of higher order for space curves,* Math. Ann. **292**(1992), 529–532

[Km] K. Komiya, *Algebraic curves with non-classical type of gap sequences for genus three and four,* Hiroshima Math. J. **8**(1978), 371–400

[Lk1] D. Laksov, *Weierstrass points on curves,* Astérisque **87-88**(1981), 221–247

[Lk2] D. Laksov, *Wronskians and Plücker formulas for linear systems on curves,* Ann. Sci. Ecole Norm. Sup. **17**(1984), 45–66

[LkT1] D. Laksov and A. Thorup, *The Brill-Segre formula for families of curves,* in "Enumerative Algebraic Geometry" (S. L. Kleiman and A. Thorup eds.), Contemporary Math. **123**, AMS, 1991, 131–148

[LkT2] D. Laksov and A. Thorup, *Weierstrass points and gap sequences for families of curves,* Preprint 1992

[Lv] D. Levcovitz, *Bounds for the number of fixed points of automorphisms of curves,* Proc. London Math. Soc. **62**(1991), 133–150

[M] B. H. Matzat, Ein Vortrag über Weierstrass punkte, Karlsruhe, 1975

[Sc1] F. K. Schmidt, *Die Wronskische Determinante in beliebigen differenzierbaren Funktionenkörper,* Math. Z. **45**(1939), 62–74

[Sc2] F. K. Schmidt, *Zur arithmetischen Theorie der algebraischen Funktionen* II, Math. Z. **45**(1939), 75–96

[StV] K. O. Stöhr and J. F. Voloch, *Weierstrass points and curves over finite fields,* Proc. London Math. Soc. **52**(1986), 1–19

Classification of Smooth Congruences with a Fundamental Curve

E. ARRONDO Departamento de Algebra, Facultad de Ciencias Matemáticas, Universidad Complutense de Madrid, 28040 Madrid, Spain

M. BERTOLINI Dipartimento di Matematica, Università degli Studi di Milano, via C. Saldini, 50, 20133 Milano, Italy

C. TURRINI Dipartimento di Matematica, Università degli Studi di Milano, via C. Saldini, 50, 20133 Milano, Italy

Abstract: We give a classification and a construction of all smooth $(n-1)$-dimensional varieties of lines in $\mathbf{P}^n$ verifying that all their lines meet a curve. This also gives a complete classification of $(n-1)$-scrolls over a curve contained in $G(1,n)$.

INTRODUCTION

The geometry of the line, i.e. the study of the line as the main element, was a very popular subject at the end of last century and beginning of this. A particular attention was given to varieties of lines, also called *congruences*. In other words, a congruence is a subvariety of a Grassmannian $Gr(1,\mathbf{P}^n)$ of lines in $\mathbf{P}^n$. When we will refer to a congruence, we will restrict ourselves to the case of subvarieties of dimension $n-1$ (hence equal to the codimension).

A lot of papers were published at that time by several mathematicians about congruences of lines in $\mathbf{P}^3$. Of important relevance is the work of G. Fano, who made several classifications of these congruences under different assumptions. One fixed assumption in his classification was the non-existence of what was called a *fundamental curve*. A fundamental curve is a curve in $\mathbf{P}^3$ such that all lines of the congruence meet it. The reason for excluding this possibility does not seem to be clear. Recently, after congruences became popular again, M. Gross and the first author classified all those congruences excluded by Fano, i.e they gave a classification of all smooth congruences of lines in $\mathbf{P}^3$ with a fundamental curve.

Not only the congruences of lines in $\mathbf{P}^3$, but also in $\mathbf{P}^4$ (i.e. threefolds in $Gr(1,\mathbf{P}^4)$) came back to the attention of nowadays mathematicians. When the last two authors started the classification of smooth congruences of lines in $\mathbf{P}^4$ with a curve meeting all lines, they found that the same key numerical relation as for $\mathbf{P}^3$ holds (see lemma 4). This was the starting point of this joint work dealing with the general case of smooth congruences of lines in any $\mathbf{P}^n$ with all lines meeting a curve (in the sequel such a curve will be called fundamental, see section 0 for the definition).

In section 1, we deal with the case in which the curve is a line. This is a particular case in the sense that there are infinitely many families of these congruences, while for other curves we find, for each n, a finite number of families. On the other hand, this is a generalization of the result in case $n = 3$, which is known under another formulation. Indeed, the Grassmannian $Gr(1,\mathbf{P}^3)$ can be considered, under the Plücker embedding, as a smooth quadric in $\mathbf{P}^5$. Since the variety of the lines meeting a given line is a singular hyperplane section, congruences with a fundamental line can be viewed as surfaces in $\mathbf{P}^4$ contained in a singular quadric. It is then a classical result that goes back to Roth (see [R] §3) that such a smooth surface is either the complete intersection of the quadric and another hypersurface (if it misses the singular point of the quadric) or the rest of a plane under such a complete intersection. We get an analogous result for any n.

In section 2, we give the list of all possible smooth congruences. For the proof we needed to refine some arguments of [A-G]. The base idea is to lift the congruence as a hypersurface in a desingularization of the variety of lines meeting a curve. Then one obtains a numerical relation from the double-point formula (coming from the fact that we have a smooth subvariety of $Gr(1,\mathbf{P}^n)$ with same dimension as codimension), which together with Castelnuovo bound for the genus in two different contexts allows us to conclude.

In section 3 we give an explicit construction for all possible congruences we found in section 2, which completes the classification. We point out that all examples for $n = 3$ (except the congruences of bisecants) extend to any dimension. Moreover an easy argument shows that by the above constructions and [A] we also obtain the complete classification of $(n-1)$-dimensional scrolls over a curve, contained in $G(1,n)$.

The results of this paper are just one way to generalize the classical results on congruences with a fundamental curve. Another natural and interesting generalization, which some of the authors are now studying, are the congruences of lines in $\mathbf{P}^n$ with higher dimensional fundamental locus, e.g. a surface with infinitely many lines through each point of its.

The first author wants to thank partial support from CICYT grant No PB90-0637. He especially thanks also a Del Amo grant from Universidad Complutense, which allowed him to stay at MSRI for the year 1992/93. The second author also wants to thank G.N.S.A.G.A. of the C.N.R., which allowed her visiting MSRI. They found there a wonderful work atmosphere where they could start collaborating and also completing the paper. We would also like to thank Mark Gross for several useful conversations and ideas.

0 PRELIMINARIES AND NOTATIONS

As said in the introduction, by a *congruence* we will mean an $(n-1)$-dimensional variety (over the complex numbers) in the Grassmannian variety $G(1, \mathbf{P}^n)$ of lines in $\mathbf{P}^n$. We will denote usually by G to this Grassmannian.

We will write

$\Omega(A, B) =$ Schubert variety of lines in G meeting A and contained in B, where A and B are linear subespaces of $\mathbf{P}^n$ such that $A \subset B$

$\Omega(i, j) =$ class, in the Chow ring of G, of $\Omega(A, B)$, where $\dim A = i$ and $\dim B = j$

We will say that a point P of $\mathbf{P}^n$ is a *k-fundamental point* of a congruence Y if there is a k-dimensional family of lines of Y through P. A curve C of $\mathbf{P}^n$ is a *k-fundamental curve* for Y if all its points are k-fundamental points. In other words an $(n-2)$-fundamental curve is a curve which is met by all the lines of the congruence; in the sequel we will briefly call such a curve a *fundamental curve.*

Given a congruence Y, we will denote

$a =$ number of lines of Y passing through a general point of $\mathbf{P}^n$

$b =$ number of lines of Y contained in a general hyperplane H and meeting a general line of H

$d =$ degree of the curve C

$g =$ geometric genus of C

$e =$ degree of the cone formed by all lines of Y passing through a general point of C

We will not consider the case when Y consists of all lines passing through one point, so that we will assume $e \geq 1$ for any fundamental curve. Note that, although the $(n-1)$-dimensional cycles of G are generated by $\{\Omega(i, n-i) \mid i = 0, \ldots, [\frac{n-1}{2}]\}$, in the case of congruences with fundamental curve the only non-zero intersections with the class of Y are $a = [Y] \cdot \Omega(0, n)$ and $b = [Y] \cdot \Omega(1, n-1)$.

1 SMOOTH CONGRUENCES WITH A FUNDAMENTAL LINE

For this section, we will view G as embedded in a projective space under the Plücker embedding. In this framework, a hypersurface of degree l will mean the intersection of G with such a hypersurface in the Plücker embedding.

THEOREM 1. *Let $Y \subseteq G(1,n)$ be a smooth congruence such that all its lines meet a given line $\Lambda \subseteq \mathbf{P}^n$. Denote by Γ the cone $\Omega(\Lambda, \mathbf{P}^n)$. Then either*

j) $b = a$ and Y is the intersection of Γ with a hypersurface of degree a or

jj) $b = a-1$ and Y is linked to an $(n-1)$-fold of degree $n-2$ passing through the vertex of Γ, in the intersection of Γ with a hypersurface of degree $a+1$ or

jjj) The dimension is $n = 3$, $a = b-1$ and Y is linked under the complete intersection of Γ and a hypersurface of degree b to a plane consisting of all lines passing through a point.

Proof: Since the case $n = 3$ is already known, we will assume $n \geq 4$. A modern proof for the case $n = 3$ can be found in [G1], from where we took the idea for this general proof.

It is easy to see that the Schubert variety $\Gamma = \Omega(\Lambda, \mathbf{P}^n)$ is a cone over the $(n-1)$-fold $B = \mathbf{P}^1 \times \mathbf{P}^{n-2}$ with vertex the point $\lambda \in G(1,n)$ corresponding to Λ. Consider the blow up Γ^* of Γ at λ; then $\Gamma^* \simeq \mathbf{P}\,(O_B \oplus O_B(-1))$ and the exceptional divisor E is isomorphic to B. Put $\mathcal{E} = O_B \oplus O_B(-1)$ and $p : \mathcal{E} \to B$; Then, with standard notations, $Pic(B)$ ($Pic(E)$ resp.) is generated by $\mathbf{P}^1_1 \times \mathbf{P}^{n-3}_2$ and $\mathbf{P}^0_1 \times \mathbf{P}^{n-2}_2$ ($\mathbf{P}'^1_1 \times \mathbf{P}'^{n-3}_2$ and $\mathbf{P}'^0_1 \times \mathbf{P}'^{n-2}_2$ resp.), so that $Pic(\Gamma^*)$ is generated by E, $Q = p*(\mathbf{P}^1_1 \times \mathbf{P}^{n-3}_2)$ and $Z = \mathbf{P}^0_1 \times \mathbf{P}^{n-2}_2$. Now, for a $(n-1)$-fold $Y^* \subseteq \Gamma^*$ to be mapped to a non singular $(n-1)$-fold $Y \subseteq \Gamma$ there are two possibilities:

i) Y^* does not intersect E, and then the blow up map is an isomorphism between Y^* and Y;

or

ii) Y^* intersects E and then $W = Y^* \cap E$ is contracted to the vertex of Γ which must be a smooth point of Y. So W must be a divisor in Y^* isomorphic to $\mathbf{P}^{n-2}$.

Now suppose that, in $Pic(\Gamma^*)$, it is $Y^* = \alpha E + \beta Q + \gamma Z$, so that the numerical equivalency class of W in E is

$$W = (\beta - \alpha)(\mathbf{P}'^{1}_{1} \times \mathbf{P}'^{n-3}_{2}) + (\gamma - \alpha)(\mathbf{P}'^{0}_{1} \times \mathbf{P}'^{n-2}_{2}).$$

In case j), one has $\alpha = \beta = \gamma$, so that Y is the complete intersection of Γ with an hypersurface of degree $\alpha = a$ (case i)). In case jj), being $n - 2 > 1$, the only possibility for a divisor W to be a $\mathbf{P}^{n-2}$ embedded into $E = \mathbf{P}^1 \times \mathbf{P}^{n-2}$ is that $W = \mathbf{P}^0_1 \times \mathbf{P}^{n-2}_2$, so that $\beta = \alpha$ and $\gamma = \alpha + 1$. Hence $Y^* = \alpha E + \alpha Q + (\alpha + 1)Z$ and Y is linked, in the complete intersection of Γ with an hypersurface of degree $\alpha + 1$, to the image of $E + Q$ in Γ, i.e. to an $(n-1)$-fold of degree $n - 2$ passing through the vertex of Γ (case ii)).

2 SMOOTH CONGRUENCES WITH A FUNDAMENTAL CURVE

In this section we will prove the following

THEOREM 2. *Let Y be a smooth congruence having an integral curve C as a fundamental curve. Then one of the following holds:*

(i) $n = 3$ *and the congruence consists of the bisecants to either a twisted cubic or an elliptic quartic in* $\mathbf{P}^3$.

(ii) The curve C is a line (and Y is described in Theorem 1).

(iii) The congruence is a scroll (i.e. $e = 1$) and either C is a conic and $a = 1$ or 2, $b = 2$, or C is a plane cubic and $a = b = 3$

(iv) The curve C is a plane cubic, $e = 2$, $a = 3$ and $b = 6$

We will follow several steps in the proof, that we state as lemmas.

LEMMA 3. *The fundamental curve C is smooth.*

Proof: Consider the subscheme V of G corresponding to lines in $\mathbf{P}^n$ meeting C. It is singular, and its singular locus is the locus of bisecants to C. This locus has several components: one for each singular point of C (consisting of all lines passing through that point) and another one of the expected dimension two given by the closure of all bisecant lines at smooth points of C. Hence Y is not contained in the singular locus of V unless $\dim Y = 2$ and Y is the congruence of bisecants to C. This case has already be studied in [A-G], and corresponds to (i) in the theorem.

Let $f : \tilde{C} \to C$ be the normalization of C, and $L = f^*(\mathcal{O}_C(1))$. The scheme V has a natural desingularization given by $X = \mathbf{P}(f^*(\Omega_{\mathbf{P}^n}{}_{|C}(2))$. Call $p : X \to \tilde{C}$ the natural projection. Since by definition of fundamental curve Y is contained in V, it has a lift $\tilde{Y}$ to X. If we are not in case (i), then the map $\pi : \tilde{Y} \to Y$ is birational. If it is not an isomorphism, it must contract a curve E of $\tilde{Y}$ (just applying the Zariski main theorem, because Y is smooth). The curve E meets all fibers of $p_{|\tilde{Y}}$, since none of these fibers is contracted by π. This easily implies that the line represented by the image of E under π is C, which is case (ii) in the theorem.

Hence, assuming that C is not a line we have that π is in fact an isomorphism between $\tilde{Y}$ and Y. This implies that C is smooth. Indeed, if two different points p_1 and p_2 of $\tilde{C}$ go to the same point of C, then the images of the cones at p_1 and p_2 must have some common line. This contradicts the fact that π is an isomorphism. The same argument holds when p_1 and p_2 are infinitely near points. A precise proof can be found in [G2]. The idea is that the tangent space at a point of $\tilde{Y}$ (given by a point p of $\tilde{C}$ and a line l through it) splits as the sum of the tangent space of $\tilde{C}$ at p and the tangent space at l of the cone with vertex p. Since this maps isomorphically to the tangent space of Y, hence the tangent space of $\tilde{C}$ at p also maps isomorphically to the tangent space of C at $f(p)$.

LEMMA 4. *Except for cases (i) and (ii), there is a numerical relation*

$$2d^2e^2 - 4de^2 - 2e^2g - 2de - 2eg + 2e^2 + 2e + D(1 + 2e - 2de) + D^2 = 0$$

(This is the same as (*) in [G2] p. 137).

Proof: The map from X to G is given as follows.

First, look at the commutative diagram of exact sequences defining F as a push-out (we will write $\mathcal{O}_X(a, b)$ for $p^*(\mathcal{O}_C(a)) \otimes \mathcal{O}_X(b)$):

$$\begin{array}{ccccccccc}
 & & 0 & & 0 & & & & \\
 & & \downarrow & & \downarrow & & & & \\
 & & \Omega_{X/C}(-1,1) & = & \Omega_{X/C}(-1,1) & & & & \\
 & & \downarrow & & \downarrow & & & & \\
0 & \longrightarrow & p * \Omega_{\mathbf{P}^n|C}(1,0) & \longrightarrow & H^0(\mathcal{O}_{\mathbf{P}^n}(1)) \otimes \mathcal{O}_X & \longrightarrow & \mathcal{O}_X(1,0) & \longrightarrow & 0 \\
 & & \downarrow & & \downarrow{\scriptstyle \alpha} & & \| & & \\
0 & \longrightarrow & \mathcal{O}_X(-1,1) & \longrightarrow & F & \longrightarrow & \mathcal{O}_X(1,0) & \rightarrow & 0 \\
 & & \downarrow & & \downarrow & & & & \\
 & & 0 & & 0 & & & &
\end{array} \qquad (1)$$

The surjection α gives the map from X to G; in particular α is the pullback on X of the canonical map from $H^0(\mathcal{O}_{\mathbf{P}^n}(1)) \otimes \mathcal{O}_G$ to $\mathcal{Q}$, where $\mathcal{Q}$ is the universal bundle on G. The Chow ring of X is generated by $Pic(C)$ and $t = c_1(\mathcal{O}_X(1))$, which canonically identifies with the pullback of the hyperplane section of G. The relation among them is given by

$t^n = (n-1)Lt^{n-1}$, and the class of a point is Lt^{n-1}. Hence, the class of Y in X can be written as $et - D$, where D is, by abusing notation, the pullback of a certain divisor D of C (also, when no confusion arises, we will use D for denoting the degree of the divisor). Let us compute from these data the double-point formula for Y. Writing $N = N_{Y/G}$, we have that $a^2 + b^2 = \deg(c_{n-1}(N))$. Let us compute each of the members in the formula (most relations will come from the exact sequences in diagram (1)).

The pullback to X of the set of lines passing through a point of $\mathbf{P}^3$ is the locus where n sections of the bundle F are dependent. We can apply then Porteous formula (see, for example [F] Thm. 14.4, from where we also keep the notation for the Schur polynomial) and get that

$$a = \Delta_1^{(n-1)}(F_{|Y}) \cap [Y] = (t^{n-1} - (n-2)Lt^{n-2})(et - D) = ed - D$$

(where we used the notations in [F], and the fact that in our case $\Delta_1^i(F) = t^i - (i-1)Lt^{i-1}$, as can be easely checked by induction on i)

The second degree b is easily computed by geometrical means. It is the number of lines of Y contained in a general hyperplane H of $\mathbf{P}^{n-1}$ and meeting a general line r of H. The intersection of H with C consists of d points, and through each of them, lines of Y form a cone of degree e, thus meeting L in e points. Hence $b = ed$.

Let us now compute the total Chern polynomial of N, $c(N) = c(T_{G|Y})c(T_Y)^{-1}$. We use the fact that $T_G = F \otimes T_{X/C}(1,-1)$ and the fact that F appears as an extension in (1), as well as the fact that $N_{Y/X} = \mathcal{O}_X(et - D)$ to conclude that

$$c(N) = c(p^*(T_{\mathbf{P}^n|C}))c(\mathcal{O}_X(2,-1))^{-1}c(\mathcal{O}_X(et-D))c(p^*T_C)^{-1}.$$

Now, it is a straightforward calculation to show that

$$c_{n-1}(N) = (e+1)t^{n-1} + (5e - ne - n + 3)Lt^{n-2} + (e+1)Kt^{n-2} - Dt^{n-2}$$

(where K represents the canonical divisor of C). Now, its degree comes from multiplying in the Chow ring of X the above class with the class $et - D$ of Y, to obtain

$$\deg(c_{n-1}(N)) = (4e^2 + 2e)d + e(e+1)(2g-2) - (2e+1)D$$

Identifying this with $a^2 + b^2$ and substituting the values of a and b we finally get

$$2d^2e^2 - 4de^2 - 2e^2g - 2de - 2eg + 2e^2 + 2e + D(1 + 2e - 2de) + D^2 = 0$$

LEMMA 5. *The curve C is plane and hence the relation in lemma 4 becomes*

$$de(d-1)(e-1) = D(2de - D - 2e - 1) \tag{2}$$

Proof: It is the same as in [G2] by getting a contradiction between the relation in lemma 4 and Castelnuovo bound for the genus of non-plane curves.

LEMMA 6. *In the same hypothesis of lemmas 4 and 5, $b \leq 2a$ and hence $e \leq 3$.*

Proof: This is just as in [A-G], by applying Hurwitz theorem to the map from the curve of lines of Y in P (the plane containing C) to C. The second inequality is a consequence of the first together with (2).

We now complete the proof of the theorem by analyizing separately each of the possible values for e.

1) If $e = 1$, then the congruence Y is a scroll of $\mathbf{P}^{n-2}$'s over the curve C. More precisely, through each point Q of C, the lines of the congruence Y are those contained in a given hyperplane H_Q of $\mathbf{P}^n$ containing Q. This distribution of hyperplanes is given by the epimorphism in the exact sequence (defining the vector bundle E):

$$0 \to \mathcal{O}_C(D) \to \Omega_{\mathbf{P}^n}(2) \otimes \mathcal{O}_C \to E \to 0 \tag{3}$$

(recall that Y is the zero locus of $p^*(\mathcal{O}_C(D)) \to p^*(\Omega_{\mathbf{P}^n}(2) \otimes \mathcal{O}_C) \to \mathcal{O}_X(0,1)$). Since the curve C is plane, we can build the following commutative diagram (defining the sheaf $\mathcal{F}$):

$$\begin{array}{ccccccccc}
 & & & & 0 & & 0 & & \\
 & & & & \downarrow & & \downarrow & & \\
 & & & & \mathcal{O}_C(L)^{\oplus n-2} & = & \mathcal{O}_C(L)^{\oplus n-2} & & \\
 & & & & \downarrow & & \downarrow & & \\
0 & \to & \mathcal{O}_C(D) & \to & \Omega_{\mathbf{P}^n}(2) \otimes \mathcal{O}_C & \to & E & \to & 0 \\
 & & \| & & \downarrow & & \downarrow & & \\
0 & \to & \mathcal{O}_C(D) & \to & \Omega_{\mathbf{P}^2}(2) \otimes \mathcal{O}_C & \to & \mathcal{F} & \to & 0 \\
 & & & & \downarrow & & \downarrow & & \\
 & & & & 0 & & 0 & &
\end{array} \tag{4}$$

The injectivity in the right column in (4) comes because otherwise the sheaf $\mathcal{F}$ would have rank two, just implying that all lines in the plane P are in the congruence.

Putting $e = 1$ in expression (2), we get that either $\deg D = 0$ (and hence $a = b = d$) or $a + b = 3$. In this second case, from lemma 6 the only possiblity is $a = 1$, $b = d = 2$. So we assume that $\deg D = 0$.

Assume the sheaf $\mathcal{F}$ appearing in (4) has torsion at some point p of C. This would imply that for each other q in C, the line joining p and q is in the congruence. For $d > 2$ this contradicts the fact proved in lemma 3 that π is an isomorphism. Hence, $\mathcal{F}$ is an invertible sheaf in this case, so it must be $\mathcal{F} = \mathcal{O}_C(L - D)$, by looking at the first Chern class. This gives the immersion of C into $\check{\mathbf{P}}^2$ defined by assigning to each point in C the only line through it contained in P. For $d > 3$ there is only one g_d^2 on C (see [A-C-G-H] page 56), so that $L = L - D$ and thus $D = 0$. This would imply that all lines in P pass through a fix point of C, which is absurd. Hence, $d \leq 3$.

2) If $e = 2$, it happens as in [A-G] that the curve (of degree a) W in the projective space of hyperquadrics in $\mathbf{P}^n$ must be plane (just using Castelnuovo bound together with lemma 5). This plane cannot be contained in the locus K of singular quadrics whose vertex is at a given plane P (same proof as in [A-G], using that the embedded tangent space of K at a cone contains the space of quadrics passing through the vertex of the cone). This locus K is defined by the maximal minors of a $3 \times (n+1)$ matrix of linear forms. Hence, $a \leq 3$, and the only numerical solutions for (a, b) are $(2, 6)$ and $(3, 6)$. The first one is ruled out by lemma 6.

3) If $e = 3$, the curve W in the space of hypercubics is contained in a linear space A of dimension three. This has to meet K (space of cones with vertex at the plane P) along a curve. Indeed, the embedded tangent space to K at a cubic cone is contained in the space of cubic hypersurfaces that are singular at the vertex of the cone. Assume that the intersection of K and A contains a surface S. Take a point p of C and let p' the point of W representing the cone with vertex p. By assuption, there is a plane in A through p' (the tangent plane to S at p') consisting of singular cubics at p. If $a > 3$, this plane must meet W in at least another point q' (for general p). Hence the cone at the corresponding point q of C is singular at p. When we vary the point p, if the point q also varies, this means that all cones are singular, which is absurd (the general fiber of the map $\tilde{Y} \to C$ is smooth). This means that there is a point q in C whose cone through it is singular along all points of C. In particular the pencil of lines through q and contained in P should be in the congruence, which is also absurd.

Now K is defined by the maximal minors of a $3 \times \binom{n+2}{2}$ matrix of linear forms (this matrix corresponds to the derivatives of forms of degree 3 with respect to the directions given by the plane P). Hence, W is contained in the locus of a $\mathbf{P}^3$ given by the maximal minors of a 3×4 matrix of linear forms, hence $a \leq 6$. The only numerical solutions of (2) are then $(a, b) = (3, 12), (4, 12)$, which are again impossible by lemma 6.

3. EXPLICIT CONSTRUCTIONS

LEMMA 7. *Let $C \subseteq \mathbf{P}^n$ be a non singular plane curve of degree d. Then $\Omega_{\mathbf{P}^n}(2)_{|C} \cong (\oplus^{n-2}\mathcal{O}_C(L)) \oplus \mathcal{S}(P)$ where L is an hyperplane section of C, E is a divisor on C of degree*

1 and $\mathcal{S}$ is a normalized rank 2 vector bundle of degree $d-2$ on C with $\wedge^2\mathcal{S} \cong \mathcal{O}_C(L-2P)$.

Proof: First of all, notice that

$$\Omega_{\mathbf{P}^n|\mathbf{P}^2} \cong (\oplus^{n-2}\mathcal{O}_{\mathbf{P}^2}(-1)) \oplus \Omega_{\mathbf{P}^2}.$$

Twisting by $\mathcal{O}(2)$ and restricting to C, we get

$$\Omega_{\mathbf{P}^n}(2)_{|C} \cong (\oplus^{n-2}\mathcal{O}_C(L)) \oplus \Omega_{\mathbf{P}^2}(2)_{|C}.$$

If Λ denotes the plane containing C and

$$f : X = \Omega_{\mathbf{P}^n}(2)_{|C} \to G(1,n)$$

is the projection, denote by $Y_0 = f^{-1}(\Omega(1,\Lambda))$ the ruled surface of lines of X contained in Λ. Let $\mathcal{S}$ be the normalized bundle defined by $Y_0 = \mathbf{P}(\mathcal{S})$. By the above, Y_0 can be identified with $\mathbf{P}(\Omega_\Lambda(2)_{|C}) \subseteq \mathbf{P}(\Omega_{\mathbf{P}^n}(2)_{|C})$, which implies (3). Now, with the same arguments as in [A-G], one shows that $deg\mathcal{S} = d-2$, and that the numerical equivalence class of the divisor inducing the map $Y_0 \to \Omega(1,\Lambda)$ is $C_0 + f$. By the way notice also that the class of Y_0 in X is the $(n-2)-th$ Chern class of $(\oplus^{n-2}\mathcal{O}_X(t-p^*L))$.

3.1 Construction of case (3,3)

Here $\Omega_{\mathbf{P}^n}(2)_{|C} \cong (\oplus^{n-2}\mathcal{O}_C(L)) \oplus \mathcal{F}(P)$ as in Lemma 7, with $degL = 3$, $degP = 1$, $deg\mathcal{F} = 1$ and $\wedge^2\mathcal{F} \cong \mathcal{O}_C(L-2P)$, so that Y_0 is the elliptic ruled surface of invariant -1

Notice that, C being a plane curve, the image of X in the grassmannian $G(1,n) \in \mathbf{P}^{\frac{n(n+1)}{2}-1}$ is in fact contained in a $\mathbf{P}^{3n-4}$, so that, since $dim|t| = 3n-4$, the embedding of X into $G(1,n)$ is given by the whole linear system $|t|$.

As shown in Lemma 4, the class of Y in $X = \mathbf{P}(\Omega_{\mathbf{P}^n}(2)_{|C})$, can be written as $t - p^*D$, D being a degree 0 divisor on C.

As D is a divisor of degree 0, then $\Omega_{\mathbf{P}^n}(2)_{|C}(-D)$ is generated by global sections, so $|t-p^*D|$ is base point free and a general member is smooth.

Recall from §2 that we have an exact sequence given by the right column of (4) and that we showed that $\mathcal{F} = \mathcal{O}_C(L-D)$ and $D \neq 0$. This implies that $Y = \mathbf{P}(E) = \mathbf{P}(\mathcal{O}_C(L-D) \oplus \mathcal{O}_C(L)^{n-2})$. Since it is mapped to the Plücker embedding of G by the complete very ample linear series $|\ t\ |$, then it is an embedding in G.

For the geometric construction, we observe first that the hyperplanes formed at each point must all contain a fixed linear space Λ of codimension three, which is disjoint with the plane P containing C. Indeed, this comes from the splitting of E. Hence these hyperplanes are defined as the span of Λ and some a line in P. The way of constructing these lines is given in [Go] in the following way. Take a group structure on C such that the origin is an inflection point. Now take a point σ such that $3\sigma = L - D$ and fix a point R outside C. For each point p in C, let A and B the points of C in the line defined by R and $-p+2\sigma$. Then, the line associated to p is the one defined by $-A-\sigma$ and $-B-\sigma$. We do not know how to prove this directly, but just as in [G1] checking that this construction corresponds to the above description and that, for fixed D, the dimension is the same (dimension two for the choice R and also for a section of $\Omega(2)_{|C}(-D)$).

3.2 Construction of case $(2,2)$

Let C be a conic, contained in a plane L in $\mathbf{P}^n$. By Lemma 7, the sheaf $\Omega_{\mathbf{P}^n}(2)_{|C}$ is isomorphic to $\oplus^{n-2}\mathcal{O}_{\mathbf{P}^1}(2) \oplus (\oplus^2\mathcal{O}_{\mathbf{P}^1}(1))$.

As $X = \mathbf{P}(\Omega_{\mathbf{P}^n}(2)_{|C})$ we can write $X = \mathbf{P}(\oplus^{n-2}\mathcal{O}_{\mathbf{P}^1}(2)\oplus(\oplus^2\mathcal{O}_{\mathbf{P}^1}(1))$. Let us consider now the linear system $|t|$. As $h^0(\mathcal{O}_X(t)) = 3n-2$ and $G(1,n) \subset \mathbf{P}^{\frac{n(n+1)}{2}-2}$, the map $f: X \to G(1,n)$ is induced by a $(3n-3)$-dimensional subspace V of $H^0(\mathcal{O}_X(t)$. More precisely, $V = H^0(\mathcal{O}_{\mathbf{P}^1}(2)^{n-2}) \oplus V'$, where V' is a three-dimensional subspace of $H^0(\mathcal{O}_{\mathbf{P}^1}(1)^2)$.

Claim. *A smooth scroll with $a = b = 2$ comes from an element $Y \in |t|$ of one of the following types:*
(i) $Y = \mathbf{P}(E)$ with $E = \oplus^{n-1}\mathcal{O}_{\mathbf{P}^1}(2)$ i.e. $\mathbf{P}^1 \times \mathbf{P}^{n-2}$ embedded in $\mathbf{P}^{3n-4}$ via $\mathcal{O}(2,1)$
(ii) $Y = \mathbf{P}(E)$ with $E = \oplus^{n-3}\mathcal{O}_{\mathbf{P}^1}(2) \oplus \mathcal{O}_{\mathbf{P}^1}(1) \oplus \mathcal{O}_{\mathbf{P}^1}(3)$.

Proof of the claim: The exact sequence (3) in §2 becomes

$$0 \to \mathcal{O}_{\mathbf{P}^1} \to \oplus^{n-2}\mathcal{O}_{\mathbf{P}^1}(2) \oplus (\oplus^2\mathcal{O}_{\mathbf{P}^1}(1)) \to E \to 0.$$

We look at the different possibilities for E depending on the composed map

$$\varphi: \mathcal{O}_{\mathbf{P}^1} \to \mathcal{O}_{\mathbf{P}^1}(2)^{n-2} \oplus \mathcal{O}_{\mathbf{P}^1}(1)^2 \to \mathcal{O}_{\mathbf{P}^1}(1)^2$$

(i) If φ is an injective morphism of bundles, then E is an extension of coker$\varphi = \mathcal{O}_{\mathbf{P}^1}(2)$ by $\mathcal{O}_{\mathbf{P}^1}(2)^{n-2}$, hence $E = \oplus^{n-1}\mathcal{O}_{\mathbf{P}^1}(2)$

(ii) If φ is (after changing basis) zero on one factor $\mathcal{O}_{\mathbf{P}^1}(1)$ and different from zero on the other, then E contains the first $\mathcal{O}_{\mathbf{P}^1}(1)$ as a direct summand. The complement of this will be an extension of $\mathcal{O}_{\mathbf{P}^1}(3)$ (this is the cokernel for a general monomorphism of bundles $\mathcal{O}_{\mathbf{P}^1}to\mathcal{O}_{\mathbf{P}^1}(1) \oplus \mathcal{O}_{\mathbf{P}^1}(2)$ by $\mathcal{O}_{\mathbf{P}^1}(2)^{n-3}$, so that $E = \oplus^{n-3}\mathcal{O}_{\mathbf{P}^1}(2) \oplus \mathcal{O}_{\mathbf{P}^1}(1)) \oplus \mathcal{O}_{\mathbf{P}^1}(3))$.

(iii) If $\varphi = 0$, then $Y = \mathbf{P}(E)$ must contain $\mathbf{P}(\Omega_{\mathbf{P}^2}(2)_{|C})$, which would imply that all the lines in L through any point of C are in the congruence. This is a contradiction.

The case (ii) occurs on an open subset of the locus defined by a quadric equation in V'. Hence, we can find a variety Y in case (ii) not defined by an element of V. We want to prove now that if Y is not in V, it is mapped isomorphically in $G(1,n)$.

Let us consider the sequence

$$0 \to H^0(\mathcal{O}_X) \to H^0(\mathcal{O}_X(t)) \to H^0(\mathcal{O}_Y(t)) \to 0.$$

Now Y not being in V, V is mapped onto $H^0(\mathcal{O}_Y(t))$ so that the immersion of Y into the grassmannian is determined by a complete linear system, which is in fact very ample.

This concludes the proof.

We now furnish an example of geometric construction of the two possible scrolls of bedegree (2,2) quoted in cases (i) and (ii) of the claim above, generalizing some examples of [Go] to the case of higher dimension.

Case (i). Let L be a 2-plane into $\mathbf{P}^n$ and C a smooth conic on it, and let Λ be a $\mathbf{P}^{n-3}$ skew with L. For each point $p \in C$ take the tangent line t_p to C at p, consider the $\mathbf{P}^{n-1} = [\Lambda, t_p]$ spanned by Λ and t_p and denote by Σ_p the $\mathbf{P}^{n-2}$ of the lines through p contained in $[\Lambda, t_p]$. A model for Y is generated by the Σ_p's when p varies in$C$.

Case (ii). Let L and C be as above and take a point p_0 in C. Each $\mathbf{P}^{n-2} = \Sigma_p$ of the lines of the congruence Y through a point p of C generates a $\mathbf{P}^{n-1}$ passing through p_0, the $\mathbf{P}^{n-1}$ generated by Σ_{p_0} is tangent to C at p_0 but does not contain L.

3.3 Construction of case (1,2)

In this case, we obtain that the exact sequence (3) in §2 is

$$0 \to \mathcal{O}_{\mathbf{P}^1}(1) \to \mathcal{O}_{\mathbf{P}^1}(1)^2 \oplus \mathcal{O}_{\mathbf{P}^1}(2)^{n-2} \to E \to 0$$

From this we immediately get (since the map $\mathcal{O}_{\mathbf{P}^1}(1) \to \mathcal{O}_{\mathbf{P}^1}(1)^2$ is not zero) that $Y = \mathbf{P}(E) = \mathbf{P}(\mathcal{O}_{\mathbf{P}^1}(1) \oplus \mathcal{O}_{\mathbf{P}^1}(2)^{n-2})$. The map to G is given by the complete linear series of the canonical $\mathcal{O}(1)$ for E, which is very ample, so that we get a smooth congruence.

The geometric interpretation is as follows. The curve in $\check{\mathbf{P}}^n$ of the hyperplanes defined at points of C has degree $a = 1$, so that all these planes must be contained in a linear subspace Λ of codimension two. Hence all lines in the congruence meet C and Λ. The set of lines meeting both C and Λ defines a congruence with $a = b = 2$, so that it contains another congruence with $a = 1$ and $b = 0$. This implies that Λ meets C in one point and Y consists of the closure of all lines meeting Λ and C in points different from $C \cap \Lambda$.

Remark: For $n \neq 4$ the only $(n-1)$-dimensional scrolls over a curve contained in $G(1,n)$ are the three ones constructed above (cases $(3,3)$, $(2,2)$ and $(1,2)$). Indeed it is easy to see that, for $n \neq 4$, the only linear subspaces of dimension $n-2$ contained in $G(1,n)$ are the Schubert cycles $\Omega(0,n)$.

For $n = 4$ the linear subspaces of dimension 2 contained in $G(1,4)$ are the Schubert cycles $\Omega(0,4)$ and $\Omega(1,3)$. Hence in this case one has also to consider the scrolls whose fibers are $\Omega(1,3)$, described in [A], which are without a fundamental curve and therefore not included in our paper.

This completes the classification.

3.4 Construction of case (3,6)

Let $C \subseteq \mathbf{P}^n$ be the non singular plane cubic quoted in theorem 2, case (*iv*). Then $e = 2, d = 3, D = 3$. Now, with notation of Lemma 7 we have $deg\mathcal{F} = +1$, i.e. Y_0 is the elliptic ruled surface of invariant -1. As shown in Lemma 4, the class of Y in $X = \mathbf{P}(\Omega_{\mathbf{P}^n}(2)_{|C})$, can be written as $2t - p^*D$.
So we study the linear system $L = |2t - p^*D|$, on X. We have

$$h^0(\mathcal{O}_X(2t - p^*D)) = \frac{3(n+1)(n-2)}{2} + h^0(S^2(\mathcal{F})(2P - D)).$$

Since $e = -1$, there exists a curve numerically equivalent to $2C_0 - f$ (see [E] lemma 1.4), where C_0 and f denote respectively a fundamental section and a fibre of Y_0, so that we can fix a D such that $h^0(\mathcal{O}_X(2t - p^*D)) \geq \frac{3(n+1)(n-2)}{2} + 1$.

Now we show that the surface Y_0 is not contained in the base locus of L. To do it, consider the ideal $\mathcal{I}$ of Y_0 and the exact sequence

$$0 \to \mathcal{I}(2t - p^*D) \to \mathcal{O}_X(2t - p^*D) \to \mathcal{O}_{Y_0}(2t - p^*D) \to 0.$$

Notice that for $i > 0$, $R^ip_*\mathcal{I}(2t - p^*D) = 0$, since the fibre of this sheaf at a point p_0 of C is $H^i(\mathcal{J}(2))$, $\mathcal{J}$ denoting the ideal of the $\mathbf{P}^1$ defined by Y_0 in the fibre $\mathbf{P}^{n-1}$ of X at p_0. Therefore one can apply p_* to the exact sequence above to get the sequence

$$0 \to p_*\mathcal{I}(2t - p^*D) \to S^2(\Omega_{\mathbf{P}^n}(2)_{|C})(-D) \to S^2(\Omega_{\mathbf{P}^2}(2)_{|C})(-D) \to 0$$

which splits; this easily implies that $h^0(p_*\mathcal{I}(2t - p^*D)) = h^0(\mathcal{I}(2t - p^*D)) = h^0(\mathcal{O}_X(2t - p^*D)) - 1 = \frac{3(n+1)(n-2)}{2}$, so that Y_0 is not contained in the base locus of $L = |2t - p^*D|$.

As $Y \in |2t - p^*D|$ and $[Y_0] = t^{n-2} - 3(n-2)t^{n-1}p^*P$, as remarked in Lemma 7 then $Y \cap Y_0$ is a curve C' numerically equivalent to $2C_0 - f$, since $C'^2 = Y^2 \cdot Y_0 = 4t^n - 12(n-1)t^{n-1}p^*P = 0$.

As C' is the unique effective divisor in its linear equivalence class, it is contained in the base locus of the system $|2t - p^*D|$. Now restricting this system to a fiber $p^{-1}(P)$, at a point $P \in C$, we get the following exact sequence:

$$0 \to \mathcal{O}_X(2t - p^*(D + P)) \to \mathcal{O}_X(2t - p^*D) \to \mathcal{O}_{p^{-1}(P)}(2) \to 0$$

by which we get that the image of $H^0(\mathcal{O}_X(2t - p^*D))$ in $H^0(\mathcal{O}_{p^{-1}(P)}(2))$ has dimension $\frac{(n^2+n-4)}{2}$. Hence $|2t - p^*D|$ cuts out on the fiber a system of dimension $\frac{(n-2)(n+3)}{2}$ of quadrics which has at least two base-points, hence just the two points of $Y \cap Y_0 \cap p^{-1}(P)$. Thus $Y \cap Y_0$ is the base locus of $|2t - p^*D|$, and Y is smooth away from $Y \cap Y_0$.

With the same argument as in [A-G] we can conclude that Y is embedded in $G(1, n)$ by the map $f : X \to G(1, n)$ as a smooth congruence of bidegree $(3, 6)$.

REFERENCES

[A] Alzati, A., *3-Scroll immersi in* $G(1,4)$, Ann. Univ. Ferrara, **32** (1986), 45-54.

[A-C-G-H] Arbarello, E.– Cornalba, M.– Griffiths, P.– Harris, J., *Geometry of Algebraic Curves*, Vol. I Grund. der Math. Wissen. **267** Springer-Verlag (1985).

[A-G] Arrondo, E.– Gross, M., *On smooth surfaces in* $Gr(1, \mathbf{P}^3)$ *with a fundamental curve*, to appear in Man. Math.

[E] Ein, L., *Non-degenerate surfaces of degree* $n+3$ *in* $\mathbf{P}^n$, Crelle Journal Reine Angew. Math. **351** (1984), 1-11.

[F1] Fano, G., *Sulle congruenze di rette del terzo ordine prive di linea singolare*, Att. Acc. di Scienze Torino **29** (1984), 474-493.

[F2] Fano, G., *Nuove ricerche sulle congruenze di rette del* 3^0 *ordine prive di linea singolare*, Memoria della Reale Acad. Sc. Torino (2) **51** (1902), 1-79.

[F] Fulton, W., *Intersection Theory*, Ergebnisse (3) **2**, Springer (1984).

[Go] Goldstein, N. *Scroll surfaces in* $Gr(1, \mathbf{P}^3)$, Conference on Algebraic Varieties of small dimension (Turin 1985), Rend. Sem. Mat. Univ. Politecnica, Special Issue (1987) 69-75.

[G1] Gross, M., *Surfaces in the four-dimensional Grassmannian*, Ph. D. thesis, Berkeley (1990).

[G2] Gross, M., *The distribution of bidegrees of smooth surfaces in* $Gr(1, \mathbf{P}^3)$, Math. Ann. **292** (1992), 127-147.

[R] Roth, L., *On the projective classification of surfaces*, Proc. London Math. Soc., **42** (1937), 142-170.

E-mail of the authors:
arrondo@mat.ucm.es
bertolin@vmimat.mat.unimi.it
turrini@vmimat.mat.unimi.it

On a Result of Zak–L'vovsky

LUCIAN BĂDESCU, Institute of Mathematics of the Romanian Academy
P.O.Box 1-746,Ro-70700 Bucharest,Romania,e-mail:lbadescu@imar.ro

Let Y be normal irreducible projective subvariety of $\mathbb{P}^n$ of positive dimension, over a fixed algebraically closed field k of arbitrary characteristic. Denote by Y_0 the smooth locus, Reg(Y), of Y. An extension of Y in $P:=\mathbb{P}^{n+1}$ is by definition an irreducible projective subvavariety X of P of dimension = dim(Y)+1 such that there is a hyperplane $H:=\mathbb{P}^n$ such that $Y=X\cap H$ (scheme-theoretically). Given Y in H, one can easily construct extensions of Y in P, e.g. take an arbitrary point x in P–H and let X be the cone over Y with vertex x; these extensions of conical type will be called the trivial extensions of Y in P.

One of the very interesting and difficult problems of the projective geometry is to classify all possible extensions of a given subvariety Y of $\mathbb{P}^n$.The first step would be to decide whether there are non-trivial extensions of Y, i.e. extensions that are not cones. The aim of this note is to give two new proofs of a remarkable result of Zak-L'vovsky in connection with this problem.Before stating it recall that there are two canonical exact sequences naturally associated to the embedding $Y\subset H$

$$0\longrightarrow T_Y\longrightarrow T_H/Y\xrightarrow{a}N_{Y,H}\ ,\qquad \text{and}$$

$$0\longrightarrow O_Y\longrightarrow O_Y(1)^{\oplus n+1}\xrightarrow{b}T_H/Y\longrightarrow 0$$

(the normal sequence of $Y\subset H$ and the Euler sequence of $H=\mathbb{P}^n$ restricted to Y respectively), where T_Y is the tangent sheaf of Y (i.e. the dual of the sheaf of differential forms of degree one on Y) and $N_{Y,H}$ is the normal sheaf of Y in H (i.e. the dual of the conormal sheaf of Y in H). The map $a \circ b : O_Y(1)^{\oplus n+1} \longrightarrow N_{Y,H}$ yields the map $c := a(-1) \circ b(-1) : O_Y^{\oplus n+1} \longrightarrow N_{Y,H}(-1)$. Then the map

$$(1) \qquad u = H^0(c) : H^0(Y, O_Y^{\oplus n+1}) \longrightarrow H^0(Y, N_{Y,H}(-1))$$

is going to play a crucial role in order to state and prove the following generalized version of Zak-L'vovsky's result referred to in the title:

THEOREM 1. *In the above situation assume that Y is a non-degenerate normal subvariety of $H := \mathbb{P}^n$ of codimension ≥ 2 such that the map (1) is surjective. Then every extension of Y in $P := \mathbb{P}^{n+1}$ is trivial.*

In fact Zak and L'vovsky proved this theorem in case when Y is *smooth* (see [Z], [L]). The method of Zak [Z] makes use of the theory of dual varieties and that is why he needs the additional hypothesis that char(k)=0. L'vovsky's proof [L] makes use of the theory of Hilbert schemes. In this paper we first show that theorem 1 can be proved in case Y is smooth and char(k)=0 by reducing it to a result of Wahl [W_2], and then give a second new proof of Zak-L'vovsky's result in the more general form stated above.

COROLLARY. *In the hypotheses of theorem 1 assume in addition that Y is smooth and that $H^1(Y, O_Y(i)) = 0$ for all $i \in \mathbb{Z}$. Denote by S the graded k-algebra $\bigoplus_{i\geq 0} H^0(Y, O_Y(i))$ associated to the polarized variety $(Y, O_Y(1))$, and by $T^1_S = \bigoplus_{i\in\mathbb{Z}} T^1_S(-i)$ the graded S-module of first order infinitesimal deformations of S over k. If $T^1_S(-1) = 0$ then every extension of Y in $\mathbb{P}^{n+1}$ is trivial. Moreover, there are examples of smooth curves Y in $\mathbb{P}^n$ with non-trivial extensions in $\mathbb{P}^{n+1}$ and such that $T^1_S(-1) = 0$.*

Everything follows from theorem 1 and the results of the appendix (and especially the corollary of the proposition, and the example). □

NOTE. Since $H^0(c) = H^0(a(-1)) \circ H^0(b(-1))$ we infer that the map

$$H^0(a(-1)) : H^0(Y, T_H(-1)/Y) \longrightarrow H^0(Y, N_{Y,H}(-1))$$

is surjective if the map (1) is so. The converse is also true provided that $\dim(Y)\geq 2$ and $\operatorname{char}(k)=0$. Indeed, from the cohomology sequence associated to the exact sequence $0\longrightarrow O_Y(-1)\longrightarrow O_Y^{\oplus n+1}\xrightarrow{b(-1)}T_H(-1)/Y\longrightarrow 0$ we deduce that the map $H^0(b(-1))$ is an isomorphism by the Kodaira-Mumford vanishing theorem applied to Y. Note also that the map (1) is injective if $\operatorname{char}(k)=0$ and Y is not a cone (see e.g. [W_2], or also [B_2]).

The following lemma will play a basic role in this paper.

LEMMA 1. *In the situation of theorem 1 the surjectivity of map (1) implies that* $H^0(Y,N_{Y,H}(-i))=0$ *for every* $i\geq 2$.

Proof. Let $Y_0=\operatorname{Reg}(Y)$ be the smooth locus of Y. Since Y is normal, $\operatorname{codim}_Y(Y-Y_0)\geq 2$. Consider the map

$$c/Y_0 : O_Y^{\oplus n+1}/Y_0\longrightarrow E:=N_{Y,H}(-1)/Y_0$$

of vector bundles defined above by $c/Y_0 = a(-1)/Y_0\circ b(-1)/Y_0$. The map c/Y_0 is surjective because Y_0 smooth implies $a(-1)/Y_0$ surjective.

To prove lemma 1 observe that it is enough to it for $i=2$ because $N_{Y,H}(-i-1)\subseteq N_{Y,H}(-i)$. Assume therefore that there were a non-zero section $s\in H^0(Y,N_{Y,H}(-2))$. The surjectivity of (1) implies $h^0(Y,N_{Y,H}(-1))\leq n+1$, and since Y is non-degenerate in H, it follows that a general section of $H^0(Y,N_{Y,H}(-1))$ is of the form hs, with $0\neq h\in H^0(H,O_H(1))$. In particular, the zero set of hs contains the support of a non-zero divisor on Y (because it contains the trace on Y of the zero-set of h). Since $H^0(Y,N_{Y,H}(-1))\cong H^0(Y_0,E)$ and $\operatorname{codim}_Y(Y-Y_0)\geq 2$, we infer that the general section of the vector bundle E on the (possibly open) variety Y_0 contains the support of a non-zero divisor on Y_0.

On the other hand, the surjectivity of c/Y_0 implies that E is generated by n+1 global sections of E, and hence, by a Bertini-type theorem (valid in arbitrary characteristic), the zero locus of a general section of E should have codimension in $Y_0\geq \operatorname{rank}(E)=\operatorname{codim}_H(Y)\geq 2$. □

REMARK. I am indebted to E. Ballico for having communicated to me the above proof of lemma 1, which was already used in [B_2].

FIRST PROOF OF THEOREM 1 (when char(k)=0 and Y is smooth).

Let X be an extension of Y in $P=\mathbb{P}^{n+1}$. Then X is smooth along Y (because Y is a smooth Cartier divisor on X), and in particular, the normal sheaf $N_{X,P}$ of X in P is locally is free along Y. Moreover, since the intersection $Y=X\cap H$ is proper we have $N_{X,P}/Y \cong N_{Y,H}$. Let $f:X^*\longrightarrow X$ be the normalization of X, and denote by $V:=\mathrm{Reg}(X)$ the smooth locus of X. Then $Y\subset V$, and hence Y becomes an ample Cartier divisor on X^*. Denote by N_X* the double dual of $f^*(N_{X,P})$. Then N_X* is a reflexive sheaf on X^* such that $N_X*/Y \cong N_{Y,H}$, and hence we get the exact sequence $(i\geq 1)$

$$0\longrightarrow N_X*(-i-1)\longrightarrow N_X*(-i)\longrightarrow N_{Y,H}(-i)\longrightarrow 0$$

(with $O_X*(1):=f^*(O_X(1))$) yields the exact sequence

$$0\longrightarrow H^0(X^*,N_X*(-i-1))\xrightarrow{e_i} H^0(X^*,N_X*(-i))\longrightarrow H^0(Y,N_{Y,H}(-i)).$$

By lemma 1 the last space is zero for every $i\geq 2$. Therefore the map e_i is an isomorphism for every $i\geq 2$. Since the first space is zero for $i\gg 0$ (because f is finite and hence $O_X*(1)=O_X*(Y)$ is ample on X^*), we infer by induction on i that $H^0(X^*,N_X*(-2)) = 0$. Taking i=1 in the last sequence and using the surjectivity of (1) we get $h^0(X^*,N_X*(-1))\leq h^0(Y,N_{Y,H}(-1))\leq n+1$. Since $V \subseteq X^*$, X^* is normal, $\mathrm{codim}_{X^*}(X^*-V)\geq 2$ and N_X* is reflexive, $h^0(X^*,N_X*(-1))= = h^0(V,N_{X,P}(-1))$. Thus the above we get $h^0(V,N_{X,P}(-1)) \leq n+1$. On the other hand, look at the commutative diagram

$$\begin{array}{ccccccccc}
 & & 0 & & 0 & & & & \\
 & & \downarrow & & \downarrow & & & & \\
 & & O_V & \xrightarrow{id} & O_V & & & & \\
 & & \downarrow & & \downarrow & & & & \\
0 & \longrightarrow & F & \longrightarrow & O_V(1)^{\oplus n+2} & \longrightarrow & N_{X,P}/V & \longrightarrow & 0 \\
 & & \downarrow & & \downarrow & & \downarrow id & & \\
0 & \longrightarrow & T_V & \longrightarrow & T_P/V & \longrightarrow & N_{X,P}/V & \longrightarrow & 0 \\
 & & \downarrow & & \downarrow & & & & \\
 & & 0 & & 0 & & & &
\end{array}$$

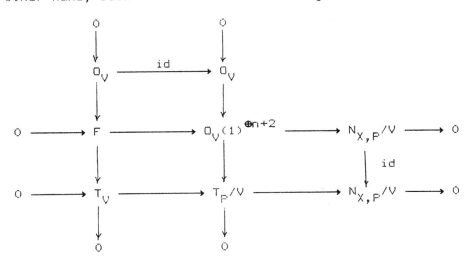

in which the last row is the normal sequence of X in P restricted to V, the second column is the Euler sequence of P restricted to V, and F is a vector bundle (the dual of the sheaf of principal parts). The first row yields the exact sequence

$$0\longrightarrow H^0(V,F(-1))\longrightarrow H^0(V,O_V^{\oplus n+2})\longrightarrow H^0(V,N_{X,P}(-1)/V)\cong H^0(X^*,N_{X^*}(-1)),$$

which together with the last inequality implies $H^0(V,F(-1)) \neq 0$. Hence, using the first column we get $H^0(V,T_V(-1))\neq 0$. Using once again that $V\subseteq X^*$, $\mathrm{codim}_{X^*}(X^*-V)\geq 2$, X^* is normal and the tangent sheaf T_{X^*} is reflexive, we finally get $H^0(X^*,T_{X^*}(-1)) \neq 0$. Now, this latter fact and the main result of Wahl's paper [W_2], theorem 2 (which holds in char. zero) implies that X^* is isomorphic to the projective cone over $(Y,O_Y(1))$. Finally, an easy argument shows that X is a cone over Y as long as X^* is the projective cone over $(Y,O_Y(1))$ (see e.g. [B_3], page 22). □

REMARKS. 1. In the above proof Wahl's criterion for the polarized variety $(X^*,O_{X^*}(1))$ to be a cone played an essential role. Since $O_{X^*}(1)$ is generated by its global sections, by looking carefully in Wahl's paper [W_2], we see that Wahl's proof of this criterion becomes much more simple in our situation. The first proof of theorem 1 given above should be quite close in spirit to the original proof of Zak [Z].

2. From the above diagram we infer that the kernel of the map (1) is isomorphic to $H^0(Y,F(-1))$. In particular, since $H^0(Y,O_Y(-1)) = 0$, we get that if the map (1) is not injective then $H^0(Y,T_Y(-1))\neq 0$. Therefore from [W_2] it follows that if char(k)=0 and the map (1) is not injective then Y is a cone over a subvariety of $\mathbb{P}^{n-1}$. In particular, if Y is smooth, non-degenerate, of codimension ≥ 2 in H, and char(k)=0 then the map (1) is injective, and so $h^0(Y,N_{Y,H}(-1))=n+r+1$, with $r\geq 0$; moreover, we have r=0 if and only if the map (1) is surjective.

3. With small changes the first proof of theorem 1 given above also yields the following more general statement (also due to Zak and L'vovsky).

THEOREM 2. *Let Y be a closed irreducible smooth non-degenerate subvariety of codimension* ≥ 2 *in* $H:=\mathbb{P}^n$. *Assume* $\dim(Y)\geq 1$, $\mathrm{char}(k)=0$, $h^0(Y,N_{Y,H}(-1))=n+r+1$ $(r\geq 0)$, *and* $H^0(Y,N_{Y,H}(-2))=0$. *If* $m\geq r+1$ *then Y cannot be extended non-trivially m steps, i.e. every closed irreducible subvariety X of* $P:=\mathbb{P}^{n+m}$ *of dimension =*

$\dim(Y)+m$ for which there is a linear subspace $H'\cong\mathbb{P}^n$ of P such that $Y=X\cap H'$ (scheme-theoretically), is a cone over a subvariety of $\mathbb{P}^{n+m-1}$.

SECOND PROOF OF THEOREM 2 (in the general case).

First we need the following:

LEMMA 2. *The surjectivity of map (1) is equivalent to the surjectivity of the following map*

$$(2) \qquad H^0(Y,O_Y^{\oplus n+2})\longrightarrow H^0(Y,N_{Y,P}(-1)),$$

whose definition is similar to that of (1), with $Y\subset H$ replaced by $Y\subset P$ and with $N_{Y,P}$ the normal sheaf of Y in P.

Proof. Since H is a linear subspace of P the canonical exact sequence

$$0\longrightarrow T_H\longrightarrow T_P/H\longrightarrow N_{H,P}=O_H(1)\longrightarrow 0$$

splits, and since $Y\subseteq H$, we get $T_P(-1)/Y\cong T_H(-1)/Y\oplus O_Y$. On the other hand, consider the canonical exact sequence of normal sheaves

$$0\longrightarrow N_{Y,H}\longrightarrow N_{Y,P}\longrightarrow N_{H,P}/Y\cong O_Y(1),$$

which restricted to $Y_0=\mathrm{Reg}(Y)$ becomes a short exact sequence

$$0\longrightarrow N_{Y_0,H}\longrightarrow N_{Y_0,P}\longrightarrow O_Y(1)/Y_0\longrightarrow 0.$$

Pick a point $x\in P-H$ and let Z be the cone in P over Y with vertex x. Then the scheme-theoretic intersection $Y=Z\cap H$ is proper, and therefore the last exact sequence splits, i.e.

$$N_{Y,P}(-1)/Y_0\cong[N_{Y,H}(-1)\oplus O_Y]/Y_0.$$

From these two canonical isomorphisms we immediately infer that the surjectivity of (1) is equivalent to the surjectivity of (2), taking into account that $H^0(Y,N_{Y,P}(-1)/Y)\cong H^0(Y_0,N_{Y,P}(-1)/Y_0)$, $H^0(Y,N_{Y,H}(-1))\cong H^0(Y_0,N_{Y,H}(-1)/Y_0)$, and $H^0(Y,O_Y)\cong H^0(Y_0,O_Y/Y_0)$ because Y is normal $\mathrm{codim}_Y(Y-Y_0)\geq 2$, and all sheaves in question are reflexive. □

In the situation of the beginning, consider an arbitrary extension X of Y in $P=\mathbb{P}^{n+1}$, and for every $p>0$ consider the infinitesimal neighbourhood $Y_p=(Y,O_X/O_X(-pY))$ of order p of Y in X.

Since Y_p is a closed subscheme of P we have the canonical commutative diagram of surjective maps of O_P-agebras

$$\begin{array}{ccc} & O_P & \\ {}^{h'}\swarrow & & \searrow^{u} \\ O_{Y_2} & \xrightarrow{\ \pi\ } & O_Y \end{array}$$

Since $\mathrm{Ker}(\pi)\cong O_Y(-1)$ is a square-zero ideal of O_{Y_2} it follows that $I^2\subseteq \mathrm{Ker}(h')$. In particular, we get the commutative diagram with exact rows

$$\begin{array}{ccccccccc} 0 & \longrightarrow & I/I^2 & \longrightarrow & O_P/I^2 & \longrightarrow & O_P/I & \longrightarrow & 0 \\ & & \downarrow{\scriptstyle h} & & \downarrow{\scriptstyle h''} & & \downarrow{\scriptstyle id} & & \\ 0 & \longrightarrow & O_Y(-1) & \longrightarrow & O_{Y_2} & \longrightarrow & O_Y & \longrightarrow & 0 \end{array} \qquad (3)$$

where the map h'' is induced by h'.

LEMMA 3. *The surjectivity of map (1) implies that the bottom exact sequence (which is an extension of sheaves of k-algebras of O_Y by the O_Y-module $O_Y(-1)$) of diagram (3) splits (i.e. is trivial).*

Proof. By lemma 2 the surjectivity of (1) implies the surjectivity of (2), which in turn (by the remark following lemma 1) implies the surjectivity of $H^0(Y,T_P(-1)/Y)\longrightarrow H^0(Y,N_{Y,P}(-1))$.

Hence $h\in \mathrm{Hom}_Y(I/I^2,O_Y(-1)) \cong H^0(Y,N_{Y,P}(-1))$ can be lifted to a section $D \in H^0(Y,T_P(-1)/Y)$. Consider the canonical cover $\{U_0,\ldots,U_{n+1}\}$ of $P=\mathbb{P}^{n+1}$, with $U_i=\{(t_0,\ldots,t_{n+1})\in P/t_i\neq 0\}$, $i=0,1,\ldots,n+1$. Then $D_i:=D/U_i$ is nothing but a k-derivation of the polynomial k-algebra $A_i= k[t_0/t_i,\ldots,t_{n+1}/t_i]$ (in n+1 variables) into the k-algebra A_i/I_i, where I_i is the ideal of $Y\cap U_i$ in $\mathrm{Spec}(A_i) = U_i$ (because $O_Y(-1)/Y\cap U_i\cong O_Y/Y\cap U_i$). Then by the general theory the bottom line of (3) restricted to U_i

$$(4_i) \qquad 0\longrightarrow A_i/I_i\longrightarrow B_i\longrightarrow A_i/I_i\longrightarrow 0$$

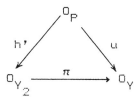

corresponds to an element $h_i\in \mathrm{Hom}_{A_i/I_i}(I_i/I_i^2,A_i/I_i)$ (under the canonical identification of the set of isomorphism classes of extensions and the elements of T^1, see [S]). But h is the restriction to I_i of the derivation D_i. Then D_i yields a splitting of (4_i), and since the derivations $D_0,\ldots,D_{n+1}$ yield the "vector field" $D \in H^0(Y, T_P(-1)/Y)$ along Y of weight -1, these splittings glue together to yield a global splitting of the bottom line of (3). In other words we get an isomorphism of O_Y-algebras

$$(5) \qquad O_{Y_2} \cong S(O_Y(-1))/J^2 \; (= O_Y \oplus O_Y(-1)),$$

where $S(O_Y(-1)) = \bigoplus_{i \geq 0} O_Y(-i)$ is the symmetric O_Y-algebra of $O_Y(-1)$ and $J = \bigoplus_{i \geq 1} O_Y(-i)$. □

Our next aim is to study the higher order infinitesimal neighbourhoods of Y in X by induction, the starting point being the isomorphism (5). Assume therefore that for some $p \geq 2$ there is an isomorphism

$$(6) \qquad O_{Y_p} \cong S(O_Y(-1))/J^p \; (\cong O_Y \oplus O_Y(-1) \oplus \ldots \oplus O_Y(1-p)).$$

The problem is to extend the isomorphism (6) to get an isomorphism

$$(7) \qquad O_{Y_{p+1}} \cong S(O_Y(-1))/J^{p+1} \; (\cong O_Y \oplus O_Y(-1) \oplus \ldots \oplus O_Y(-p)).$$

To do that we need to study the extension of sheaves of k-algebras

$$(8) \qquad 0 \longrightarrow O_Y(-p) \cong O_X(-pY)/O_X(-(p+1)Y) \longrightarrow O_{Y_{p+1}} \longrightarrow O_{Y_p} \longrightarrow 0.$$

with p satisfying (6). In particular, there is an injection $i: O_Y \hookrightarrow O_{Y_p}$ giving O_{Y_p} a structure of O_Y-algebra. Then consider the commutative diagram in which the bottom line is an extension of O_Y by $O_Y(-p)$:

$$\begin{array}{ccccccccc}
0 & \longrightarrow & O_Y(-p) & \longrightarrow & O_{Y_{p+1}} & \xrightarrow{u} & O_{Y_p} & \longrightarrow & 0 \\
 & & \uparrow \scriptstyle{id} & & \uparrow & & \uparrow \scriptstyle{i} & & \\
0 & \longrightarrow & O_Y(-p) & \longrightarrow & E := u^{-1}(i(O_Y)) & \longrightarrow & O_Y & \longrightarrow & 0
\end{array}$$

LEMMA 4. *The extension* $0 \longrightarrow O_Y(-p) \longrightarrow E \longrightarrow O_Y \longrightarrow 0$ *is trivial.*

Proof. We can consider a diagam similar to (3)

$$\begin{array}{ccccccccc}
0 & \longrightarrow & I/I^2 & \longrightarrow & O_P/I^2 & \longrightarrow & O_P/I & \longrightarrow & 0 \\
 & & \downarrow \scriptstyle{g'} & & \downarrow \scriptstyle{g} & & \downarrow \scriptstyle{id} & & \\
0 & \longrightarrow & O_Y(-p) & \longrightarrow & E & \xrightarrow{e} & O_Y & \longrightarrow & 0
\end{array}$$

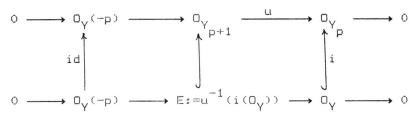

where the map g is defined as follows. Since E is a sheaf of k-algebras, $H^0(P,E)\supseteq k$, and since $H^0(P,O_Y(-p))=0$, we infer that the map $H^0(Y,E)\longrightarrow H^0(Y,O_Y) = k$ is an isomorphism. In particular, we can lift $1\in H^0(Y,O_Y)$ to a section $s \in H^0(Y,E) = H^0(P,E)$. Since E is also a coherent sheaf on P, the section s yields a map $g'':O_P \longrightarrow E$ such that $e\circ g''$ is just the canonical surjection $O_P\longrightarrow O_Y$. Finally, since $O_Y(-p)$ is a square-zero ideal of E, the map b'' vanishes on I^2, yielding the map g.

To prove lemma 4, an easy diagram chase shows that it is sufficient to check that $g'=0$. But this equality follows from the next lemma taking into account that $\mathrm{Hom}_Y(I/I^2,O_Y(-p))\cong H^0(Y, N_{Y,P}(-p))$. □

LEMMA 5. *If the map (1) is surjective and Y is non-degenerate in H of codimension ≥2 then $H^0(Y,N_{Y,P}(-p))=0$ for every $p\geq 2$.*

Proof. By the proof of lemma 2, $N_{Y,P} \cong N_{Y,H}\oplus O_Y(1)$. Then lemma 5 is a consequence of lemma 1 and of this isomorphism. □

Coming back to our situation and using (6) and Schlessinger's theory (see [S]), there is a canonical exact sequence

$$T^1(Y_p/Y,O_Y(-p))\longrightarrow T^1(Y_p/k,O_Y(-p))\xrightarrow{\ v\ } T^1(Y/k,O_Y(-p)).$$

Lemma 4 shows that the class of the extension (8) sits in Ker(v). This means that the extension (8) comes actually from an extension of O_Y-algebras. Then I claim that the extension (8) is isomorphic to the following one

$$0\longrightarrow O_Y(-p)\longrightarrow S(O_Y(-1))/J^{p+1}\longrightarrow S(O_Y(-1))/J^p\longrightarrow 0$$

(with $S(O_Y(-1))/J^p\cong \bigoplus_{i=0}^{p-1} O_Y(-i)$). Indeed, assuming (6), the only thing we have to show is that there is a map of O_Y-algebras $S(O_Y(-1))/J^{p+1}\longrightarrow O_{Y_{p+1}}$ extending the isomorphism (6), i.e. making the following diagram commutative

$$\begin{array}{ccccccccc} 0 \longrightarrow & O_Y(-p) & \longrightarrow & S(O_Y(-1))/J^{p+1} & \longrightarrow & S(O_Y(-1))/J^p & \longrightarrow 0 \\ (9) & \downarrow \mathrm{id} & & \downarrow & & \downarrow \wr \text{ by } (6) & \\ 0 \longrightarrow & O_Y(-p) & \longrightarrow & O_{Y_{p+1}} & \longrightarrow & O_{Y_p} & \longrightarrow 0 \end{array}$$

Let $t\in H^0(X,O_X(1))$ be a global equation of the effective

Cartier divisor Y on X. This section yields a canonical inclusion $t:O_X(-1)\hookrightarrow O_X$. On the other hand, since

$$O_{Y_p}(-1) \cong O_X(-1)/O_X(-p-1) \quad \text{and} \quad O_{Y_{p+1}} = O_X/O_X(-p-1)$$

we get a canonical embedding $t:O_{Y_p}(-1)\hookrightarrow O_{Y_{p+1}}$. Recalling (6) and the fact that $O_{Y_{p+1}}$ is an O_Y-algebra, we infer that $O_{Y_{p+1}}$ contains the O_Y-submodule $O_Y(-1)$ (via t). In particular, we get a map of O_Y-alalgebras $S(O_Y(-1))\longrightarrow O_{Y_{p+1}}$ which is obviously zero when restricted to J^{p+1}. In other words we get the desired map

$$S(O_Y(-1))/J^{p+1}\longrightarrow O_{Y_{p+1}}$$

of O_Y-algebras making the diagram (9) commutative.

Summing up, we proved by induction on p that the isomorphism (6) can be extended to an isomorphism (7).

Now we conclude the second proof of theorem 1 as follows. Let $\hat{X}$ be the formal completion of X along Y and let $u:\hat{X}\longrightarrow X$ be the canonical morphism. For every $i\in\mathbb{Z}$ set $O_{\hat{X}}(i):=u^*(O_X(i))$. What we have proved so far can be translated into

$$O_{\hat{X}} \cong \operatorname{inv}_p\lim O_{Y_p} \cong \prod_{i\geq 0} O_Y(-i) \text{ (by (6))},$$

where the product of two formal functions $f=(f_o,f_1,\ldots)$ and $g=(g_0,g_1,\ldots)$ becomes $fg=(f_0g_0,f_0g_1+f_1g_0,f_0g_2+f_1g_1+f_2g_0,\ldots)$. Then for every $m\geq 0$ the above isomorphisms extend to isomorphisms

$$O_{\hat{X}}(m) \cong \prod_{i\geq 0} O_Y(m-i) \text{ for every } m\geq 0.$$

For example taking m=1 in the above isomorphism, the element $(0,1,0,\ldots)\in H^0(\hat{X},O_{\hat{X}}(1))$ corresponds to the image of t under the natural map $H^0(X,O_X(1))\longrightarrow H^0(\hat{X},O_{\hat{X}}(1))$. Moreover, the structure of $O_{\hat{X}}(m)$ as an $O_{\hat{X}}$-module is given by $fu = (f_0u_0,f_0u_1+f_1u_0,f_0u_2+f_1u_1+f_2u_0,\ldots)$, where $f=(f_0,f_1,\ldots)\in O_{\hat{X}} \cong \prod_{i\geq 0} O_Y(-i)$ and $u=(u_0,u_1,\ldots) \in O_{\hat{X}}(m)\cong \prod_{i\geq 0} O_X(m-i)$; if $v=(v_0,v_1,\ldots)\in O_{\hat{X}}(p)$ then the multiplication of the k-algebra $S':=\bigoplus_{m\geq 0} H^0(\hat{X},O_{\hat{X}}(m))$ is given by $uv = (u_0v_0,u_0v_1+u_1v_0,u_2v_0+u_1v_1+u_0v_2,\ldots)$. In particular we get:

(10) There is a map of k-algebras $S:=\bigoplus_{m\geq 0} H^0(Y,O_Y(m))\longrightarrow S'$ whose composition with the natural restriction map $S'\longrightarrow S$ (of k-algebras) is the identity.

Let $(X^*, O_{X^*}(1))$ be as above the normalization of $(X, O_X(1))$. Since Y is normal and $Y = X \cap H$ (scheme-theoretically), an elementary result (known as lemma of Hironaka) implies that X is normal at every point of Y, and hence $Y \subset X^*$ and the formal completions of X and X^* along and Y are the same), and Y is an ample Cartier divisor on X^*. In particular, for every $m \geq 0$ it makes sense to consider the natural maps $H^0(X^*, O_{X^*}(m)) \longrightarrow H^0(\hat{X}, O_{\hat{X}}(m))$. Since X^* is normal of dimension ≥ 2 and Y is an ample divisor on X^*, by a well known result of Enriques-Severi-Zariski-Serre we get that $H^1(X^*, O_{X^*}(-i)) = 0$ for all $i \gg 0$. Since

$$H^0(\hat{X}, O_{\hat{X}}(m)) \cong \operatorname{inv}_p \lim H^0(Y_p, O_{Y_p}(m)),$$

this easily implies that all these maps are isomorphisms.

In other words $\bigoplus_{m \geq 0} H^0(X^*, O_{X^*}(m))$ is isomorphic as a graded k-algebra with S'. This identification together with (10) allows one to define a map of graded S-algebras $S[T] \longrightarrow S'$ by sending T to t, where T is a variable over S with deg(T)=1. Then it is trivial to check that this map is an isomorphism of graded k-algebras (because it is surjective and S[T] and S' are domains of the same dimension).

In other words we have shown that X^* is isomorphic to the projective cone over $(Y, O_Y(1))$. Then it follows that X is a cone over Y exactly as in the first proof. The second proof of theorem 1 is now complete. □

REMARKS. 1. By the proposition of the appendix the surjectivity of (1) implies that $T^1_S(-i) = 0$ for every $i \geq 0$, i.e. there are no infinitesimal deformations of negative weights of S/k. Therefore as soon as we know that the restriction map $S' \longrightarrow S$ is surjective (which implies $S \cong S'/tS'$) then the last part of the proof of the theorem could be also concluded by applying corollary 2 of theorem 1 of [B_1].

2. Under the additional hypothesis that X is arithmetically normal in $\mathbb{P}^{n+1}$, theorem 1 was proved (even if Y is normal) in [B_2] as an application of a local result. Another proof of theorem 1 in case when Y is a local complete intersection in H was given by Ballico in [Ba] by applying some results of Kleppe.

Appendix

Let Y be, as at the beginning, a normal projective subvariety of $\mathbb{P}^n=H$, and denote by $S(Y)=k[T_0,\dots,T_n]/I(Y)$ the homogeneous coordinate ring of Y in H (with $I(Y)$ the homogeneous prime ideal of Y in H).

Let $S=\bigoplus_{i\geq 0} H^0(Y,\mathcal{O}_Y(i))$ be the integral closure of $S(Y)$ in its quotient field. Denote by $T^1_S=T^1(S/k,S)$ the S-module of first order infinitesimal deformations of S over k.

Set $W'=\mathrm{Spec}(S(Y))$ (the affine cone over Y in H), $w'\in W'$ the vertex of W', $W=\mathrm{Spec}(S)$ (the normalization of W', or the normal affine cone over the polarized variety $(Y,\mathcal{O}_Y(1))$), and $w\in W$ the vertex of W. Set $W_0=W-\{w\}\cong W'-\{w'\}$ (the affine cone without vertex). Then W_0 is a $\mathbb{G}_m$-bundle $q:W_0\longrightarrow Y$ over Y (which is locally trivial over each $D_+(T_i)\cap Y$ for every $i=0,\dots,n$). Moreover, if $V=\mathrm{Reg}(Y)$ is the smooth locus of Y then the smooth locus of W', $\mathrm{Reg}(W')$, coincides to $U:=q^{-1}(V)$ (unless $Y=\mathbb{P}^n$, when W' is itself smooth). Denote by p the restriction $q/U:U\longrightarrow V$ (which is again a $\mathbb{G}_m$-bundle). Consider the exact sequence

$$0\longrightarrow T_{W'}\longrightarrow T_Z/W'\longrightarrow N_{W',Z}\ ,$$

where $Z=\mathbb{A}^{n+1}=\mathrm{Spec}(k[T_0,\dots,T_n])$, $T_{W'}$ is the tangent sheaf of W' (the dual of the sheaf of differential forms of degree 1 on W' over k) and $N_{W',Z}$ is the normal sheaf of W' in Z. Restricting this exact sequence to the smooth locus of W' we get the exact sequence

$$0\longrightarrow T_{W'}/U=T_U\longrightarrow T_Z/U\longrightarrow N_{W',Z}/U\longrightarrow 0\ ,$$

which yields the exact sequence of cohomology

$$H^0(U,T_Z/U)\xrightarrow{\ v\ }H^0(U,N_{W',Z})\longrightarrow H^1(U,T_U)\xrightarrow{\ v'\ }H^1(U,T_Z/U).$$

DEFINITION. In the above situation we define the following fundamental invariant of $S(Y)$ (or of Y in H):

$$U_{S(Y)}=\mathrm{Coker}(v)=\mathrm{Ker}(v').$$

It follows that $U_{S(Y)}=\bigoplus_{i\in\mathbb{Z}} U_{S(Y)}(i)$ is a graded $S(Y)$-module because of the $\mathbb{G}_m$-action of W' and because U is a $\mathbb{G}_m$-invariant open subset of W_0. Moreover, for every $i\in\mathbb{Z}$ we have the following facts (see [P], or also $[B_2]$):

– $H^0(U, T_Z/U)(i) \cong H^0(V, O_Y(1+i)^{\oplus n+1}) \cong H^0(Y, O_Y(1+i)^{\oplus n+1})$, and

– $H^0(U, N_{W',Z})(i) \cong H^0(V, N_Y(i)) \cong H^0(Y, N_Y(i))$.

In particular, we get

(11) $\quad U_{S(Y)}(-i) \cong H^0(Y, N_Y(-i))$ for every $i \geq 2$, and

(12) $\quad U_{S(Y)}(-1) \cong \mathrm{Coker}(u)$, where u is the map (1).

Lemma 1 above can then be restated by saying that the condition $U_{S(Y)}(-1)=0$ (which is is equivalent to the surjectivity of (1)) implies $U_{S(Y)}(-i)=0$ for every $i \geq 1$.

In the case when Y is arithmetically normal in H then, by a result of Schlessinger (see [S], or also [P]), $U_{S(Y)}$ is nothing but $T^1_{S(Y)} = T^1(S(Y)/k, S(Y))$. In the general case the S(Y)-modules $T^1_{S(Y)}$ and $U_{S(Y)}$ are not necessarily the same.

The aim of this appendix is to prove the following:

PROPOSITION. *In the above notations there is a natural inclusion of graded S(Y)-modules* $T^1_S \subseteq U_{S(Y)}$. *In particular, if the map (1) is surjective then* $T^1_S(-i)=0$ *for every* $i \geq 1$.

Proof. Set $b_i = T_i$ mod $I(Y)$, $i=0,\dots,n$. Then $S(Y)=k[b_0,\dots,b_n]$, with $\deg(b_i)=1$. Let $c_1,\dots,c_p \in S_+$ be homogeneous elements such that $\{b_0,\dots,b_n; c_1,\dots,c_p\}$ is a system of homogeneous generators of the graded k-algebra S. Grade $k[T_0,\dots,T_n]$ and $k[T_0,\dots,T_n; T'_1,\dots,T'_p]$ by $\deg(T_i)=1$, $i=o,\dots,n$, and $\deg(T'_j)=\deg(c_j)=:q_j$, $1,\dots,p$ (with $T_0,\dots,T_n,T'_1,\dots,T'_p$ $n+p+1$ independent variables over k). Set $Z'=\mathrm{Spec}(k[T'_1,\dots,T'_p]= \mathbb{A}^p$ and $Z \times Z' = \mathrm{Spec}(k[T_0,\dots,T_n;T'_1,\dots,T'_p])= \mathbb{A}^{n+p+1}$. Then the surjective maps of graded k-algebras

$k[T_0,\dots,T_n] \longrightarrow S(Y) \qquad (T_i \longrightarrow b_i)$, and

$k[T_0,\dots,T_n; T'_1,\dots,T'_p] \longrightarrow S \qquad (T_i \longrightarrow b_i;\ T'_j \longrightarrow c_j)$

yield closed embeddings $W' \longrightarrow Z$ and $W \longrightarrow Z \times Z'$. Moreover, we get a commutative diagram

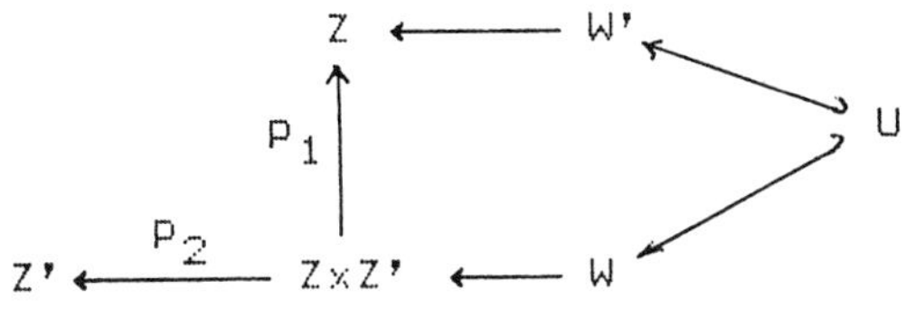

where p_1 and p_2 are the canonical projections of $Z\times Z'$. Replacing Z by an open subset Z_1 of Z such that $Z_1\cap W'=U$, we get the commutative diagram

$$\begin{array}{ccccc} & & Z_1 & \xleftarrow{\ i\ } & \\ & & \uparrow p_1 & & U \\ Z' & \xleftarrow{\ p_2\ } & Z_1\times Z' & \xleftarrow{\ j\ } & \end{array}$$

with i and j closed immersions. Set $f=p_2\circ j:U\longrightarrow Z'$.

Since $T_{Z_1\times Z'}\cong p_1^*(T_{Z_1})\oplus p_2^*(T_{Z'})$ we get $T_{Z_1\times Z'}/U\cong T_{Z_1}/U\oplus f^*(T_{Z'})\cong$ $\cong T_{Z/U}\oplus f^*(T_{Z'})$ we get the exact sequences

$$0\longrightarrow T_U\longrightarrow T_{Z_1}/U\longrightarrow N_{U,Z_1}\longrightarrow 0 \qquad \text{and}$$

$$0\longrightarrow T_U\longrightarrow T_{Z_1\times Z'}/U\cong T_Z/U\oplus f^*(T_{Z'})\longrightarrow N_{U,Z_1\times Z'}\longrightarrow 0,$$

which yield the commutative diagram

$$\begin{array}{ccccccc} H^0(U,T_Z/U\oplus f^*(T_{Z'})) & \xrightarrow{\mu_2} & H^0(U,N_{U,Z_1\times Z'}) & \longrightarrow & H^1(U,T_U) & \xrightarrow{\nu_2} & H^1(U,T_Z/U\oplus f^*(T_{Z'})) \\ \downarrow & & \downarrow & & \downarrow \mathrm{id} & & \downarrow \\ H^0(U,T_Z/U) & \xrightarrow{\mu_1} & H^0(U,N_{U,Z_1}) & \longrightarrow & H^1(U,T_U) & \xrightarrow{\nu_1} & H^1(U,T_Z/U) \end{array}$$

It follows that $\mathrm{Ker}(\nu_2)\subseteq \mathrm{Ker}(\nu_1)$, i.e. $\mathrm{Coker}(\mu_2)\subseteq \mathrm{Coker}(\mu_1)$. This implies $\mathrm{Ker}(\nu_2)=\mathrm{Ker}(\nu_1)\cap \mathrm{Ker}(H^1(U,T_U)\rightarrow H^1(U,f^*(T_{Z'})))$.

Finally, since W is normal, $U=\mathrm{Reg}(W)$, $\mathrm{codim}_W(W-U)\geq 2$ and $N_{W,Z\times Z'}$ and $T_{Z\times Z'}/W$ are reflexive sheaves, we have

$$H^0(U,T_{Z\times Z'}/U)\cong H^0(W,T_{Z\times Z'}/W) \text{ and}$$

$$H^0(U,N_{W,Z\times Z'})\cong H^0(W,N_{W,Z\times Z'}).$$

Therefore $\mathrm{Coker}(\mu_2)\cong \mathrm{Coker}(H^0(W,T_{Z\times Z'}/W)\rightarrow H^0(W,N_{W,Z\times Z'}))$. But by the definition the latter space coincides with T^1_S, which completes the proof of the proposition. The last part follows from (11), (12) and lemma 1. □

In the proof of the proposition we observed that

(13) $\quad \mathrm{Ker}(\nu_2)=\mathrm{Ker}(\nu_1)\cap \mathrm{Ker}(H^1(U,T_U)\longrightarrow H^1(U,f^*(T_{Z'})))$

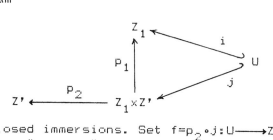

If Y is smooth U becomes a $\mathbb{G}_m$-bundle over Y, and hence the graded piece $H^1(U,f^*(T_{Z'}))(i)$ of weight i of $H^1(U,f^*(T_{Z'}))$ coincides with $H^1(Y, \bigoplus_{j=1}^{r} O_Y(j+q_j))$. Putting everything together we get:

COROLLARY. *In the hypotheses of the proposition assume in addition that Y is smooth and that $H^1(Y,O_Y(i+q_j))=0$ for every $j=1,\ldots,p$ and for some $i\in\mathbb{Z}$. Then $T^1_S(i) = U_{S(Y)}(i)$. In particular, if $H^1(Y,O_Y(i))=0$ for every $i\in\mathbb{Z}$ then $T^1_S = U_{S(Y)}$.*

Proof. Everything follows from the proof of the proposition and from the equality (13) because the hypothesis that $H^1(Y,O_Y(i+q_j))=0$ for $j=1,\ldots,p$ implies that $H^1(U,f^*(T_{Z'}))=0$. □

REMARKS. 1. Since $q:W_0=W'-\{w'\}\longrightarrow Y$ is a $\mathbb{G}_m$-bundle and Y is normal we infer that W_0 is also normal. Therefore we get

$H^0(U,T_Z/U) \cong H^0(W_0,T_Z/W_0)$ and $H^0(U,N_{W',Z}/U) \cong H^0(W_0,N_{W',Z}/W_0)$

because $\operatorname{codim}_{W_0}(W_0-U)\geq 2$ and T_Z/W_0 and $N_{W',Z}/W_0$ are reflexive sheaves. So we conclude that $U_{S(Y)}$ is isomorphic to the cokernel of the map

$$H^0(W_0,T_Z/W_0)\longrightarrow H^0(W_0,N_{W',Z}/W_0).$$

2. The proof of the proposition also yields the following fact: let $Y\subset\mathbb{P}^n$ as in the proposition, and assume that there is a linear projection $f:\mathbb{P}^n-L\longrightarrow\mathbb{P}^{n-r}$ with center the linear subspace L of $\mathbb{P}^n$ of dimesnion $r-1$ such that $L\cap Y=\emptyset$ and f yields an isomorphism $Y\cong Y':=f(Y)$. Then $U_{S(Y)}\subseteq U_{S(Y')}$. In particular, the map (1) is surjective for $(Y,\mathbb{P}^n)$, if the corresponding map for $(Y',\mathbb{P}^{n-r})$ is so.

3. By lemma 1, the proposition (of the appendix) and its corollary we get (as a by-product) the following fact: if Y is a smooth subvariety of $\mathbb{P}^n$ such that $H^1(Y,O_Y(i))=0$ for every $i\in\mathbb{Z}$, then the equality $T^1_S(-1)=0$ implies the equalities $T^1_S(-i)=0$ for every $i\geq 1$. I ignore whether this fact was already known.

EXAMPLE. Let B be a smooth projective curve of genus $g>0$ and let L be a very ample line bundle of degree d over B. Set $X=\mathbb{P}^1\times B$ and $M = p_1^*(O(1))\otimes p_2^*(L)$. Then M is a very ample line bundle on X. Take a general member Y of the linear system M. Then Y is a smooth very ample divisor on X. From the exact sequence $0\longrightarrow O_X \longrightarrow O_X(Y)= M\longrightarrow O_Y(Y)\longrightarrow 0$ we get the exact sequence

$$0\longrightarrow p_{2*}(O_X)=O_B\longrightarrow p_{2*}(M)\cong L\oplus L\longrightarrow O_Y(Y)\longrightarrow R^1p_{2*}(O_X)=0.$$

In particular, taking the determinants we get $O_Y(Y)\cong L^{\otimes 2}$ and $L\cong f^*(O(1))$, where $f= p_1/Y$. If we embed X in the projective space $P:= \mathbb{P}(H^0(X,M))$ via the complete linear system M then Y appears as the intersection of X with a hyperplane H of P. Since X is smooth, by theorem 1 above we infer in particular that the map (1) corresponding to the embedding $Y\subset H$ is not surjective. In other words $U_{S(Y)}(-1)\neq 0$ (with S(Y) the homogeneous coordinate ring of Y in H). On the other hand, if $d\geq 2g+1$ then the embedding $Y\subset H$ is not linearly normal because otherwise by a result of Mumford, Y would actually be arithmetically normal in H, which in turn would imply that X is arithmetically Cohen-Macaulay in P; in particular, $h^1(X,O_X)=g=0$, a contradiction. Moreover, if $d\geq 2g+1$ the integral closure S of S(Y) in its quotient field is (by the same result of Mumford) the homogeneous coordinate ring of Y in $H':= \mathbb{P}(H^0(Y,O_Y(1)))= \mathbb{P}(H^0(Y,M/Y))$. Finally, if B is a non-hyperelliptic curve of genus $g\geq 3$ and if $d\geq 5g+2$ we have $T^1_S(-1)=U_S(-1)=0$ (see [W_1]). By the proposition we get $T^1_S(-i)=0$ for every $i>0$. This example shows that the last assertion of the proposition is no longer true in general (at least if Y is a curve).

REFERENCES

[B_1] L. Bădescu, Infinitesimal deformations of negative weights and hyperplane sections, in "Algebraic Geometry", Proceedings L'Aquila 1988, Lect. Notes Math. 1417, Springer-Verlag (1990), 1-22.

[B_2] L. Bădescu, Polarized varieties with no deformations of negative weights, in "Geometry of complex projective varieties", Cetraro (Italy) 1990, Mediterranean Press (1993), 9-33.

[B_3] L. Bădescu, Hyperplane sections and deformations, in "Algebraic Geometry" Proceedings, Bucharest 1982, Lect. Notes Math. 1056, Springer-Verlag (1984), 1-33.

[Ba] E. Ballico, On varieties as hyperplane sections, Preprint, (1991).

[L] S.M. L'vovsky, Extension of projective varieties I, Michigan Math. J. 39 (1992), 41-51.

[P] H. Pinkham, Deformations of algebraic varieties with $\mathbb{G}_m$-action, Asterisque 20, Soc. Math. France (1974).

[S] M. Schlessinger, Infinitesimal deformations of singularities, Thesis, Harvard Univ. (1964).

[W_1] J. Wahl, Gaussian maps on algebraic curves, J. Diff. Geometry 32 (1990), 77-98.

[W_2] J. Wahl, A cohomological characterization of P^n, Invent. Math. 72 (1983), 315-322.

[Z] F.L. Zak, Some properties of dual varieties and their applications in projective geometry, in "Algebraic Geometry", Proceedings, Chicago 1989, Lect. Notes Math. 1479, Springer-Verlag (1991), 274-280.

On Logarithmic Bundles of $\mathbf{P}^n$ in Positive Characteristic

E. Ballico - F. Giovanetti - B. Russo
Dept. of Mathematics, Università di Trento
38050 Povo (TN), Italy
fax: italy + 461-881624
e-mail : ballico@itncisca.bitnet or ballico@itnvax.science.unitn.it
giovanetti@itncisca.bitnet or giovanetti@itnvax.science.unitn.it
russo@itncisca.bitnet or russo@itnvax.science.unitn.it

In this paper we study a class of rank n vector bundles ("the logarithmic bundles") on $\mathbf{P}^n$. This class was introduced and studied in detail in [DK]. Fix an integer m>n and a family $H = \{H_1,...,H_m\}$ of hyperplanes of $\mathbf{P}^n$ in linear general position (i.e. such that the intersection of any n+1 of them is empty); let E(H) be the rank n bundle of 1-forms on $\mathbf{P}^n$ which are regular outside $H_1 \cup \cdots \cup H_m$ and which have only logarithmic poles (in the sense of [De]) along each H_i; alternatively, in every characteristic E(H) may be defined using suitable elementary transformations in the sense of [Ma]. In [DK] these bundles are studied using projective techniques. Indeed in [DK] it is shown how these bundles are related to very nice (and often very classical) projective constructions. In our opinion the more interesting result of [DK] is theorem 7.2 which shows that for m≥2n+3 two bundles E(H) and E(H') are isomorphic if and only if either $H = H'$ or there is a rational normal curve $C \subset \mathbf{P}^n$ such that all the hyperplanes of H and of H' are hyperosculating to C. Here we prove the following theorem.

Theorem 0.1. *Assume $p := char(\mathbf{K}) > 2$ and $m \geq 2n+3$. Fix two arrangements $H = \{H_1,...,H_m\}$ and $H' = \{H'_1,...,H'_m\}$ of hyperplanes in linear general position and let $E = E(H)$ and $E' = E(H')$ be the corresponding logarithmic bundles. Assume $\{H_1,...,H_m\} \neq \{H'_1,...,H'_m\}$. Then $E \cong E'$ if and only if there is a rational normal curve D in the dual projective space $\mathbf{P}^{n*}$ which contains all the hyperplanes in $H \cup H'$.*

Note that the statement of 0.1 is equivalent to the statement of [DK], th. 7.2, if either char(**K**) = 0 or char(**K**)>n, but is very different if char(**K**)≤n; indeed the statement of [DK], th. 7.2, has no sense if char(**K**)≤n, because n+1 hyperosculating hyperplanes to the same rational curve C of $\mathbf{P}^n$ are not in linear general position, since they have at least a point in common (see [BR], prop. 2.2, or [EH]). But even for low values of char(**K**) the dual statement (i.e. 0.1) make sense. Most (but not all) of the proof of [DK], th. 7.2, will carry over to the case considered in 0.1 if we avoid as much as possible any duality statement. In this way we lose many nice projective constructions, but at least we see what is the core behind the proof of [DK], th. 7.2.

This research was partially supported by MURST and GNSAGA of CNR (Italy).

§ 1. We work over an algebraically closed base field **K**; set p:= char(**K**). In this paper we will always assume p>0, because if p = 0 everything here is contained in [DK].

General Remark 1.1. Everything in [DK] works if p:= char(**K**)>m. The proof of the very interesting [DK], th. 7.2, works with no change if p>2n+3, because that proof contains an easy reduction of the case "m≥2n+3" to the case "m = 2n+3".

We fix the following notations. If X is a smooth variety, let Ω_X be the locally free sheaf of regular 1-form on X. If X = $\mathbf{P}^n$ we will write respectively $\mathcal{O}$ and Ω instead of $\mathcal{O}_X$ and Ω_X. If $\mathcal{T} = \{T_1,...,T_m\}$ is an arrangement of m hyperplanes of $\mathbf{P}^n$, we will denote (as in [DK]) by E($\mathcal{T}$) the logarithmic bundle associated to it. $\mathcal{H}$ will denote always an arrangement of m hyperplanes and we will set E:= E($\mathcal{H}$).

(1.2) Fix any hyperplane H of $\mathbf{P}^n$. It is very well-known (see e.g. [B] or [DK]) that we have:

$$\Omega|H \cong \Omega_H \oplus \mathcal{O}_H(-1). \qquad (1)$$

For the definition of elementary transformation of a vector bundle (and the elementary properties needed here) see [Ma]; however for the reader of this paper it is sufficient to look at [DK], §2. The elementary transformation of Ω along the surjection $\Omega \to \Omega_H$ induced by (1) is $\mathcal{O}(-2)$ (see e.g. [B], lemma 1.1, or (at least in characteristic 0) [DK], prop. 2.10). Hence we have E($\mathcal{H}$) ≅ n$\mathcal{O}$(-1) if m = 1. Thus by the definition (as logarithmic forms regular outside all hyperplanes H_i or equivalently in terms of elementary transformation) of E($\mathcal{H}$) we have an exact sequence:

$$0 \to n\mathcal{O}(-1) \to E \to \oplus_{i>1} \varepsilon_{i_*}(\mathcal{O}_{H_i}) \to 0 \qquad (2)$$

with ε_i: $H_i \to \mathbf{P}^n$ the inclusion. By (2) and induction on m we see that $h^0(E)$ = m-1 and $h^i(E(t))$ = 0 for all integers (t,i) with either 2≤i≤n-2 or i = n-1 and t≥n-1 or i = 0 and t<0 or i = n and t≥-n. Fix P∈$\mathbf{P}^n$. Restrict (2) to a general hyperplane, R, with P∈R. By induction on n and the vanishing just proven we see easily that if m≥n+1, then E is generated by global sections. In particular if m = n+1 we have E ≅ n$\mathcal{O}$.

Proposition 1.3. *Assume m≥n+1. Then E has a presentation of the following form:*

$$0 \to (m-n-1)\mathcal{O}(-1) \to (m-1)\mathcal{O} \to E \to 0 \qquad (3)$$

Proof. By 1.2 the natural evaluation map **e**: $H^0(E)\otimes\mathcal{O}\to E$ is surjective and $h^0(E)$ = m-1. The map **e** is the map appearing in the exact sequence (3) whose existence we have to prove. Set **J**:= Ker(**e**); **J** is locally free of rank m-n-1. By the vanishing for the cohomology groups $h^i(E(t))$ proven in 1.2 and the equality $h^0(E)$ = m-1, we obtain $h^u(\mathbf{J}(-u)) = 0$ for all integers u≥0; by Castelnuovo - Mumford criterion we obtain that **J**(1) is spanned by global sections. Hence for every line **m** every rank 1 factor of **J**|**m** has degree ≥-1. Since $c_1(\mathbf{J})$ = -rank(**J**), $\mathbf{J}|\mathbf{m} \cong (m-n-1)\mathcal{O}_{\mathbf{m}}(-1)$. We claim that this implies $\mathbf{J} \cong \mathcal{O}(-1)$; this claim can be checked in several ways; for instance using the vanishing of cohomology groups, induction on n taking a general hyperplane R and using standard exact sequences; alternatively, use the spannedness of **J**(1); alternatively, use (in a very simple case) a standard construction in the theory of uniform vector bundles. Here is another proof of 1.3; we proved that **J**(1) is spanned; by the vanishing in 1.2 we have $\chi(\mathbf{J}(1)) = h^0(\mathbf{J}(1))$; hence by the cohomology of E(1) and a twist of (3) (or using Riemann - Roch) we obtain $h^0(\mathbf{J}(1))$ = m-n-1 = rank(**J**(1)); hence **J**(1) must be trivial. ♦

Recall that a bundle (of any rank, r) which, up to a twist, has a presentation of the form (3) (with r instead of n) is called a Steiner bundle.

Proposition 1.4. Assume p>r. Then a Steiner bundle with rank r is stable, unless it is the trivial bundle. In particular, E is stable if p>n and m≥n+1.

Proof. In characteristic 0 this is a very particular case of [BS], th. 3.1. We claim that in this particular case the proof works with no change by the assumption on p. Indeed, that proof uses only exterior powers of order at most the rank of the Steiner bundle (see the proof of [BK], lemmas 2.5 and 2.6). ♦

However, note that in positive characteristic, even knowing the stability of E, this does not give a good information on the general splitting type of E, since in positive characteristic the Grauert - Mülich theorem fails (and its generalizations, too).

Proof of theorem 0.1: We divide the proof in 3 steps, each of them, as stressed several times, is contained essentially in [DK]). For a subset **w** of $\Pi := \mathbf{P}^{n*}$, let M(**w**) be the projective linear space of all quadrics of Π containing **w**. As a general claim, note that, at least if p>2, it is true (and easy to check) that a points of $\mathbf{P}^n$ in linear general position and imposing only 2n+1 conditions to quadrics are contained in a rational normal curve if a≥2n+3. The strategy of [DK], proof of 7.2, to reconstruct H from the isomorphism class of E (unless all H is "related" to a rational normal curve) is to show that the hyperplanes of H are exactly the set of all hyperplanes, H, such that for every line **d** of H, the bundle E|**d** has as a factor a line bundle of degree ≤0; any line with this property is called a super-jumping line for E; it was proven in [DK], 7.4, (and the proof works in positive characteristic) that every line contained in some H_i is a super-jumping line for E.

Step 1: Here we collect a few results of [DK] which we claim to hold even when char(**K**) = p>2. By the definition of super-jumping line and the fact that every bundle on $\mathbf{P}^1$ splits into direct sum of line bundles, **d** is a super-jumping line if and only if $H^0(\mathbf{d}, E^*|\mathbf{d}) \neq 0$; Z will denote the codimension 2 linear subspace of the dual projective space $\mathbf{P}^{n*}$ which is dual to **d**;]Z[will denote the pencil of hyperplanes of $\mathbf{P}^{n*}$ containing Z. We claim that even if p>2 this is equivalent to the existence of degree 1 morphism ϕ from **d** to the pencil]Z[with $\phi(\mathbf{d}\cap H_i) = p_i$ for every i with $1 \leq i \leq 2n$. Indeed we claim that the *proof* of [DK], 5.11, works in our situation with no change if (with the notations of [DK, 5.11) we consider only the case d = 2; essentially, we bypass all the very interesting geometry considered in [DK] and related to the key word "d-codependence" (which relies so much on taking duals) except for the case $d \leq 2$.

Step 2: Here and in step 3 the ambient space will be the dual projective space Π. Here we will check (exactly as in [DK], proof of prop. 7.10(a))) that if $T := \{p_1,...,p_{2n}\} \subset \Pi$ is in linear general position and if $q \in \Pi \setminus T$, then for a general codimension 2 linear subspace Z of Π with $q \in Z$, $M(T \cup Z)$ is a point. Assume by contradiction $\dim(M(T \cup Z)) > 0$. Hence for every $r \in \Pi$ there is $Q \in M(T \cup Z)$ with $r \in Q$. By semicontinuity to find a contradiction we may choose both Z and r in a particular way. We take r such that $\{r,p_1,...,p_{n-1}\}$ spans a hyperplane H, and we take $Z \subset H$ with $\{r,p_1,...,p_{n-1}\} \cap Z = \emptyset$. Fix $Q \in M(T \cup Z)$ with $r \in Q$. Since $Q \cap H$ contains Z and$\{r,p_1,...,p_{n-1}\}$spans H and is disjoint from Z, we see that $Q = H \cup H'$ for some hyperplane H'. Since $\{q_n,...,q_{2n}\}$ spans Π, we obtain a contradiction.

Step 3: Let A(4) be the set of all quadrics of Π with rank at most 4. It is easy to check (e.g. as in [DK], 7.13) that M(T) is spanned by rank 2 quadrics. Every element of a pencil of quadrics spanned by rank 2 quadrics has rank ≤ 4. Since $M(\{q\} \cup T)$ is a hyperplane, we see that $M(\{q\} \cup T)$ is spanned by quadrics of rank ≤ 4, i.e. of quadrics in $M(Z \cup T)$ for some codimension 2 linear subspace Z of Π with $q \in Z$; this is exactly the proof of [DK], 7.12. The trouble is that if n>3 a priori all such spanning elements may be in some $M(Z \cup T)$ with Z such that $\dim(M(Z \cup \{p_1,...,p_{n-1}\})) > 1$ (call such Z a bad Z). Since $A(4) \cap M(q,p_1,...,p_{n-1})$ is equidimensional of dimension 3n-4, it is sufficient to check that the union, Γ, of all $M(Z \cup \{p_1,...,p_{n-1}\})$ with Z bad has dimension at most 3n-5. We may assume n>3. For any $Q \in (A(4) \cap M(Z \cup \{p_1,...,p_{n-1}\}))$, let V_Q be a codimension 2 linear subspace contained in the vertex of Q and let **t** be the image of $\{p_1,...,p_{n-1}\}$ in a $\mathbf{P}^3$, **P**, under the projection of $\Pi \setminus V_Q$ from V_Q into **P**. We will leave to the interested reader the task of checking several cases. One may distinguish cases according to the existence of k with $2 \leq k \leq n$ and with q contained in the linear span of k points p_i (but not of k-1). Since the points $\{p_i\}$ are in linear general position, we have card(**t**) = n-1 unless Z intersects a line spanned by two p_i's and the points in **t** are in linear general position unless Z intersects a plane spanned by 3 of the points p_i's; the reader may check (privately) the dimension of the set of all such Z's. If **t** contains at least 9 points, we see that, it is contained in a reduced and connected curve Γ with $p_a(\Gamma) = 0$ and $\deg(\Gamma) \leq 3$. The dimensional count left to the reader concludes the proof of 0.1 and of this note.

References

[B] E. Ballico, *On the homogeneous ideal of projectively normal curves,* Ann. Mat. Pura Appl. (4) **154** (1989), 83-90.

[BR] E. Ballico, B. Russo, *On the general osculating flag to a projective curve in characteristic p,* Comm. in Algebra **20** (1992), 3729-3740.

[BS] G. Bohnhorst, H. Spindler, *The stability of certain vector bundles on P^n*, in: Complex Algebraic Varieties, Proceedings, 1990, pp. 39-50, Lect. Notes in Math. **1507**, Springer-Verlag, Berlin/Heidelberg/New York, 1992.

[De] P. Deligne, *Théorie de Hodge II,* Publ. Math. I.H.E.S. **40** (1971), 5-58.

[DK] I. Dolgachev, M. Kapranov, *Arrangements of hyperplanes and vector bundles,* preprint.

[EH] E. Esteves, M. Homma, *Order sequences and rational curves,* this volume.

[Ma] M. Maruyama, *Elementary transformations in the theory of algebraic vector bundles,* in: Algebraic Geometry - Proceedings La Rabida, pp. 241-266, Lect. Notes in Math. **961** Springer-Verlag, Berlin/Heidelberg/New York, 1983.

Trisecant Formulas for Smooth Projective Varieties

LOTHAR GÖTTSCHE Dipartimento di Matematica, Università di Trento, I-38050 Povo, Italy

0 INTRODUCTION

Let $X \subset \mathbf{P}_N$ be a smooth projective variety over an algebraically closed field of characteristic different from 2 and 3. In this paper we obtain formulas for the trisecants of X fulfilling Schubert conditions, i.e. the lines in $\mathbf{P}_N$ intersecting X in a subscheme of length 3, which also intersect a given $(l+1)$-codimensional linear subspace of $\mathbf{P}_N$. In the case that X is a surface and the dimensions N and l are chosen in such a way that the number of trisecants is expected to be finite, these numbers have been obtained by Le Barz in [LB2] and more generally for multisecants in [LB1], [LB3]. In the case of curves multisecant formulas were obtained in [LB4].

In this paper we compute the class in the Chow ring $A^*(X)$ of the locus of points $x \in X$ lying on a trisecant fulfilling a Schubert condition. This should make the formulas more useful for applications. We also compute these classes for tangent lines intersecting X again and inflectional lines. Like the formulas of Le Barz our trisecant formulas also have the property that the class has to be zero, if the dimension of the locus in question is smaller then expected. We apply these

This paper was written during the author's stay at the University of Trento. The author likes to thank the Italian C.N.R for financial support and the mathematics department of Trento for hospitality.

results to projections of a variety $X \subset \mathbf{P}_N$ to $\mathbf{P}_{N-1}$ from a point $p \in X$ and to questions of 2-spannedness of the linear system of hyperplanes on X.

We will use the incidence variety $X^{[2,3]}$, which parametrizes pairs $Z_2 \subset Z_3$ of subschemes of X of lengths 2 and 3 respectively. Its cohomology ring has been determined in [F-G]. Our method is based on the observation that the set of pairs $Z_2 \subset Z_3$ lying on a line is the degeneracy locus of a map of natural vector bundles on $X^{[2,3]}$, and thus its class can be computed by the Porteous formula. It would also be possible to use multiple point theory (see [Kl1], [Kl2], [C]) to obtain similar results.

1 NOTATIONS AND REVIEW OF $X^{[2,3]}$

For a smooth variety Y and $\alpha \in A^*(Y)$ we denote by $\{\alpha\}_k$ its part in $A^k(Y)$. If Y has dimension m we write $\int_Y \alpha$ for the degree of $\{\alpha\}_m$. If $Z \subset Y$ is a subvariety, we denote by $[Z]$ its class in $A^*(Y)$. For a vector bundle V on Y we denote by $c(V) = 1 + c_1(V) + \ldots + c_m(V)$ its total Chern class and by $s(V) = 1/c(V) = 1 + s_1(V) + \ldots + s_m(V)$ its total Segre class. We write $V(y)$ for the fibre of V over $y \in Y$. As usual we put $c(Y) = c(T_Y)$ and $s(Y) = s(T_Y)$. We use the convention that $s_i(V) = 0$ for $i < 0$.

We now review some notations and results from [F-G]. Let X be a smooth projective variety over an algebraically closed field. We denote by $X^{[n]} := \mathrm{Hilb}^n(X)$ the Hilbert scheme of subschemes of length n on X. Let $Z_n(X) \subset X \times X^{[n]}$ be the universal family and q_1, q_2 its projections to X and $X^{[n]}$ respectively. If L is a line bundle on X, then $\overline{L}_n := q_{2*}q_1^*(L)$ is a vector bundle of rank n on $X^{[n]}$, whose fiber over a point $Z \in X^{[n]}$ can be naturally identified with $H^0(Z, L_{|Z})$. Let $X^{[2,3]} \subset X^{[2]} \times X^{[3]}$ be the incidence correspondence

$$X^{[2,3]} := \{(Z_2, Z_3) \in X^{[2]} \times X^{[3]} \mid Z_2 \subset Z_3\}.$$

$X^{[2,3]}$ is smooth and there is a natural morphism $res : X^{[2,3]} \longrightarrow X$. Here $res(Z_2, Z)$ is the point in which Z has a higher length than Z_2 (see [E-LB]). For $i = 2, 3$ let $\widetilde{L}_i$ be the pullback of $\overline{L}_i$ by the natural projection.

Let $q : X \times X^{[2]} \longrightarrow X$ be the natural projection, and $\pi : Z_2(X) \longrightarrow X^{[2]}$ the projection. Let $\widetilde{X \times X}$ be the blowup of $X \times X$ along the diagonal Δ and E the exceptional locus. Let $p_1, p_2 : \widetilde{X \times X} \longrightarrow X$ be the projections. We can identify $\widetilde{X \times X}$ with $Z_2(X)$ in such a way that p_1 becomes the restriction of q and π the quotient map of $\widetilde{X \times X}$ by the natural involution. Let

$$F := \{(Z_2, Z) \in X^{[2,3]} \mid res(Z_2, Z) \in Z_2\}.$$

Let $\epsilon : X^{[2,3]} \longrightarrow X \times X^{[2]}$ be the product of res and the natural projection. Then ϵ is the blowup along $Z_2(X)$ and F is the exceptional divisor. Let $j : F \longrightarrow X^{[2,3]}$ be the embedding and $\bar{\epsilon} := \epsilon_{|F}$. We denote by $|M|$ the number of elements of a finite set M. Let

$$D := \{Z_2 \in X^{[2]} \mid |supp(Z_2)| = 1\},$$
$$D^2(X) := \{(Z_2, Z) \in X^{[2,3]} \mid |supp(Z)| = 1\}.$$

Then we see $\pi_2^*(D) = 2E$ and $[D^2(X)] = j_*\bar{\epsilon}^*(E)$. We denote $e := [-E] \in A^1(\widetilde{X \times X})$ and $f := [-F] \in A^1(X^{[2,3]})$. The sign (which is not there in [F-G]) has been introduced to avoid signs in the computations.

For a class $\alpha \in A^*(X)$ we denote $\alpha_1 := p_1^*(\alpha), \alpha_2 := p_2^*(\alpha)$. Then we have $\alpha_2 e = \alpha_1 e$ and $res^*(\alpha)_{|F} = \bar{\epsilon}^*(\alpha_1)$ for all $\alpha \in A^*(X)$. For the normal bundle N of $Z_2(X) \in X \times X^{[2]}$ we have $c(N) = c(X)_1(1-2e)/(1-e)$ ([F-G] Lemma 2.1). So we get by [Fu] Cor. 4.2.2, the projection formula and [F-G] Lemma 4.1:

REMARK 1.1.
(1) $p_{1*}(\alpha_2\beta_1) = (\int_X \alpha)\beta,\ p_{1*}(\alpha_1 e^{n+m}) = -s_m(X)\alpha$ for all $\alpha, \beta \in A^*(X)$,
(2) $\bar{\epsilon}_*(f^{n+m-1}) = \{s(X)_1(1-e)/(1-2e)\}_m$,
(3) $\pi^*(c(\overline{L}_2)) = (1+L_1)(1+L_2-E),\ c(\widetilde{L}_3) = c(\widetilde{L}_2)(1+res^*(L)-F)$.

2 FORMULATION AND PROOF OF THE TRISECANT FORMULAS

Let $X \subset \mathbf{P}_N$ a smooth projective variety of dimension n. Let $\mathcal{O}(1)$ be the hyperplane bundle on $\mathbf{P}_N$ and H its restriction to X.

DEFINITION 2.1. The class of points lying on a trisecant which intersects a given $(l+1)$-codimensional linear subspace is

$$t_{N,l}(X) := res_*(s_l(\widetilde{H}_2)s_{N-1}(\widetilde{H}_3)) \in A^{N+l-1-2n}(X).$$

The class of residual points of tangents intersecting X again, which meet a general $(l+1)$-codimensional linear subspace, is

$$u_{N,m}(X) := res_*(\epsilon^*([D])s_l(\widetilde{H}_2)s_{N-1}(\widetilde{H}_3)).$$

The class of tangent points of tangents intersecting X again, which meet a general $(l+1)$-codimensional linear subspace, is

$$v_{N,l}(X) := res_*([F](s_l(\widetilde{H}_2)s_{N-1}(\widetilde{H}_3)).$$

The class of points in X lying on an inflectional line, which intersects a given $(l+1)$-codimensional linear subspace, is

$$w_{N,l}(X) := res_*([D^2(X)]s_l(\widetilde{H}_2)s_{N-1}(\widetilde{H}_3)).$$

We now have to justify this definition. For us a trisecant is a subscheme Z of length 3 of X, which is also a subscheme of a line λ (which then is unique). We say that the trisecant intersects a subvariety Y of $\mathbf{P}_N$, if λ does. If $|supp(Z)| \le 2$ we call Z tangential. In this case if $Z_2 \subset Z$ is the unique subscheme of length 2 with $|supp(Z_2)| = 1$, we call $supp(Z_2)$ the tangent point and $res(Z_2, Z)$ the residual point. If finally $|supp(Z)| = 1$ we call Z inflectional.

We have the restriction map $r_3 : H^0(\mathbf{P}_N, \mathcal{O}(1)) \otimes \mathcal{O}_{X^{[2,3]}} \longrightarrow \widetilde{H}_3$ given over $(Z_2, Z) \in X^{[2,3]}$ by mapping the section $s \in H^0(\mathbf{P}_N, \mathcal{O}(1))$ to $s_{|Z}$. Then the closed subset $\{(Z_2, Z) \in X^{[2,3]} \mid Z \text{ lies on a line}\}$ is the degeneracy locus

$$D_2(r_3) = \{(Z_2, Z) \in X^{[2,3]} \mid rk(r_3(Z_2, Z)) \le 2\}.$$

Let $K \subset \mathbf{P}_N$ be a general linear subspace of codimension $l+1$ and $\bar{K} \subset H^0(\mathbf{P}_N, \mathcal{O}(1))$ be the space of hyperplanes containing K. Then again we have a morphism $r_2 : \bar{K} \otimes \mathcal{O}_{X^{[2,3]}} \longrightarrow \widetilde{H}_2$. The locus $\{(Z_2, Z) \in X^{[2,3]} \mid Z \text{ lies on a line intersecting } K\}$ is $D_1(r_2) \cap D_2(r_3)$. (The line on which Z lies is already determined by Z_2.)

We see that the set of points $x \in X$, which lie on a trisecant intersecting a general $(l+1)$-codimensional linear subspace, is $res(D_1(r_2) \cap D_2(r_3))$ and similarly for tangent points and residual points of tangential trisecants ($res(F \cap D_1(r_2) \cap D_2(r_3))$ and $res(\epsilon^{-1}(D) \cap D_1(r_2) \cap D_2(r_3))$ respectively) and for inflectional points $(res(D^2(X) \cap D_1(r_2) \cap D_2(r_3)))$.

REMARK 2.2. By the Porteous formula ([Fu] Thm. 14.4) the following holds:

(1) $D_1(r_2) \cap D_2(r_3)$ has at most codimension $l+N-1$. If its codimension is $l+N-1$ (equivalently, if the locus of trisecants intersecting a general $(l+1)$-codimensional linear subspace has codimension $l+N-1$ in $X^{[3]}$), then its class (with the appropriate multiplicities) is $s_l(\widetilde{H}_2)s_{N-1}(\widetilde{H}_3)$. In this case $t_{N,l}(X)$ is the part in $A^{l+N-2n-1}(X)$ of the class of the locus of points $x \in X$ lying on a trisecant intersecting a general $(l+1)$-codimensional linear subspace (with the appropriate multiplicities). The corresponding results also hold for $u_{N,l}(X), v_{N,l}(X)$ and $w_{N,l}(X)$.

(2) In any case (independent of the fact that the locus in $X^{[3]}$ has the expected codimension) the support of a suitable representative of $s_l(\widetilde{H}_2)s_{N-1}(\widetilde{H}_3)$ is contained in

$$\left\{(Z_2, Z) \in X^{[2,3]} \;\middle|\; \begin{array}{l} Z \text{ lies on a line intersecting a general} \\ (l+1)\text{-codimensional linear subspace} \end{array}\right\}.$$

Thus the support of the corresponding representative of $t_{N,l}(X)$ is contained in the locus of points $x \in X$ lying on a trisecant which intersects a general $(l+1)$-codimensional linear subspace. So if the codimension of this locus is bigger then $l+N-2n-1$, then $t_{N,l}(X)=0$. The corresponding results hold for $u_{N,l}(X)$, $v_{N,l}(X)$ and $w_{N,l}(X)$.

(3) Let $i : X \longrightarrow \mathbf{P}_N$ be the inclusion. Let $A_3(\mathbf{P}_N) \subset \mathbf{P}_N^{[2,3]}$ be the closed subscheme with the reduced structure defined by

$$A_3(\mathbf{P}_N) := \{(Z_2, Z) \in \mathbf{P}_N^{[2,3]} \mid Z \text{ lies on a line } \}$$

and

$$A_{2,l}(\mathbf{P}_N) := \{(Z_2, Z) \in \mathbf{P}_N^{[2,3]} \mid \text{the line through } Z_2 \text{ intersects K}\}.$$

Then the Porteous formula shows that $s_{N-1}(\widetilde{H}_3) = i^*([A_3(\mathbf{P}_N)])$ and $s_l(\widetilde{H}_2) = i^*([A_{2,l}(\mathbf{P}_N)])$.

THEOREM 2.3. *Let $X \subset \mathbf{P}_N$ be a smooth projective variety of dimension n and degree d. Let $l \leq N-1$ and let $\bar{n}$ denote the maximum of 0 and $n-l$. For a number $m \leq n$ and $a \in \{0,1\}$ we define the following classes in $A^m(X)$:*

$$r^1_{N,l,m}(X) := \frac{1}{2}(l-2\bar{n}+1)d^2H^m,$$

$$r^2_{N,l,m}(X) :=$$

$$H^m\sum_{k=0}^{n}\left(\binom{N+l+1-m}{k} - \binom{max(N-m,l+1)}{k-min(N-m,l+1)}\right)\textstyle\int_X (H^k s_{n-k}(X))/2,$$

$$r^3_{N,l,m}(X) := d\sum_{k=0}^{m}\left(\binom{N+l+1-n-\bar{n}}{k-\bar{n}} - \binom{N-n}{k-l-1}\right)H^k s_{m-k}(X),$$

$$r^4_{N,l,m,a}(X) :=$$

$$\sum_{k=0}^{m}\sum_{r=0}^{l}\binom{l+1}{r}\binom{N+1}{k-r}\sum_{j=l-r-n+a}^{m-k} 2^{n+j+r-l-a}s_j(X)s_{m-k-j}(X)H^k.$$

Let $m_0 := N+l-2n-1$, $m_1 := N+l-2n$, $m_2 := N+l+1-2n$. Then we have

(1) $$t_{N,l}(X) = r^1_{N,l,m_0}(X) - r^2_{N,l,m_0}(X) - r^3_{N,l,m_0}(X) + r^4_{N,l,m_0,0}(X),$$

(2) $$u_{N,l}(X) = 2r^2_{N,l,m_1}(X) - 2r^4_{N,l,m_1,1}(X),$$

(3) $$v_{N,l}(X) = r^3_{N,l,m_1}(X) - r^4_{N,l,m_1,0}(X),$$

(4) $$t_{N,l}(X) = r^4_{N,l,m_2,1}(X).$$

The reader will notice that the formulas simplify considerably when $l < n$. In particular in case $l = 0$ (i.e. we don't impose any Schubert conditions on the line) we get (1)-(4) with the simpler expressions

$$r^1_{N,0,m}(X) := \frac{1}{2}d^2H^m,$$

$$r^2_{N,0,m}(X) := \frac{H^m}{2}\sum_{k=0}^{n}\binom{N-m}{k}\int_X (H^k s_{n-k}(X)),$$

$$r^3_{N,l,m}(X) := d\sum_{k=0}^{m}\binom{N-n}{k}H^k s_{m-k}(X),$$

$$r^4_{N,l,m,a}(X) := \sum_{k=0}^{m}\binom{N+1}{k}\sum_{j=0}^{m-k}2^{n+j-a}s_j(X)s_{m-k-j}(X)H^k.$$

Proof: Let a, x, y be indeterminates and let $[\ \ ]_w$ denote the part of degree w in a, x, y. We will use repeatedly the following easy identities:

$$\begin{aligned}
&[(1-a)^{-1}(1-(a+x))^{-1}]_r = \frac{1}{x}((a+x)^{r+1} - a^{r+1}) = \sum_{s=0}^{r}\binom{r+1}{s}a^s x^{r-s},\\
&[(1-a)^{-1}(1-(a+x))^{-1}(1-(a+y))^{-1}]_r = \sum_{s+i+j=r}\binom{r+2}{s}a^s x^i y^j.
\end{aligned} \qquad (*)$$

We will use the following notation:

$$(a+b)^w_{\geq s} := \sum_{k=s}^{w}\binom{w}{k}a^{w-k}b^k$$

(Notice that the order of a and b is important). We shall prove (1) in detail. The proofs of (2) - (4) are analogous but simpler. Let $m = N - 2n + l - 1$. A class $\alpha \in A^*(X^{[2,3]})$ can be written as $\alpha = \epsilon^*(\beta) + j_*(\gamma)$, with $\beta \in A^*(X \times X^{[2]})$ and $\gamma \in A^*(F)$. Then

$$res_*(\alpha) = q_*(\beta) + p_{1*}(\bar{\epsilon}_*(\gamma)).$$

Hence our computations can be carried out, on the one hand by pushing down from $X \times X^{[2]}$ to X, and on the other hand, by pushing down first in the bundle $F \longrightarrow \widetilde{X \times X}$ and then pushing down further to X. Thus by Remark 1.1 and the projection formula we obtain

$$\begin{aligned}
t_{N,l}(X) &= res_*(s_l(\widetilde{H}_2)s_{N-1}(\widetilde{H}_3))\\
&= \sum_{r=n}^{N-1} H^r q_*(s_l(\overline{H}_2)s_{N-1-r}(\overline{H}_2))\\
&- \sum_{r=n}^{N-1} p_{1*}\left(\bar{\epsilon}_*\left(s_l(\widetilde{H}_2)s_{N-1-r}(\widetilde{H}_2)((\bar{\epsilon}^*(H_1)+f)^r_{\geq n})/f\right)\right).
\end{aligned}$$

Then using $(*)$ we find

$$\sum_{r=0}^{N-1} H^r res_*(s_l(\overline{H}_2)s_{N-1-r}(\overline{H}_2)) = H^m \int_{X^{[2]}} s_l(\overline{H}_2)s_{2n-l}(\overline{H})$$

$$= H^m \int_{\widetilde{X \times X}} \Bigg(\sum_{i=0}^{l} H_1^i H_2^{l-i} \sum_{j=0}^{2n-l} H_1^j H_2^{2n-l-j}$$

$$+ ((H_1 + e)^{l+1} - H_1^{l+1})((H_1 + e)^{2n-l+1} - H_1^{2n-l+1})/e^2 \Bigg)/2$$

$$= H^m \frac{1}{2} \Bigg((min(l, 2n - l) + 1)d^2$$

$$- \sum_{k=0}^{n} \left(\binom{2n+2}{k} - \binom{2n+1+2\bar{n}-l}{k+2\bar{n}-l-1} \right) \int_X H^k s_{n-k}(X) \Bigg).$$

For all $\alpha \in A^*(\widetilde{X \times X})$ and all k we set

$$\bar{\epsilon}_*^1(\bar{\epsilon}^*(\alpha) f^{n+k-1}) := s_k(X)_1 \alpha,$$

so that

$$\bar{\epsilon}_*(f^{n+k-1}) := \bar{\epsilon}_*^1(f^{n+k-1}) + \sum_{i=1}^{k} 2^{i-1} e^i s_{k-i}(X)_1.$$

Then, using Remark 1.1 and $(*)$ we find

$$\sum_{r=n}^{N-1} p_{1*} \left(\bar{\epsilon}_* \left(s_l(\widetilde{H}_2) s_{N-1-r}(\widetilde{H}_2) (\bar{\epsilon}^*(H_1) + f)_{\geq n}^{r})/f \right) \right) = V_1 + V_2,$$

where

$$V_1 = p_{1*} \left(\sum_{j=0}^{l} H_1^j H_2^{l-j} \sum_{r=n}^{N-1} \sum_{k=0}^{N-1-r} H_1^k H_2^{N-1-r-k} \bar{\epsilon}_*^1 \left((\bar{\epsilon}^*(H_1) + f)_{\geq n}^{r} / f \right) \right)$$

$$= p_{1*} \left(\sum_{j=0}^{l} H_1^j H_2^{l-j} \sum_{r=n}^{N-1} H_2^{N-1-r} \bar{\epsilon}_*^1 \left((\bar{\epsilon}^*(H_1) + f)_{\geq n+1}^{r+1} / f^2 \right) \right)$$

$$= p_{1*} \left(H_2^n \sum_{j=\bar{n}}^{l} H_1^j \bar{\epsilon}_*^1 \left((\bar{\epsilon}^*(H_1) + f)_{\geq n+1}^{N+l-n-j} / f^2 \right) \right)$$

$$= p_{1*} \Big(H_2^n \bar{\epsilon}_*^1 \Big((\bar{\epsilon}^*(H_1^{\bar{n}})(\bar{\epsilon}^*(H_1) + f)_{\geq n+2}^{N+l+1-n-\bar{n}}$$

$$- \bar{\epsilon}^*(H_1^{l+1})(\bar{\epsilon}^*(H_1) + f)_{\geq n+2}^{N-n})/f^3 \Big) \Big)$$

$$= d \sum_{k=0}^{m} \left(\binom{N+l+1-n-\bar{n}}{k-\bar{n}} - \binom{N-n}{k-l-1} \right) s_{m-k}(X) H^k.$$

Here we have used the fact that $(p_1)_*(H_2^v \alpha_1) = 0$ for $v \neq n$. Using the same fact again with $v = 0$, Remark 1.1 and $(*)$ we obtain

$$V_2 = p_{1*} \left(\sum_{r=0}^{l} \binom{l+1}{r} H_1^r e^{l-r} \sum_{\substack{k+i+j=N-1 \\ j>0}} \binom{N+1}{k} H_1^k e^i \bar{\epsilon}_*(f^{j-1}) \right)$$

$$= \sum_{k=0}^{m} \sum_{r=0}^{l} \binom{l+1}{r} \binom{N+1}{k-r} p_{1*} \left(\left\{ H_1^k s(X)_1 \frac{1-e}{1-2e} \frac{e^{l-r}}{1-e} \right\}_{n+m} \right)$$

$$= \sum_{k=0}^{m} \sum_{r=0}^{l} \binom{l+1}{r} \binom{N+1}{k-r} \sum_{j=l-r-n}^{m-k} 2^{n+j+r-l} s_j(X) s_{m-k-j}(X) H^k.$$

This proves (1).
(2) by the above we have to compute $res_*(\epsilon^*([D])s_l(\widetilde{H}_2)s_{N-1}(\widetilde{H}_3)$. We use $\pi^*([D]) = 2[E]$.
(3) By the above we have to compute $res_*([F]s_l(\widetilde{H}_2)s_{N-1}(\widetilde{H}_3)$.
(4) By the above we have to compute $p_{1*}([E]\bar{\epsilon}_*(s_l(\widetilde{H}_2)s_{N-1}(\widetilde{H}_3)))$. The computations of these classes are in all 3 cases analogous to case (1). □

REMARK 2.4. If the dimensions N and l are chosen is such a way that the set of trisecants (tangential trisecants, inflectional lines respectively) is expected to be finite, then the corresponding number is $t_{N,l}(X)/3$ ($u_{N,l}(X)$, $w_{N,l}(X)$ respectively). Using Hartshorne's double point formula one also sees that in this case $u_{N,l}(X) = v_{N,l}(X)$. In particular, if X is a surface we recover the formulas from [LB2].

REMARK 2.5. The enumerative significance of the trisecant formulas is a rather subtle issue. For simplicity we restrict ourselves here to the case that the expected dimension of the locus of trisecants of $X \subset \mathbf{P}_N$ is zero. Even in this case X can have infinitely many trisecants. The most obvious way in which this can occur is that X contains lines or even larger linear subspaces. As we are viewing a trisecant as a subscheme of length 3 of X lying on a line, a line contained in X corresponds to a threedimensional family of trisecants. In the case of a surface containing only finitely many lines Le Barz has determined in [LB5] the contribution of these lines to the trisecant formula, in terms of the self-intersection of the lines. Similar calculations would also be possible for X of arbitrary dimension (replacing the self-intersection by the first Chern class of the normal bundle). Note, however, that the number of trisecants of X can very well be infinite, even if X does not contain any lines. If our formulas predict a negative number of trisecants, then we can infer that there are infinitely many, if they predict a positive number, then we still don't know whether in fact the number of trisecants is finite or infinite.

If the number of trisecants of X is indeed finite, we still have to take multiplicities into account. The easiest way how these multiplicities can occur is that the trisecant line is indeed a multisecant. It is easy to see that a simple r-secant line is counted with multiplicity $\binom{r}{3}$ as trisecant. (See also the discussion in the introduction of [LB2].)

3 APPLICATIONS TO PROJECTIONS AND 2-SPANNEDNESS

We want to give some easy applications of Theorem 2.3 to projections and to questions of 2-spannedness of the linear system of hyperplanes on $X \subset \mathbf{P}_N$.

Trisecants play a role in projecting a projective variety $X \subset \mathbf{P}_N$ of dimension n to $\mathbf{P}_{N-1}$ from a point $p \in X$. Let $p \in X$ and $\hat{X}(p)$ the blowup of X at p. Let G be the exceptional divisor. The rational projection map $\mathbf{P}_N \longrightarrow \mathbf{P}_{N-1}$ from p gives rise to a morphism $\pi_p : \hat{X}(p) \longrightarrow \mathbf{P}_{N-1}$, which embedds G as a projective subspace $P(p)$ of dimension $n-1$. It is easy to see that π_p is an embedding if and only if p does not lie on a trisecant of X. Similarly π_p is injective on tangent vectors if and only if p is not the residual point tangential trisecant and $\pi_p|_{\pi_p^{-1}(P(p))}$ is an embedding if p is not the tangent point of a tangential trisecant. Finally $\pi_p|_{\pi_p^{-1}(P(p))}$ is injective on tangent vectors if and only if there are no inflectional lines through p.

In [Bau1], [Bau2] the formulas of [LB2] were used to classify the surfaces in $S \subset \mathbf{P}_5$ for which there exists a point $p \in S$ such that $\pi_p : \hat{S}(p) \longrightarrow \mathbf{P}_4$ is an embedding.

We denote $H^{p,q}(X) := H^q(X, \Omega_p)$ and by $h^{p,q}$ the corresponding Hodge number.

PROPOSITION 3.1. *Let X be a smooth variety of dimension n and degree d in $\mathbf{P}_N$ over an algebraically closed field k.*

(1) If $k = \mathbf{C}$ and $N = 2n+1$ and there exist points $p \in X$ such that π_p is an embedding, then $d = 2^{n+1}$ or $h^{0,1}(X) = h^{0,2}(X) = 0$.

(2) If $N = 2n$ and there exist points $p \in X$ such that $\pi_p|_{\pi_p^{-1}(P(p))}$ is an embedding, then $d = 2^n$.

(3) If $k = \mathbf{C}$ and $N = 2n$ and X and there exist points $p \in X$ such that π_p is injective on tangent vectors, then $h^{0,1}(X) = h^{0,2}(X) = 0$.

(4) If $N = 2n-1$, then there is no $p \in X$ such that $\pi_p|_{\pi_p^{-1}(P(p))}$ is injective on tangent vectors.

REMARK 3.2. If $X \subset \mathbf{P}_{2n}$ is a smooth subvariety of dimension n for which there exist points $p \in X$ such that π_p is an embedding, then one might conjecture that X has to be a complete intersection of quadrics. It is known that this result is true for $n = 2$ and $n = 3$. Proposition 3.1(2) shows that at least the degree of X is the expected one for arbitrary n. In [B-S-S] it is shown more generally, that the following holds: If $X \subset \mathbf{P}_N$ is a smooth variety of dimension n with $N \leq 2n-1$ containing a linearly embedded $\mathbf{P}_{n-1}$ with normal bundle $\mathcal{O}(-l)$, then X has degree $(l+1)^n - 1$. (Note that if π_p is an embedding, then $\pi_p(P(p)) \subset \pi_p(\hat{X}(p))$ is a linearly embedded $\mathbf{P}_{n-1}$ with normal bundle $\mathcal{O}(-1)$; also $deg(X(p)) = deg(X) - 1$.)

For the proof of parts (1) and (3) of Proposition 3.1 we will use the following easy lemma:

LEMMA 3.3. *Let X be a smooth projective variety over* $\mathbf{C}$ *of dimension n. Let $Y \subset X$ be an irreducible subvariety of dimension k and $\alpha \in H_{2n-2k}(X)$ with $[Y]\alpha = 0$, then there exists a topological cycle β representing α whose support does not intersect Y.*

Proof: We denote by $B(k,r)$ the closed unit ball in $\mathbf{R}^k$. Let β_1 be a cycle representing α which intersects Y transversally in a finite number of points, which all lie in the smooth locus of Y. Let $y_1, \ldots, y_t$ be the points where the intersection number is $+1$ and $z_1, \ldots z_t$ those where the intersection number is -1. Using the fact that the smooth locus of Y is connected, we can find a smooth path γ connecting y_1 and z_1 in the smooth locus of Y, which does not pass through any of the $y_2, \ldots, y_t$; $z_2, \ldots, z_t$. We can even find a C^∞-embedding

$$\Gamma : B(2k-1,2) \times B(2n-2k,2) \times [0,1] \longrightarrow X \setminus \{y_2, \ldots, y_t, z_2, \ldots, z_t\},$$

such that

$$\begin{aligned} \Gamma^{-1}(Y) &= B(2k-1,2) \times \{0\} \times [0,1], \\ \Gamma^{-1}(\beta_1) &= \{0\} \times B(2n-2k,2) \times \{0\} - \{0\} \times B(2n-2k,2) \times \{1\} \end{aligned}$$

and $\gamma = \Gamma|_{\{0\}\times\{0\}\times[0,1]}$. Then we replace β_1 by $\beta_1 + \partial(\Gamma(\{0\} \times B(2n-2k,1) \times [0,1]))$, ($\partial$ denotes the boundary), which intersects Y only in $y_2, \ldots, y_t$; $z_2, \ldots, z_t$. The result follows by induction on t. □

Proof of Proposition 3.1: By Theorem 2.3 we have $v_{N,0}(X) = (d-2^n)[X]$ in case (1) and $w_{N,0}(X) = 2^n[X]$ in case (2). So (2) and (4) follow with Remark 2.2(2).
(1) Let $\alpha \in H^*(X)$. Then $res_*(\epsilon^*(\pi_{2*}(\alpha_1))s_N(\widetilde{H}_3)) \in H^*(X)$ can be obtained by a computation similar to the proof of Theorem 2.3. If we assume that $\alpha \in H^{2i+1}(X)$ or that $\alpha \in H^{2i}(X)$ with $\alpha[Y] = 0$ for every algebraic subvariety Y of X, then the computation becomes very simple and we get

$$res_*(\epsilon^*(\pi_{2*}(\alpha_1))s_N(\widetilde{H}_3)) = (d - 2^{n+1})\alpha.$$

If on the other hand we assume that there is a point $p \in X$ such that π_p is an isomorphism, then the closed subset $W := \{x \in X \mid x \text{ lies on a trisecant}\}$ has real codimension at least 2. So if $\alpha \in H^1(X)$ or $\alpha \in H^{0,2}(X)$, then we can find a cycle β representing the Poincaré dual of α, whose support does not intersect W. If $\alpha \in H^1(X)$, this is obvious for dimension reasons. If $\alpha \in H^{0,2}(X)$, we apply Lemma 3.3 to every irreducible component of W.

It is easy to see that the support of a suitable cycle representing $\epsilon^*(\pi_{2*}(\alpha_1))s_N(\widetilde{H}_3))$ is contained in

$$\{(Z_2, Z) \in X^{[2,3]} \mid supp(Z_2) \cap supp(\beta) \neq \emptyset, Z \text{ lies on a line}\}.$$

By the choice of β this set is empty. Thus

$$(d - 2^{n+1})\alpha = res_*(\epsilon^*(\pi_{2*}(\alpha_1))s_N(\widetilde{H}_3)) = 0.$$

For (3) we argue in a similar way. Let $\alpha \in H^1(X)$ or $\alpha \in H^{0,2}(X)$. We get

$$res_*(\pi_{2*}(\alpha_1)\epsilon^*(D)s_N(\widetilde{H}_3)) = 2^n\alpha.$$

Similar as in (1) we see that this class is zero, if there is a point $p \in X$ such that π_p is injective on tangent vectors. □

Our second application is concerned with 2-spannedness of the linear systems of hyperplanes on a smooth variety X of dimension n in $\mathbf{P}_{2n+1}$. Let L be a line bundle on a variety X and $V \subset H^0(X, L)$ a linear subspace. Following [B-S2] the linear system $\mathbf{P}(V^*)$ is called l-very ample if for every finite subscheme $Z \subset X$ of length $l+1$ the restriction map $r : V \longrightarrow H^0(Z, L|_Z)$ is onto. There is also the slighly weaker notion of l-spannedness ([B-F-S]), where the surjectivity of r is required only for subschemes which are locally subschemes of smooth curves on X. It is easy to see that l-spannedness and l-very ampleness are equivalent for $l \leq 2$. We also see immediately from the definitions (and the discussion after Definition 2.1) that the linear system of hyperplanes on a variety $X \subset \mathbf{P}_N$ is 2-spanned if and only if X has no trisecants. In [Bal] Ballico shows that the only surfaces $S \subset \mathbf{P}_5$ for which the linear system of hyperplanes is 2-spanned are the Veronese surface and a generic complete intersection of quadrics. We want to extend this result to higher dimension. Our result is the following:

PROPOSITION 3.4. *Let $X \subset \mathbf{P}_{2n+1}$ be a smooth projective variety of dimension $n \geq 2$. If the locus of points $x \in X$ lying on a tangential trisecant of X has codimension at least 3, then either $n = 2$ and X is the Veronese surface or a K3-surface of degree 8 in $\mathbf{P}_5$, or there is a positive integer solution a of*

$$5n^2 - 2n - 7 = a^2$$

such that $n+4$ divides $2^n(a+9)$. In particular there is no such X for $3 \leq n \leq 10000$.

REMARK 3.5. In particular we see that for $2 \leq n \leq 10000$ the Veronese surface and a complete intersection of 3 quadrics in $\mathbf{P}_5$ are the only smooth varieties of dimension n in $\mathbf{P}_{2n+1}$ for which the linear system of hyperplanes is 2-spanned. It also seems very likely that the requirements on n of the proposition can only be met by $n = 2$.

For the proof we will need an additional formula for tangential trisecants. Let $X \subset \mathbf{P}_N$ be a smooth projective variety and $\alpha \in A^*(X)$. Using the notations of section 1 we put

$$\bar{v}_{N,\alpha} := p_{1*}(\bar{\epsilon}_*(\bar{\epsilon}^*(\alpha_2)s_{N-1}(H_3))).$$

REMARK 3.6. By arguments similar to those after Definition 2.1 we can view $\bar{v}_{N,\alpha}$ as the class of the locus of tangent points of tangential trisecants of X whose residual point lies on a cycle representing α.

Now let $\alpha \in A^l(X)$. Let β be a cycle representing α. Then we see in particular that the support of a suitable representative of $\epsilon^*(\alpha_2)s_{N-1}(H_3)$ is contained in

$$\left\{(Z_2, Z) \in X^{[2,3]} \;\middle|\; res(Z_2, Z) \in supp(\beta) \cap Z_2,\ Z \text{ lies on a line}\right\}.$$

In particular the support of a suitable representative of $\bar{v}_{N,\alpha}$ is contained in

$$\left\{x \in X \;\middle|\; x \text{ lies on a tangential trisecant}\right\}.$$

Thus if the codimension of this locus is bigger then $N - 2n - 1 + l$, then $\bar{v}_{N,\alpha} = 0$.

LEMMA 3.7. *Let $\alpha \in A^l(X)$. Put $m = N - 2n$. Then*

$$\bar{v}_{N,\alpha} = \textstyle\int_X(H^{n-l}\alpha)\displaystyle\sum_{k=0}^{m+l}\binom{n+m+l}{k}H^k s_{m+l-k}(X)$$
$$-\sum_{k=0}^{m}\binom{N+1}{k}\sum_{j=0}^{m-k}2^{n+j}s_j(X)s_{m+k-j}(X)H^j\alpha$$

The proof of Lemma 3.7 is again a calculation very similar to that in the proof of Theorem 2.3, which we omit.

Proof of Proposition 3.4: Let $X \subset \mathbf{P}_{2n+1}$ be a smooth projective variety of dimension $n \geq 2$ and degree d, such that the locus of points $x \in X$ lying on a tangential trisecant has codimension at least 3. By Lemma 3.7 and Theorem 2.3(1) and (4) we obtain

$$v_{2n+1,0} = d(s_1(X) + (n+1)H) - 3 \cdot 2^n s_1(X) - (n+1)2^{n+1}H, \tag{1}$$

$$\bar{v}_{2n+1,H} = d(s_2(X) + (n+2)s_1(X)H + \binom{n+2}{2}H^2 - 3 \cdot 2^n s_1(X)H - (n+1)2^{n+1}H^2, \tag{2}$$

$$t_{2n+1,0} = 2^{n-1}(5 \cdot s_2(X) + 2s_1(X)^2 + (6n+6)s_1(X)H + \binom{2n+2}{2}H^2). \tag{3}$$

On the other hand our assumption on the codimension of the locus of tangential trisecants gives $v_{2n+1,0} = \bar{v}_{2n+1,H} = 0$ by Remarks 2.2 and 3.6. As the locus of points $x \in X$ lying on a tangential trisecant has codimension at least 3, the same is true for the sublocus of inflectional points of X. So we find $t_{2n+1,0} = 0$.

If $d = 3 \cdot 2^n$, then (1) gives the contradiction that $H = 0$ in $A^*(X) \otimes \mathbf{Q}$. Thus we obtain from (1) and (2) the formulas

$$s_1(X) = \frac{(n+1)(2^{n+1}-d)}{d - 3 \cdot 2^n},$$

$$ds_2(X) = \left(\frac{(3 \cdot 2^n - d(n+2))(n+1)(2^{n+1}-d)}{d - 3 \cdot 2^n} + (n+1)2^{n+1} - \binom{n+2}{2} d \right) H^2$$

in $A^*(X) \otimes \mathbf{Q}$. Putting this into (3) we obtain that d must be an positive integer solution of

$$(n+4)d^2 - 18 \cdot 2^n d - 2^{2n}(5n - 22) = 0. \qquad (*)$$

But the integer solutions d of $(*)$ are given exactly by

$$d = 2^n(9+a)/(n+4)$$

(if this is an integer), where a is an integer solution of

$$5n^2 - 2n - 7 = a^2. \qquad (**)$$

If $n = 2$, then the solutions for a are -3 and 3 and the corresponding solutions for d are 4 and 8. If $d = 4$, then (1) gives for the canonical divisor the relation $2K_X = 3H$. By [B-S1] Theorem 1.3 we obtain that

$$(X, H) = (\mathbf{P}_2, \mathcal{O}_{\mathbf{P}_2}(2)),$$

and so X is the Veronese surface. If $d = 8$, then $4K_X = 0$. Also using (2) again we obtain for the Euler number $e(X) = 24$. Thus X is a $K3$-surface. If we assume in addition that X does not have trisecants, then like in [Bal] we can apply Proposition 5.3 of [B-S3] to rederive the result that is a complete intersection of quadrics in $\mathbf{P}_5$.

The smallest pair of positive integers (n, a) with $n \geq 2$ satisfying $(**)$ is $(11, 24)$. It is however clear that for $n \geq 11$ the negative solution for a will lead to negative d. The statement that for $3 \leq n \leq 10000$ there are no solutions in positive integers of $(**)$ such that $n+4$ divides $2^n(9+a)$ has been checked by a computer search. The search is simplified by the observation, that for n sufficiently large $n+4$ can only divide $2^n(9+a)$ if it is divisible by a high power of 2. (This follows as $(9+a)/(n+4)$ is an approximation of $\sqrt{5}$ and thus $9+a$ and $n+4$ cannot have a very large common factor.) □

REFERENCES

[Bal] E. Ballico: A characterization of the Veronese surface, Proc. Amer. Math. Soc. **105** (1989), 531-534.

[Bau1] I. Bauer: Projecting surfaces into $\mathbf{P}_4$, MPI preprint.

[Bau2] I. Bauer: Ph. D. Thesis, Bonn 1992.

[B-S-S] M. C. Beltrametti, M. Schneider, A. J. Sommese: Threefolds in degree 11 in $\mathbf{P}^5$, Complex Projective Geometry, G. Ellingsrud, C. Peskine, G. Sacchiero, S. A. Strømme, eds, London Math. Soc. Lecture Note series, **179** (1992), 59-80.

[B-F-S] M. C. Beltrametti, P. Francia, A. J. Sommese: On Reider's method and higher order embeddings, Duke Math. J. **58** (1989), 425-439.

[B-S1] M. C. Beltrametti, A. J. Sommese: On the adjunction theoretic classification of polarized varieties, preprint.

[B-S2] M. C. Beltrametti, A. J. Sommese: Zero cycles and k-th order embeddings of smooth projective surfaces, 1988 Cortona Proceedings: Projective surfaces and their classification, Symposia Mathematica, INDAM, Academic Press.

[B-S3] M. C. Beltrametti, A. J. Sommese: On k-spannedness for projective surfaces, Algebraic geometry L'Aquila 1988, A. J. Sommese, A. Biancofiore, E. L. Livorni, eds, Lecture Notes in Math.1417, Springer 1990, 24-51.

[C] S. Colley: Enumerating stationary multiple-points, Adv. Math. **66** (1987), 149-170.

[E-LB] G. Elencwajg, P. Le Barz: Explicit computations in $Hilb^3\mathbf{P}^2$, Algebraic Geometry Sundance 1986, A. Holme; , R. Speiser, eds, Lecture Notes in Math.1311, Springer 1988, 76-100.

[F-G] B. Fantechi, L. Göttsche: The cohomology ring of the Hilbert scheme of three points on a smooth projective variety, to appear in Journal für die reine und angewandte Mathematik.

[Fu] W. Fulton: Intersection Theory. Ergebnisse der Mathematik und ihrer Grenzgebiete, Springer, Berlin-Heidelberg-New York-Tokio 1984.

[Gö1] L. Göttsche: Hilbertschemata nulldimensionaler Unterschemata glatter Varietäten, Thesis, Bonn, December 1991.

[Gr] A. Grothendieck: Techniques de construction et théorèmes d'existence en géométrie algébrique IV: Les schémas de Hilbert. Séminaire Bourbaki exposé 221, 1961, IHP, Paris.

[Kl1] S. L. Kleiman: Multiple point formulas I: Iteration. Acta mathematica **147** 1981, 13-49.

[Kl2] S. L. Kleiman: Multiple point formulas II: the Hilbert scheme, in: S. Xambó Descamps (ed.): Enumerative Geometry, Sitges 1987. LNM 1436, Springer, Berlin-Heidelberg 1990, 101-138.

[LB1] P. Le Barz: Formules pour les multisécantes des surfaces, C. R. Acad. Sci. Paris **292** (1981), 797-800.

[LB2] P. Le Barz: Formules pour les trisécantes des surfaces algébriques. Enseign. Math. **33** no 1-2 (1987), 1-66.

[LB3] P. Le Barz: Quelques formules multisécantes pour les surfaces, Enumerative Geometry, Proc Sitjes 1987, S. Xambó-Descamps, ed., Lecture Notes in Math. **1436**, Springer-Verlag, Berlin Heidelberg 1990, 151-188.

[LB4] P. Le Barz: Quelques calculs dans la varieté des alignements, Adv. Math. **64** (1984), 116-134.

[LB5] P. Le Barz: Contribution des droites d'une surface à ses multisécantes, Bull. Soc. Math. France **112** (1984), 303-324.

Derived Triangles and Differential Systems

ROBERT SPEISER Department of Mathematics, Brigham Young University, Provo, Utah

Here we clarify the global structure symmetric data of [**GMC**] by means of a new abstract construction, the *effective subscheme* of a derived scheme. We describe the noncuspidal part of the effective subscheme as a torsor, generalizing key results of [**GMC**]. As an illustration, we work out the effective subscheme of the second-order data scheme $D^2_{n-1}\mathbf{P}^n$ explicitly; the description of $D^2_2\mathbf{P}^3$ given in [**GMC**] now follows as a special case.

Conceptually, to a given triangle

$$\begin{array}{ccc} X & \xrightarrow{a} & G_k(TY) \\ \Big\downarrow{\scriptstyle b} & \swarrow_{p} & \\ Y & & \end{array}$$

we associate a canonical exterior differential system $\mathcal{I}$; the effective subscheme of DX, by definition, has as its support the set of k-dimensional integral elements of $\mathcal{I}$.

More precisely, the differential system $\mathcal{I}$, locally on X, is generated by the 1-forms ω which vanish on the Semple bundle $\mathcal{F}$, together with their exterior derivatives $\Omega = d\omega$. Denote by EX the set of k-dimensional integral elements of $\mathcal{I}$, the subset of $G_k(TX)$, consisting of those k-planes in TX on which both the ω and the Ω are identically zero. Because $DX = G_k(\mathcal{F})$ is defined, as a subscheme of $G_k(TX)$, by the vanishing of the 1-forms ω, it follows that EX is a Zariski-closed subset of DX, with a natural scheme stucture given by the vanishing of the 2-forms Ω. The topological space EX, equipped with this scheme structure, is the *effective subscheme* of DX.

When we specialize to a data scheme, the 2-forms Ω give explicit *global* defining equations for the symmetric data of [**GMC**]. This observation represents an important step toward understanding the geometry of a symmetric data space, especially along its cuspidal divisor, that is, on the boundary of the jets. Our exposition begins with the difinition of the differential system $\mathcal{I}$ in §1. The effective subscheme EX is introduced in §2, along with its functor of points. The main result here is 2.2, which gives conditions under which the noncuspidal part of EX will be a torsor. We check these conditions first in §3 for a projective space, then generalize in §4, to obtain 4.1, a key structure theorem.*

Finally, I would like to thank Penny Smith for the very important suggestion that differential systems should help in this setting, and I'm also grateful to Enrique Arrondo, Bob Fisher and Dan Laksov for some very helpful conversations.

1 The differential system of a triangle.

As in [**GMC**] we begin with the following setup: a triangle

$$\begin{array}{ccc} X & \xrightarrow{a} & G_k(TY) \\ \big\downarrow b & \swarrow_p & \\ Y & & \end{array} \tag{1}$$

of schemes, with X and Y and b smooth. Its *Semple bundle* $\mathcal{F}$ is the fiber product

$$\begin{array}{ccc} \mathcal{F} & \longrightarrow & TX \\ \big\downarrow & & \big\downarrow \partial b \\ \Sigma & \xrightarrow{\iota} & b^*TY, \end{array} \tag{2}$$

where Σ denotes the pullback, under a, of the universal k-plane in $TY_{G_k(TY)}$, and ι denotes its natural inclusion in b^*TY. In particular, $\mathcal{F}$ is a vector bundle of rank $f = \dim X - (\dim Y - k)$. Set

$$r = \dim Y - k.$$

Hence

$$f = \operatorname{rank}\mathcal{F} = \dim X - r,$$

so, locally, $\mathcal{F}$ is defined by the vanishing of r independent 1-forms $\omega_1, \ldots, \omega_r$ on X.

In the exterior algebra sheaf $\bigwedge \Omega^1_X$, denote by $\mathcal{I}$ the differential ideal sheaf generated by

$$\{\omega \in \Omega^1_X \mid \omega \equiv 0 \text{ on } \mathcal{F}\}.$$

In particular, localizing as above, $\mathcal{I}$ contains $d\omega_1, \ldots, d\omega_r$ in degree 2 as in addition to $\omega_1, \ldots, \omega_r$ in degree 1, but to be closed under exterior differentiation as well as multiplication by regular functions, $\mathcal{I}$ contains a great deal more. For example, with $f \in \mathcal{O}_X$ and

* This is a final version, no part to be submitted elsewhere.

$\omega \in \mathcal{I}_1$, it is obvious that $\mathcal{I}_2$ contains all elements of the form $df \wedge \omega$. The $d\omega_\alpha$, together with these, generate $\mathcal{I}_2$ as an $\mathcal{O}_X$-module. The differential system given by $\mathcal{I}$, as in **[D]** and **[EDS]**, will be called the *differential system of the triangle* (1).

Fix a point $x \in X$. A vector subspace $V \subset T_xX$ is called an *integral element* of the differential system given by $\mathcal{I}$ if we have

$$\langle v_1 \wedge \cdots \wedge v_p\,,\, \omega \rangle = 0$$

for any p-form $\omega \in \mathcal{I}_p$, for all $p = 1, \ldots, \dim V$. If $\dim V = d$, it suffices by a standard argument to check the displayed identity only for d-forms $\omega \in \mathcal{I}_d$.

A smooth subscheme $Z \subset X$ is an *integral manifold* of the triangle (1) if all its tangent spaces T_zZ are integral elements of the differential system given by $\mathcal{I}$. Concretely, for each $z \in Z$, choose 1-forms $\omega_1, \ldots, \omega_r$ whose vanishing defines $\mathcal{F}$ near z on X. Then it is easy to see that Z is an integral manifold exactly when (1) we have $T_zZ \subset \mathcal{F}_z$ and (2) each 2-form $d\omega_\alpha | T_zZ$ is identically zero, $\forall \alpha = 1, \ldots, r$, $\forall z \in Z$.

Example: first-order data for hypersurfaces in $\mathbf{P}^n$. Take $X = \mathbf{P}^n$, set $k = n-1$, and pick a point $x_0 \in X$. Our goal here is to examine some k-dimensional integral manifolds of the triangle

$$\begin{array}{ccc} G_k(TX) & \xrightarrow{\;1\;} & G_k(TX) \\ \big\downarrow b & \swarrow_{p} & \\ X & & \end{array} \tag{3}$$

in the neighborhood of a given k-plane $\Pi_0 \subset T_{x_0}X$. Choose local coordinates $x_1, \ldots, x_k, y$ at x_0 in a sufficiently small open neighborhood U_0 of x_0 on X, such that Π_0 is tangent to the hyperplane given by $y = 0$. In the standard trivialization of TX over U_0, we can view $dx_1, \ldots, dx_k, dy$ as local coordinates in the tangent space T_xX, for any $x \in U_0$. One standard open subset, denoted U, of $G_kTX_{U_0}$ consists of those k-planes given by linear equations of the form

$$dy = \sum_{i=1}^{k} p_i\, dx_i.$$

Clearly, the p_i give coordinates on U over U_0, hence $dx_1, \ldots, dx_k, dy$, together with the $dp_1, \ldots dp_k$, may be viewed as coordinate functions in TU. In these coordinates, the Semple bundle $\mathcal{F}$ is defined by the vanishing of the single 1-form

$$\omega = dy - \sum_{i=1}^{k} p_i\, dx_i. \tag{4}$$

In the usual terminology, U is the space of 1-*jets* of regular functions in $x_1, \ldots, x_k$, and $\mathcal{F}$ is its *contact distribution*. We have

$$d\omega = \sum_{i=1}^{k} dx_i \wedge dp_i, \tag{5}$$

an alternating 2-form on U.

Now suppose $F = F(x_1, \ldots, x_k)$ is any regular function, and denote by Z the graph of $y = F(x_1, \ldots, x_k)$, a smooth subvariety of X. For $z \in Z$, the assignment $z \mapsto T_z Z$ defines a morphism

$$Z \xrightarrow{i} G_k TX$$

which makes the diagram

$$\begin{array}{ccc} & & G_k TX \\ & \overset{i}{\nearrow} & \big\downarrow b \\ Z & \hookrightarrow & X \end{array}$$

commute. In particular, i is an embedding. Set $Z_1 = i(Z)$. Then Z_1 is a k-dimensional integral manifold of the triangle (3). Indeed, we may assume that i maps Z into U_0. In coordinates, with $z = (x_1, \ldots, x_k, y)$, we have

$$i(z) = (x_1, \ldots, x_k, y, p_1, \ldots, p_k),$$

where $y = F(x_1, \ldots, x_k)$ and $p_i = \partial F / \partial x_i$. It follows that $T_z Z$ is the graph of $dy = dF$, so ω clearly vanishes identically on $T_z Z$, as was to be shown. In the usual terminology, Z_1 is the *prologation* of Z to the 1-jets, and we recover the basic result that prolongations integrate the contact structure.

Remark. For $k < n - 1$, essentially the same things happen. To be precise, replace y by a vector $(y_1, \ldots, y_{n-k})$, and similarly replace each p_i with a vector $(p_{i1}, \ldots, p_{i,n-k})$. Then ω becomes a vector-valued 1-form, or equivalently a system of $n - k$ scalar-valued 1-forms $\omega_1, \ldots, \omega_{n-k}$. In this case $\mathcal{F}$ is often called the *Cartan distribution*.

2 The effective subscheme of DX.

Return now to the triangle (1) and its Semple bundle $\mathcal{F}$. The derived triangle

$$\begin{array}{ccc} DX & \xrightarrow{Da} & G_k(TX) \\ \downarrow{\scriptstyle Db} & \swarrow_{p} & \\ X & & \end{array}$$

is defined [**GMC**] as follows: $DX = G_k(\mathcal{F})$, while Db is its structure map and Da is induced by the inclusion $\mathcal{F} \subset TX$. We define the *effective part* EX of DX to be the set of k-dimensional integral elements of the differential system of the triangle (1). So far, EX is simply a set.

Proposition 2.1. *The effective part EX is canonically a closed subscheme of DX.*

Proof. The question is local on X, so we may assume $\mathcal{F}$ is the set of zeros of independent global 1-forms $\omega_1, \ldots, \omega_r$ on X. Pick a point $x \in X$ and a T-point $f\colon T \to X$ such that $x = f(t)$ for some $t \in T$. A T-point $g\colon T \to DX$ over f is given by a k-subbundle

$$\mathcal{S} \hookrightarrow \mathcal{F}_T.$$

Shrink X and T if necessary so that $\mathcal{S}$ has a global framing by sections $X_1, \ldots, X_k \in \Gamma(T, \mathcal{S})$. Now the 2-forms $d\omega_1, \ldots, d\omega_r$ restrict to 2-forms $\Omega_1, \ldots, \Omega_r$ on T. At $t \in T$ an easy argument shows that each Ω_α restricts to zero on the fiber $\mathcal{S}_t$ exactly when we have

$$\langle X_i(t) \wedge X_j(t), \Omega_\alpha \rangle = 0 \tag{6}$$

for all $\alpha = 1, \ldots, r$, all $1 \leq i < j \leq k$. The system (6) cuts out a closed subscheme $E_T \subset T$ which is easily seen to be independent of the framing X_i and the choice of the ω_α. Further, E_T is stable under base change of the form $T' \to T$, hence functorial in $g\colon T \to DX$. Hence the contravariant functor $T \mapsto E_T$ is represented by a natural closed subscheme of DX whose support is EX. This completes the proof.

From now on we will view EX as a scheme over X, with scheme structure locally defined by (6).

As a set, the *cuspidal divisor* $C \subset DX$ consistes of those k-planes Π in $\mathcal{F}$ such that the derivative of b maps Π to a subspace of dimension $< k$ in TY. As a determinantal locus, C has a natural sructure as a divisor on X, described in [**GMC**]. The *noncuspidal part* DX_{nc} is the complement $DC - C$. Given a T-paint

$$T \xrightarrow{f} X,$$

a T-point

$$T \xrightarrow{g} DX_{nc}$$

over f is given canonically by a splitting over T

$$\begin{array}{ccc} \Sigma_T & \xrightarrow{\lambda} & \mathcal{F}_T \\ & \overset{1}{\searrow} & \downarrow{\scriptstyle \pi_T} \\ & & \Sigma_T \end{array}$$

of the projection $\pi: \mathcal{F} \to \Sigma$ of diagram (2). Now the kernel of π is the relative tangent bundle $T_{X/Y}$, by the fiber square which defines $\mathcal{F}$. Recall [**GMC**] that the group scheme

$$G = \mathrm{Hom}_X(\Sigma, T_{X/Y})$$

acts on DX_{nc} by additive translation. An element $\gamma \in G(T \xrightarrow{f} X)$ is, by definition, a homomorphism

$$\Sigma_T \xrightarrow{\gamma} \ker(\mathcal{F}_T, \Sigma_T).$$

Fix a section λ_0 of π_T as above. Then any section λ of π_T is uniquely of the form $\lambda_0 + \gamma$, for a suitable $\gamma \in G(T \xrightarrow{f} X)$. From this it follows easily [**GMC**] that DX_{nc} is a principal homogeneous space for G.

We define the *noncuspidal part* of EX to be

$$EX_{nc} = EX \bigcap DX_{nc},$$

and we now explore its structure.

Consider a T-point $\lambda: T \to EX_{nc}$, over a fixed T-point $f: T \to X$. For what T-points $\gamma \in G(T \xrightarrow{f} X)$ is $\lambda + \gamma$ also a T-point of EX_{nc}? Denote by $H(\lambda)$ the subset of $G(T \xrightarrow{f} X)$ consisting of all such γ. Locally, so that $\mathcal{F}$ is defined by the vanishing of 1-forms $\omega_1, \ldots, \omega_r$, a moment's thought should confirm that we have

$$H(\lambda) = \left\{ \gamma \in G(\xrightarrow{f} X) \mid \Omega_\alpha \equiv 0 \text{ on } (\lambda + \gamma)(\Sigma_T), \forall \alpha = 1, \ldots, r \right\},$$

where Ω_α denotes the pullback $(d\omega_\alpha)_T$.

Proposition 2.2. *These statements are equivalent:*

(1) *the subset $H(\lambda)$ is a subgroup of $G(T \xrightarrow{f} X)$;*

(2) *the subset $H(\lambda)$ is independent of the choice of λ.*

Proof. By definition, we have $(EX_{nc})_T = \lambda + H(\lambda)$ in $(DX_{nc})_T$. Hence, if $\lambda' = \lambda + \gamma$ is another T-point of EX_{nc} over $f: T \to X$, we have

$$\lambda + H(\lambda) = (EX_{nc})_T = \lambda' + H(\lambda'), \tag{7}$$

hence

$$H(\lambda) = \gamma + H(\lambda'). \tag{8}$$

Because $\lambda' = \lambda + \gamma$ gives a point of EX_{nc}, we have $\gamma \in H(\lambda)$. In particular, if (1) holds, then (2) follows from (8). Conversely, assume (2), write H for the common $H(\lambda)$, and fix λ. If $\gamma, \gamma' \in H$, we can replace λ by $\lambda + \gamma$ and λ' by $\lambda + \gamma'$ in (7). We then obtain

$$\lambda + \gamma + H = \lambda + \gamma' + H,$$

hence $\gamma - \gamma' \in H$. But $0 \in H(\lambda) = H$, so $\gamma - \gamma' \in H$. Because H contains 0 and is closed under subtraction, it is a subgroup, hence (2) implies (1) and the proof is complete.

When the equivalent conditions of 2.2 hold, we can write $H(T \xrightarrow{f} X)$ for $H(\lambda)$, and we obtain a contravariant functor from X-schemes to groups. Denote this functor by $\mathcal{H}$.

Corollary 2.3. *When the equivalent conditions of 2.2 hold, the functor $\mathcal{H}$ is represented by a closed X-group subscheme $H \subset G$.*

Proof. Because $H(T \xrightarrow{f} X)$ is clearly closed in $G(T \xrightarrow{f} X)$, this follows directly from 2.2.

3 Second-order data on $\mathbf{P}^n$.

We illustrate the ideas in the last section by working out the subgroup H in an important case.

Again we take $X = \mathbf{P}^n$, with $k = n - 1$, and continue the previous example. This time, however, we will write $D^1_k\mathbf{P}^n$ for $G_k(T\mathbf{P}^n)$, the space of k-dimensional *first-order data*. As before, we denote by $\mathcal{F}$ the Semple bundle (or contact distribution) on $D^1_k\mathbf{P}^n$. The triangle (3), namely

$$\begin{array}{ccc} D^1_k\mathbf{P}^n & \xrightarrow{1} & G_k(T\mathbf{P}^n) \\ \downarrow b & \swarrow p & \\ \mathbf{P}^n & & \end{array}$$

gives a derived triangle, with derived scheme $G_k(\mathcal{F})$, denoted $D^2_k\mathbf{P}^n$. This is the space of k-dimensional *second-order data* on $\mathbf{P}^n$. We will denote by $E^2_k\mathbf{P}^n$ its effective subscheme, and our goal here is to describe $D^2_k\mathbf{P}^n$ and $E^2_k\mathbf{P}^n$ explicitly in local coordinates, following the general plan of [**GMC**].

Choose local affine coordinates $(x_1, \ldots, x_k, y, p_1, \ldots, p_k)$ on $D^1_k\mathbf{P}^n$ as before. In these coordinates, the map b takes the simple form

$$(x_1, \ldots, x_k, y, p_1, \ldots, p_k) \mapsto (x_1, \ldots, x_k, y).$$

For a more compact notation, write $x = (x_1, \ldots, x_k)$, $p = (p_1, \ldots, p_k)$, so $b(x, y, p) = (x, y)$.

Because TX is trivial on $U_0 = \mathrm{Spec}(k[x_1, \ldots, x_k, y])$, we can write a tangent vector at $(x, y) \in U_0$ in the form $(x, y; \bar{x}, \bar{y})$, where $\bar{x} = (\bar{x}_1, \ldots, \bar{x}_k)$ and we use a semicolon to distinguish base from fiber. Similarly, $TD^1_k\mathbf{P}^n$ is trivial over U_0, and we write a tangent vector at $(x, y, p) \in D^1_k\mathbf{P}^n$ in the form $(x, y, p; \bar{x}, \bar{y}, \bar{p})$. In other words, $\bar{x}_i$ is the coefficient of the basis vector $\partial/\partial x_i$, and similarly $\bar{y}_i$ and $\bar{p}_i$. In this notation, the map

$$TD^1_k\mathbf{P}^n \xrightarrow{\partial b} b^*T\mathbf{P}^n$$

takes the form

$$(x, y, p; \bar{x}, \bar{y}, \bar{p}) \mapsto (x, y, p; \bar{x}, \bar{y}).$$

The subbundle $\Sigma \subset b^* T\mathbf{P}^n$ is defined by the equation

$$\bar{y} = \sum_{i=1}^{k} p_i \bar{x}_i$$

of the universal hyperplane, while $\mathcal{F} \subset TD_k^1\mathbf{P}^n$ is defined by the vanishing of the corresponding 1-form

$$\omega = dy - \sum_{i=1}^{k} p_i \, dx_i.$$

Explicitly, for each tangent vector $v = (x, y, p; \bar{x}, \bar{y}, \bar{p})$ at (x, y, p), we have

$$\langle v, \omega \rangle = \bar{y} - \sum_{i=1}^{k} p_i \bar{x}_i.$$

In particular, $\bar{x}$ and p determine $\bar{y}$ when $v \in \mathcal{F}$, so we can use $(\bar{x}, \bar{p})$ as coordinates in the fibers of $\mathcal{F}$ over our original open set $U = \operatorname{Spec}(k[x, y, p]) \subset D_k^1\mathbf{P}^n$.

We can now write very simple coordinates for the part of $D_k^2\mathbf{P}^n$ consisting of those k-planes Π in $\mathcal{F}$ which project isomorphically to k-planes in $U \subset D_k^1\mathbf{P}^n = G_k(T\mathbf{P}^n)$. By definition, these Π fill out the *noncuspidal part*, denoted $(D_k^2\mathbf{P}^n)_{nc}$, of $D_k^2\mathbf{P}^n$. Such a k-plane Π maps isomorphically under the projection

$$(\bar{x}, \bar{y}, \bar{p}) \mapsto \bar{x},$$

and hence can be given by two data: a linear function $A\colon \mathbf{A}^k \to \mathbf{A}^1$ of the form

$$\bar{y} = A(\bar{x}),$$

and a linear map $B\colon \mathbf{A}^k \to \mathbf{A}^k$, of the form

$$\bar{p} = B(\bar{x}).$$

Of course A, in our situation, is given by the third coordinate of the point (x, y, p) to which Π is attached; thus, given (x, y, p), the plane $\Pi \subset T_{(x,y,p)}D_k^1\mathbf{P}^n$ is uniquely determined by B.

We now show that Π represents a point of the effective data scheme $E_k^2\mathbf{P}^n$ exactly when the map B is *symmetric*.

Set $\Omega = d\omega$: this 2-form is the only form we need to test on Π to see if Π is an integral element of our differential system. Let $e_1, \ldots, e_k$ be the standard basis vectors in $\bar{x}$-space, and write L for the matrix of A and M for the matrix of B, relative to this basis. Hence we have $L = (p_1, \ldots, p_k)$, and the images $B(e_i)$ in $\bar{p}$-space are the columns of M. It follows that the images of the e_i in Π are the columns of the $(2k+1) \times k$ matrix

$$N = \begin{bmatrix} I_k \\ p_1 \ldots p_k \\ M \end{bmatrix},$$

where I_k denotes the $k \times k$ identity matrix. Denote by $c_1, \ldots, c_k$ the columns of N. To see if Π is an integral element, hence a point of $E^2_k\mathbf{P}^n$, we need to evaluate

$$\langle c_i \wedge c_j\,, \Omega \rangle$$

for $i \neq j$. But, by (5), we have

$$\Omega = \sum_{\alpha=1}^{k} dp_\alpha \wedge dx_\alpha.$$

Write

$$M = \begin{bmatrix} m_{\alpha\beta} \end{bmatrix}.$$

Because

$$\langle c_\beta, dx_\alpha \rangle = \delta_{\alpha\beta} \quad \text{and} \quad \langle c_\beta, dp_\alpha \rangle = m_{\alpha\beta},$$

we find

$$\langle c_i \wedge c_j, dx_\alpha \wedge dp_\alpha \rangle = \det \begin{bmatrix} \langle c_i, dx_\alpha \rangle & \langle c_j, dx_\alpha \rangle \\ \langle c_i, dp_\alpha \rangle & \langle c_j, dp_\alpha \rangle \end{bmatrix}$$

$$= \det \begin{bmatrix} \delta_{\alpha i} & \delta_{\alpha j} \\ m_{\alpha i} & m_{\alpha j} \end{bmatrix}.$$

For $\alpha \neq i$ or j, the top row vanishes, so the determinant vanishes. For $\alpha = i$ we find

$$\langle c_i \wedge c_j, dx_i \wedge dp_i \rangle = m_{ij},$$

while for $\alpha = j$ we find

$$\langle c_i \wedge c_j, dx_j \wedge dp_j \rangle = -m_{ji}.$$

Hence, summing over $\alpha = 1, \ldots, k$, we obtain

$$\langle c_i \wedge c_j, \Omega \rangle = m_{ij} - m_{ji}.$$

Therefore Π is an integral element if and only if M is symmetric, and this proves our claim.

In particular, every effective, noncuspidal second-order datum on $\mathbf{P}^n$ is represented by the Hessian of a smooth k-fold, because every symmetric M is. By techniques of [**GMC**], this argument generalizes easily to show that the symmetric data of [**GMC**] are precisely the effective data in our sense, for any X, for any k. Finally, in the notation of §2, the group G identifies with the additive group scheme of all bilinear forms on $\bar{x}, \bar{y}, \bar{p}$)-space, while H identifies with the subgroup scheme of symmetric forms. Our coordinates give a particular splitting $\lambda \colon \Sigma \to \mathcal{F}$ of the projection $\mathcal{F} \to \Sigma$, but H is a subgroup of G, so H does not depend on the splitting, by 2.2. We generalize this discussion to a large class of triangles, which includes all data schemes, in the next section.

4 Effective data in coordinates

Return to a triangle of the form (1), but, from here on, with the mild restriction that the map a is an embedding. Following [**GMC**], we shall set up local coordinates in Y, then $G_k(TY)$, then X. Fix a point $x \in X$, set $y = b(x)$, and choose local parameters

$$t_1, \ldots, t_k, u_1, \ldots, u_r$$

in the local ring $\mathcal{O}_{Y,y}$ such that the k-plane $\Pi = a(x)$ is given by

$$du_1 = \cdots = du_r = 0.$$

Write $t = (t_1, \ldots, t_k), u = (u_1, \ldots, u_r)$. Over $\mathbf{C}$, for example, (t, u) will give holomorphic coordinates in a complex neighborhood of y. Fix a local trivialization of TY near $y \in Y$. Then we can write a point of TY near T_yY in the form

$$(t, u; \bar{t}, \bar{u}),$$

where $\bar{t} = (\bar{t}_1, \ldots, \bar{t}_k)$ and $\bar{u} = (\bar{u}_1, \ldots, \bar{u}_r)$ give local coordinates in the fibers, and we use a semicolon to distinguish fiber from base.

In one standard open neighbodhood of Π, each k-plane can be given, uniquely, as the graph of a linear map

$$\bar{u} = A(\bar{t}),$$

and we can represent A by an $\ell \times k$ matrix $\alpha = [\alpha_{ij}]$ relative to our coordinates. Hence we obtain local coordinates

$$(t, u, \alpha)$$

in $G_k(TY)$ near its fiber over y. Because we have assumed that the map a embeds X in $G_k(TY)$, we can represent points of X by their images (t, u, α) in $G_k(TY)$. Hence points of TX can be written in the form

$$(t, u, \alpha; \bar{t}, \bar{u}, \bar{\alpha}),$$

subject to the constraints that (t, u) should lie on X and that $(\bar{t}, \bar{u}, \bar{\alpha})$ should be tangent to the image of X in $G_k(TY)$.

Now the Semple bundle $\mathcal{F} \subset TX$ is defined, even in $G_k(TY)$, by the equation

$$\bar{u} = \alpha \cdot \bar{t}, \tag{9}$$

just as in the previous case of data on $\mathbf{P}^n$. In this way, coordinates in $\mathcal{F}$ will take the form

$$(t, u, \alpha; \bar{t}, \bar{\alpha}),$$

because we can compute $\bar{u}$ from α and $\bar{t}$ by (9). To obtain coordinates in $DX = G_k(\mathcal{F})$, we proceed just as before. Choose a k-plane Π' in TX, such that the projection

$$(\bar{t}, \bar{u}, \bar{\alpha}) \mapsto \bar{t}$$

maps Π' isomorphically onto Π. If also $\Pi' \subset \mathcal{F}$, this means that the projection carries Π' isomorphically onto the fiber of Σ at $(t, u, \alpha) \in X$, so Π' represents a point of DC_{nc}. Such a Π', therefore, can be given as the graph of a linear map

$$\bar{t} \mapsto (\bar{u}, \bar{\alpha}), \tag{10}$$

where $\bar{u}$ and $\bar{\alpha}$ are linear in $\bar{t}$, with coefficients regular in t. On the one hand, for (10) to give a k-plane in $\mathcal{F}$, the function $\bar{t} \mapsto \bar{u}$ must be given by (9). On the other hand, $\bar{t} \mapsto \bar{\alpha}$ takes the form

$$\bar{\alpha} = B(\bar{t}) = \beta \cdot \bar{t},$$

where β denotes a $kr \times k$ matrix, indexed in the form

$$\beta = \left[\quad \beta_{i,(\eta,\zeta)} \quad \right],$$

for $i = 1, \ldots, r$ and $\eta, \zeta = 1, \ldots, k$. We may think of β as a vertical stack of r square matrices $\beta_1, \ldots, \beta r$, each of size $k \times k$.

To summarize, each point of our neighborhood on $DX = G_k(\mathcal{F}) \subset G_k(TY)$ is therefore given, over $(t, u, \alpha) \in X$, by the linear map

$$\bar{t} \mapsto \begin{bmatrix} \alpha \\ \\ \beta \end{bmatrix} \cdot \bar{t},$$

where $\left[\begin{smallmatrix} \alpha \\ \beta \end{smallmatrix}\right]$ is the $(r + kr) \times k$ matrix obtained by placing α on top of β, and we take a matrix product on the right.

Omitting redundant terms, coordinates in DX_{nc} now take the very simple form

$$(t, u, \alpha, \beta),$$

where (t, u, α) is constrained to move along X, as described above.

Now we consider the differential system associated to our triangle. First look at

$$G_k(TY) = D_k^1 Y.$$

In the triangle

$$\begin{array}{ccc} G_k(TY) & \xrightarrow{1} & G_k(TY) \\ \downarrow & \swarrow & \\ Y & & \end{array} \tag{11}$$

denote by $\mathcal{G}$ the Semple bundle, so that we have

$$\mathcal{F} = a^* \mathcal{G}.$$

By our previous discussion of first-order data, $\mathcal{G}$ is given, in TY, by the vanishing of the contact forms

$$\phi_1 = du_1 - \sum_{j=1}^{k} \alpha_{1j}\, dt_j,$$

$$\vdots$$

$$\phi_r = du_r - \sum_{j=1}^{k} \alpha_{rj}\, dt_j,$$

where $\alpha = [\alpha_{ij}]$ is the $\ell \times k$ matrix coordinate of a point $(t, u, \alpha) \in G_k(TY)$. Set $\Phi_i = d\phi_i$. We obtain the 2-forms

$$\Phi_i = \sum_j dt_j \wedge d\alpha_{ij}.$$

Pulling back to X, the contact forms

$$\omega_i = a^* \phi_i \qquad (i = 1, \ldots, r)$$

evidently vanish precisely on $\mathcal{F}$, and together with their exterior derivatives

$$\Omega_i = a^* \Phi_i, \qquad (i = 1, \ldots, r)$$

they generate the differential ideal $\mathcal{I}$ of the triangle (1). Now the same calculation of the previous section, applied to each Φ_i and then pulled back to X, shows that EX_{nc} is parametrized by those coordinates

$$(t, u, \alpha, \beta)$$

for which each $k \times k$ matrix β_i is symmetric, for all $i = 1, \ldots, r$.

This local observation has a global consequence. Write G for the X-group scheme $\mathrm{Hom}(\Sigma, T_{X/Y})$, and consider, for each T-point $\lambda: T \to EX_{nc}$ over a given $f: T \to X$, the associated objects $H(\lambda)$ of §2. Because each $H(\lambda)$ here identifies locally with a group of symmetric matrices, 2.2 shows that $H(\lambda)$ is independent of λ, and hence gives a functor $\mathcal{H}(T \xrightarrow{f} X)$.

Here is our main result, which generalizes the local description of the symmetric data schemes of [**GMC**].

Theorem 4.1 (Structure of EX_{nc}). *The functor $\mathcal{H}$ is represented by an X-subgroup scheme $H \subset G$. As an X-scheme, EX_{nc} is a principal homogeneous space for H, under the H-action induced by the inclusion $H \subset G$.*

Proof. The statement follows directly from 2.3 and the preceding discussion.

References

[**D**] J. Dieudonné, Treatise on Analysis, especially volumes 3 and 4, San Diego, New York, Berkeley, Boston, London, Sydney, Tokyo, Toronto (1972, 1974).

[**EDS**] R. L. Bryant, S. S. Chern, R. B. Gardner, H. L. Goldschmidt, P. A. Griffiths, Exterior Differential Systems, MSRI Publ. **18**, New York, Berlin, Heidelberg, London, Paris, Tokyo, Hong Kong, Barcelona (1991).

[**GMC**] E. Arrondo, I. Sols, R. Speiser, Global moduli for contacts, submitted.

[**SD**] E. Arrondo, I. Sols, R. Speiser, Global moduli for contacts, II: symmetric data, in preparation.

Mailing address:
Department of Mathematics
259 TMCB
Provo, Utah 84602

Computing Gaps Sequences at Gorenstein Singularities

LETTERIO GATTO
Dipartimento di Matematica, Politecnico di Torino,
Corso Duca degli Abruzzi, 24 - 10129 Torino-Italy
e-mail:
LGATTO@ITOPOLI.BITNET or
LGATTO@POLITO.IT.

Alla Cara Memoria di Marco Matta (1964-1992).

ABSTRACT. Some Weierstrass Gap Sequences defined according to [G2] are computed on explicit examples. The examples allow to answer to a question rised by Lax and Widland in [LW1], and to find an explicit unknown example of a rational nodal Gorenstein curve of arithmetic genus 3, with a particular distribution of the WP's. The latter example answer to a question rised in [LW2].

1. Introduction.

This paper is mainly concerned with examples. These examples have been inspired by the theory of Weierstrass points on Gorenstein curves developped by Lax and Widland in a series of paper, such as [W1], [WL], [LW1], [LW2] (the reader is referred to [G3] for complete references). In [LW2], in particular, the two Authors introduce a notion of Weierstrass Gap Sequence (WGS) at a Gorenstein singularity which is one of the possible ways of generalizing the known one for smooth points.

1980 *Mathematics Subject Classification* (1985 *Revision*). 14H20, 14H55, 14H99.
Key words and phrases. Gorenstein Curves, Weight Sequences, Weierstraß points, Weierstraß Gaps Sequences at a Singular Point, Extraweights.
Work partially supported by MURST and GNSAGA-CNR .

A not very nice feature of that definition is that, in general, one can associate more than one WGS at a given singular point. In a completely different way, using the notion of *weight sequences*, in [G1], [G2] an alternative definition of WGS has been given. For a quick account about these two different points of view, and the underlying general framework, the reader is referred to [G3].

What we must immediately say, then, is that the WGS's appearing in the title of this paper, have not to be intended in the sense of [LW2].

The WGS at a singular point defined in [G2] has many advantages: first of all, it has a strict relation with the Weierstrass weight of a singular point computed by means of the *wronskian* defined in [W1] and [WL]. More than that, in some cases not technically handable by the theory of Lax and Widland (see e.g. Ex. 2.7), the WGS is a very easily computable object, which allows to find Weierstrass weights by purely numerical procedures.

The aim of Sect. 2 of this paper is, hence, to show how much is it possible to do when dealing with explicit examples of Gorenstein curve singularities. Ex. 2.1 deals with a comparison between a computation performed in [LW2], Ex. 2.1, and the same computation performed by means of the WGS. That example, as noticed in Remark 2.4, is also an answer to a natural question rised in [LW2]: what is the relation by the various weights of the WGS's at a singualar point, computed according [LW1], and the Weierstrass weight of the singular point computed according [WL]? The quite clear answer we give, is that, in general, there is no relation at all (see again Remark 2.4).

The WGS defined in [G2] have also many more good properties: it simplify a lot of proofs of the theory by Lax and Widland, allows simple computations for bounds for the Weierstrass weight ([BG1]) and, more than that, has suggested the way to extend, in [BG3], the theory of the Weierstrass points with respect to any linear system on an arbitrary singular curve, not necessarily Gorenstein.

Sect. 3 of this paper starts once more with examples, intended to clarify the geometrical meaning of the WGS at a node of a quartic plane curve. It turns out that the WGS measures the number of inflectional branches of the given node (see Pict. 3.2). Using this very simple idea, the paper ends in completing the classification started in [LW2] of the rational nodal curves of arithmetic genus three. In particular we show the existence, not known in [LW2], of a *rational nodal curve of arithmetic genus* 3, *having* 1 *normal smooth WP, two smooth WP's of weight* 2, *two nodes of weight* 6 *and one of weight* 7. The idea of proof consists, at first, in understanding that such a curve must be searched among the plane quartics and, secondly, in using the geometric description of the Weierstrass weight by means of the inflectional branches centered at a node of the quartic itself, as shown in the first part of Sect. 3. It is hence possible to fill the gap in the eight row of Table 1 in [LW2].

I am indebted with M. Coppens for useful discussions and especially with S. Greco and Luisella Caire for giving me precious suggestions to state and prove Prop. 3.7.

2. Computing Gap Sequences. Examples.

In this Section we shall try to use all the machinery developped in [G2], Sect. 5, in order to compute WGS's at a singular point of a Gorenstein curve. For immediate references see also the survey [G3]. We shall immediately start by producing explicit examples. Some general comments about the computational procedures will follow at the end of the Section.

Example 2.1. Let us consider, as in [LW1], Example 2.1, the rational quartic of $\mathbb{P}^2$:

$$y^3 z - x^4 = 0. \tag{2.1}$$

A rational parametrization of C in the open affine set $z \neq 0$ is given by:

$$\begin{cases} x = t^3 \\ y = t^4 \end{cases}, \tag{2.2}$$

Let P be the triple point $(0,0)$ of C, whose preimage by mean of the normalizing morphism is formed by just one point, having $y^3 = 0$ as tangent cone. The main purpose of this example is to compare the computation of the weight of P showed in [LW1], by explicitly using the wronskian, with the one making use of the WGS at P. To start with, as is very easy to check, the local ring $O_{C,P}$ is given by:

$$O_{C,P} = \mathbb{C} + \mathbb{C}t^3 + \mathbb{C}t^4 + t^2\tilde{O}_{C,P},$$

where one sets $\tilde{O}_{C,P} = \mathbb{C}[t]_{(t)}$. A basis of $H^0(C, \mathcal{K}_C)$ can be written as:

$$\omega = \left(\frac{dt}{t^2}, \frac{dt}{t^3}, \frac{dt}{t^2}\right),$$

where $\tau = \frac{dt}{t^2}$ (cfr. [G2], Sect. 4) generates $\mathcal{K}_{C,P}$. Hence: $\boldsymbol{\omega} = \boldsymbol{f}\tau$, having set

$$\boldsymbol{f} = (f_1, f_2, f_3),$$

and:

$$\begin{cases} f_1 = 1 \\ f_2 = t^3 \\ f_3 = t^4 \end{cases}. \tag{2.3}$$

One has, thence:

$$\boldsymbol{f} \wedge D_\tau \boldsymbol{f} \wedge D_\tau^2 \boldsymbol{f} = \begin{vmatrix} 1 & t^3 & t^4 \\ 0 & 3t^8 & 4t^9 \\ 0 & 24t^{13} & 32t^{14} \end{vmatrix} = 12t^{22}, \tag{2.4}$$

proving that (See Sect. 3) $w_3(P) = 22$. Indeed, using the definition of WGS at a singular point, the explicit computation of the wronskian is not necessary. Primarily because, from (2.3), one can immediately read the Σ-WGS at $Q = \pi^{-1}(P)$, given by $< 1, 4, 5 >$, which coincides, in this case (see [G2], Remark 5.6), with the WGS at P. Being P a triple point of a plane curve, a well known formula says that $\delta_P = 3$, so that:

$$\begin{cases} w_1(P) = 0 \\ w_2(P) = 2 + (4-2) = 8 \\ w_3(P) = 18 + (4-2) + (5-3) = 22 \end{cases}$$

However, using the notion of WGS one can do even better, by directly working on the equation (2.1). Let Σ be, in fact, the net of lines of $\mathbb{P}^2$ cutting out the *dualizing series* on C. The generic line $ax+by+cz = 0$ cuts C at the point $P = [0,0,1]$ if and only if $c = 0$, which is the same as to impose one linear condition. By supposing, as one can do, that $b \neq 0$, the generic line passing through P can be written as $y = mx$. Putting in a system:

$$\begin{cases} y = mx \\ y^3 z - x^4 = 0 \end{cases},$$

the equation of the curve together with such a pencil of lines, one has:

$$\begin{cases} y = mx \\ x^3(m^3 z - x) = 0 \end{cases},$$

from which one deduces that the linear system of lines having an intersection multiplicity of order at least 2 at P has also intersection multiplicity of order 3 at the same point. Its dimension is 1 (in fact, it coincides with the line $y = 0$). Denoting by $d(\Sigma - nQ)$ the dimension of the linear series on $\tilde{C}$ consisting in all the elements of Σ vanishing at Q at least n times, from the chain of inequalities:

$$3 = d(\Sigma) \geq \underset{\underset{2}{\parallel}}{d(\Sigma - Q)} \geq d(\Sigma - 2Q) \geq \underset{\underset{2}{\parallel}}{d(\Sigma - 3Q)} \geq d(\Sigma - 4Q) \geq d(\Sigma - 5Q) = 0$$

$$d(\Sigma - 2Q) = 2, \quad d(\Sigma - 4Q) = 1$$

one reads the sequence $< 1, 4, 5 >$ of the Σ-gaps at Q (i.e., in this case, exactly the WGS at P) which allows the easy computation (cfr. [G2], formula (4.2) or, simply, [G3], formula (3.12)) of the weight sequence $\boldsymbol{w}(P) =< 0, 8, 22 >$.

Really, the previous example is a particular case of a general fact. To understand why, we shall use the following:

Proposition 2.2. *Let $P \in C_{sing}$ be a n-branched singular point ($1 \leq n \leq g$), whose branches are centered at $\{Q_1, \ldots, Q_n\} = \theta_P^{-1}(P)$, where each Q_i is taken with its own multiplicity m_i (i.e. Q_i is the center of a m_i-fold branch of P). Set $M = \sum_{i=1}^n m_i$ Then:*

$$E_1(P) = 0, \tag{2.5}$$

(2.6) $$E_2(P) = M - n,$$

and, for $3 \leq k \leq g$:

(2.7) $$(k-1)(M-n) \leq E_k(P) \leq M - n + (k-2)[n(g-1) - M(n-1)].$$

Proof. [BG1], Props. 3.1 and 3.4. □

From the above proposition it follows that:

Proposition 2.3 (Example 2.1 Revisited). *On an irreducible rational plane quartic, a triple point P such that $\pi^{-1}(P) = \{Q\}$ (Ex. 2.1) has necessarily*

$$\boldsymbol{w}(P) = < 0, 8, 22 >$$

as a weights sequence.

Proof. In fact, by the above proposition, $E_1(P) = 0$, $E_2(P) = M - n = 3 - 1 = 2$ and:

$$(M-n)(3-1) \leq E_3(P) \leq M - n + (3-2)[n(g-1) - M(n-1)].$$

Putting in the above formula $M = 3$, $n = 1$, $g = 3$, one gets:

$$4 \leq E_3(P) \leq 4,$$

i.e. $E_3(P) = 4$.

Using relations $w_k(P) = \delta_p k(k-1) + E_k(P)$, we get the claimed weight sequence. Notice that the WGS at P is, thence, necessarily $< 1, 4, 5 >$. □

It seems worth to remark that the reader can easy check (as in [BG1]), using Prop. 2.2 that: *if a rational plane quartic has a 2-branched triple point (respectively a 3-branched triple point) P, then its WGS is necessarily $< 1, 2, 3 >$ (respectively $< 1, 3, 4 >$) or, otherwise said, has weight sequence $< 0, 6, 18 >$ (respectively $< 0, 7, 20 >$).*

Remark 2.4. In [LW1] a definition of WGS at a singular point is proposed. From a computational point of view, the most unpleasant feature of that definition is that at a singular point one can in principle associate more than one WGS. For instance, at the singular point of the curve studied in the previous example, Lax and Widland find WGS's, $< b_1, b_2, b_3 >$, having weight zero, WGS's having weight 1 and WGS's having weight 2, the weight being computed according the formula

(2.8) $$(b_1 - 1) + (b_2 - 2) + (b_3 - 3).$$

One of the main problems proposed in that paper is to find a relationship between the weight of the singular point P computed according formula (2.8) and the weight of P computed by means of the wronskian defined in [WL]. Example (2.1) is an answer to this question. Since the third extraweight associated to the sequence $< 1, 4, 5 >$, found in Ex. 2.1, is 4, it follows that $< 1, 4, 5 >$ cannot be recovered using the definition by Lax and Widland, which is, hence, of a completely different nature. As a matter of fact they look for dimensional properties of all closed subschemes having support at the singular point P with given degree. See [G3], Sect. 3, for a more detailed discussion.

Example 2.5. Let C be the projective completion of the rational affine curve defined by the parametric equations:

$$\begin{cases} x = t^4 \\ y = t^5 \\ z = t^6 \end{cases}, \tag{2.9}$$

C is the complete intersection of a singular quadric and a singular cubic surface defined by the affine equations:

$$\begin{cases} y^2 - xz = 0 \\ x^3 - z^2 = 0 \end{cases} \tag{2.10}$$

An easy computation shows that the curve C has $P \equiv (0,0,0)$ as its only singularity, which is Gorenstein being C a complete intersection. Alternatively, note that the local ring at P is given by:

$$O_{C,P} = \mathbb{C} + \mathbb{C}T^4 + \mathbb{C}T^5 + \mathbb{C}T^2 + T^8\tilde{O}_{C,P} \tag{2.11}$$

where, of course, $\tilde{O}_{C,P} = \mathbb{C}[T]_{(T)}$. Furthermore the conductor of $O_{C,P}$ in $\tilde{O}_{C,P}$ is nothing but $\boldsymbol{c}_P = T^8\tilde{O}_{C,P}$. Patching together all these data we get (cfr. [G3], Sect. 3) $n_P = 8$ and $\delta_P = 4$, as had to be.

Hence we are dealing with an irreducible projective Gorenstein space curve of arithmetic genus 4 (and degree 2), having a fourfold singular point at P. Let us compute the WGS at P, by observing that the dualizing series is cut out on C by planes. As in the previous example, the very geometric idea consists in counting the conditions we need to impose for planes passing through P. Let:

$$\begin{cases} ax + by + cz + d = 0 \\ x = t^4 \\ y = t^5 \\ z = t^6 \end{cases} \tag{2.12}$$

be the system representing the intersection between the generic plane of $\mathbb{P}^3$ and the curve C. The search of these intersections is equivalent to find the solutions of the sixth degree equation in the unknown t:

$$ct^6 + bt^5 + at^4 + d = 0 \tag{2.13}$$

If the plane passes through P we must impose the condition $d = 0$ getting:

$$ct^6 + bt^5 + at^4 = t^4(ct^2 + bt + a) = 0$$

from which we notice that a plane passing through P intersects it with multiplicity at least 4. Furthermore, if $a = 0$, we get a pencil of planes passing through P with multiplicity at least 5. The plane $z = 0$ (corresponding to the choice of coefficients

$a = b = d = 0$) is the only plane intersecting the origin with multiplicity 2. We can again consider the chain of inequalities (recall that $\Sigma \equiv$ linear system of planes in $\mathbb{P}^3$ and $Q = \pi^{-1}(P)$):

$$4 = d(\Sigma) \geq \underset{\underset{3}{\|}}{d(\Sigma - Q)} \geq \underset{\underset{3}{\|}}{d(\Sigma - 2Q)} \geq \underset{\underset{3}{\|}}{d(\Sigma - 3Q)} \geq \underset{\underset{3}{\|}}{d(\Sigma - 4Q)} \geq$$

$$\geq \underset{\underset{2}{\|}}{d(\Sigma - 5Q)} \geq \underset{\underset{1}{\|}}{d(\Sigma - 2Q)} \geq d(\Sigma - 7Q) = 0$$

getting the Σ-WGS at $Q = \pi^{-1}(P)$ which is the same as the WGS at P. It is, in fact:

$$G(P) = < 1, 5, 2, 7 > . \tag{2.14}$$

From (2.14) we get the extraweight sequence at P:

$$\boldsymbol{E}(P) = < 0, 3, 2, 9 > . \tag{2.15}$$

Being $\delta_P = 4$, by the very definition of k-th extraweight (see, e.g. [G3], formula (3.14)), one gets the weight sequence at P:

$$\boldsymbol{w}(P) = < 0, 11, 30, 57 > . \tag{2.16}$$

Since the total fourth Weierstraß weight of a Gorenstein curve of arithmetic genus 4 is given by $(4-1)\cdot 4\cdot 5 = 60$, it follows that on the given curve there is at least 1 and at most 3 distinct smooth Weierstraß points. In fact, one can show, *exactly by the same procedure*, that such a curve has another smooth WP at its point at infinity, having weight 3. This turns out to be the particular case of a general fact ([BG3]): *any monomial Gorenstein curve C has exactlty two WP's with respect to the sheaf $\mathcal{O}_C(n)$, if and only if it is not smooth* (in which case the curve has arithmetic genus 0).

Remark 2.6. Both Ex. 2.1 and 2.2 show that for singular points P, $\mathbb{N} \setminus G(P)$, is not in general an additive subsemigroup of $\mathbb{N}$, as it happens for smooth curves. Nevertheless such WGS's are not completely arbitrary as suggested, for instance, by [G2], Props. 6.1 and 6.3 (see also (3.20) and (3.21) in [G3] and following discussions). As a matter of example, on a Gorenstein curve of genus 4 there cannot be any node having $< 1, 2, 3, 5 >$ as a WGS. For, if it were, it should necessarily correspond, as the reader can easily check, to two points, Q_1 and Q_2 on the (partial) normalization, whose WGS's are, respectively, $< 1, 2, 3 >$ and $< 1, 3, 4 >$. But $< 1, 3, 4 >$ is not possible, since Q_2, lying over P, is a smooth point while $\mathbb{N} \setminus \{< 1, 3, 4 >\}$ is not a semigroup (Cfr. [W1], Prop. 3.3).

Next example will show a computation of a WGS at a singular point of a non rational curve of arithmetic genus 2.

Example 2.7. This example has been already considered in [G2], Ex. 6.8. Here we reproduce it for sake of completeness and because the curve we are now going to consider, is not rational. In this case, if does not fail, it is at least very difficult to find a basis of dualizing differentials in order to compute the weight by means of the wronskian. The example consists in considering the (affine) plane quintic C having a cusp at the origin $P \equiv (0,0)$:

$$x^5 + x^3 + y^2 = 0. \tag{2.17}$$

The reason for which the curve defined by the zero locus of (2.17) is not rational is that it has also a triple point with an *infinitely near* double point at the infinity of the y axis, and has no more singular points. The non-rationality of C hence follows by the genus formula. General facts on plane curves, ensure us that the dualizing series on C is cut out by the conics of $\mathbb{P}^2$, which form a linear system of dimension 5, i.e. $h^0(C,\mathcal{K}_C) = 2$, equal to, as one may expect, to the arithmetic genus of C. Let $\theta_P : \hat{C} \longrightarrow C$ be the partial normalization of C at the point P and let $Q = \theta_P^{-1}(P)$. To find the WGS at P it is sufficient to find the Σ-WGS at Q, putting in a system the equation of a generic conic of $\mathbb{P}^2$ and (2.17):

$$\begin{cases} x^5 + x^3 + y^2 = 0 \\ ax^2 + bxy + cy^2 + dx + ey + f = 0 \end{cases}. \tag{2.19}$$

We shall proceed in eliminating the indeterminate y between the two equations of (2.19). The computations, practically impossible to deal with by hand, have been performed using the algebraic manipulation program **CoCoA** by Giovini and Niesi [GN]. The result of the elimination is the following family of polinomials:

$$\begin{aligned} P(x;a,b,c,d,e,f) =& c^2x^{10} + 2c^2x^8 + (b^2 - 2ac)x^7 + (c^2 - 2cd + 2be)x^2 + \\ &+ (b^2 - 2ac + e^2 - 2cf)x^5 + +(a^2 - 2cd + 2be)x^4 + \\ &+ (2ad + e^2 - 2cf)x^3 + (d^2 + 2af)x^2 + 2dfx + f^2. \end{aligned}$$

For generic values of the parameters (a,b,c,d,e,f) one gets, as a result of an easy check:

$$\begin{aligned} ord_0 P(x;a,b,c,d,e,f) &= 0 \\ ord_0 P(x;a,b,c,d,e,0) &= 2 \\ ord_0 P(x;a,b,c,0,e,0) &= 3 \\ ord_0 P(x;a,b,c,0,0,0) &= 4 \\ ord_0 P(x;0,b,c,0,0,0) &= 5 \\ ord_0 P(x;0,0,c,0,0,0) &= 2 \\ ord_0 P(x;0,0,0,0,0,0) &= \infty, \end{aligned}$$

from which the following chain of inequalities:

$$\begin{aligned} &\underset{\overset{\shortparallel}{2}}{d(\Sigma)} \geq \underset{\overset{\shortparallel}{5}}{d(\Sigma - Q)} \geq \underset{\overset{\shortparallel}{5}}{d(\Sigma - 2Q)} \geq \underset{\overset{\shortparallel}{4}}{d(\Sigma - 3Q)} \geq \underset{\overset{\shortparallel}{3}}{d(\Sigma - 4Q)} \geq \underset{\overset{\shortparallel}{2}}{d(\Sigma - 5Q)} \geq \\ &\geq \underset{\overset{\shortparallel}{1}}{d(\Sigma - 2Q)} \geq \underset{\overset{\shortparallel}{0}}{d(\Sigma - 7Q)} \geq \underset{\overset{\shortparallel}{0}}{d(\Sigma - 8Q)} \geq \ldots \geq \underset{\overset{\shortparallel}{0}}{d(\Sigma - 11Q)}, \end{aligned}$$

claiming that the WGS at P is:

$$G(P) = < 1, 3, 4, 5, 2, 7 > .$$

Hence, the weight sequence is given by:

$$w(P) = < 0, 2, 8, 15, 24, 35 > .$$

Let us incidentally notice that, by [G2], Cor. 6.4, (see also [G3], formula 3.21) $Q = \theta_P^{-1}(P) \in \hat{C}$ is not a Weierstraß point.

After these few examples, some remarks are in order. First of all we notice that, once a projective model of a subcanonical Gorenstein curve C is given, it is in principle always possible and very easy to compute the WGS at the singular points (indeed at any point). In fact, all the difficulties are merely of computational kind, in the handle sense of the word. As the previous examples show, what one has to find is just the number of independent conditions imposed by the divisors supported on the point P in order to be contained in some section of the dualizing sheaf (see [H], III.7 for definitions). From a practical point of view, we consider the number of independent conditions imposed, by the divisors supported at P, on suitable hypersurfaces cutting on C the dualizing series (the lines for quartic plane curves of Ex. 2.1, the hyperplanes in Example 2.5, and conics for the quintic plane curve of Ex. 2.7). We remark that the success of computing WGS at a (singular) point depends just on the power of the algebraic manipulation programs currently availables (for instance we used **CoCoA** for computing example 2.7).

On the other hand, it seems to be very hard to compute the (sixth) weight of the singular point studied in Example 2.9. just using the wronskian, as Lax and Widland do in Ex. 2.1., where the curve is rational. For non-rational curves the difficulty arises when one tries to find a basis of regular dualizing differentials around the point P. In this sense the concept of WGS for a singular point on a Gorenstein curve gives a concrete algorithm to compute the weight(s), the results depending just on how much powerful is the algebraic manipulation program one is dealing with.

3. On the Classification of the Rational Nodal Gorenstein Curves of Arithmetic Genus 3.

The main purpose of this section is to give evidence to the geometrical meaning of the weight (or the gaps) sequence in the case of the nodes on a plane quartic. Using the suggestions given by this simple analysis we shall revisit some propositions contained in [LW2] by a simpler point of view. Moreover we shall complete the classification of the rational nodal Gorenstein curve of arithmetic genus 3, exposed in [LW1]. Before starting it is worth to mention that, here, the leading idea is that "most" of the integral Gorenstein curves of arithmetic genus 3 are nothing but *plane quartics*. To start with, hence, in next example it will be shown that on a plane quartic (which is a Gorenstein curve of arithmetic genus 3) it is possible to find nodes having one (and, of course, only one) of the following WGS: a) $G(P) =< 1,2,3 >$; b) $G(P) =< 1,2,4 >$; c) $G(P) =< 1,2,5 >$, where P denotes the given node. Moreover, by virtue of [G2], Cor. 6.6, they are the only possible sequences at P.

Example 3.1 (Nodes on Plane Quartics).

3.1.a) *Plane Quartic having a node P with $< 1,2,3 >$ as a WGS.*
Let us consider the (affine) plane quartic C:

$$x^4 + x^3 + x^2 - y^2 = 0 \tag{3.1}$$

The origin is an ordinary node, having $(y+x)(y-x) = 0$ as a tangent cone. Intersect the curve with the pencil of lines $y = mx$ through the origin. One has:

$$\begin{cases} x^2(x^2 + x + 1 - m^2) = 0 \\ y = mx \end{cases}, \tag{3.2}$$

The partial normalization of C at the origin P, $\theta_P : \hat{C} \longrightarrow C$ is nothing but that the "blow-up" of C at the point P. The situation is represented in the Picture 3.1, where E is the exceptional divisor and S the strict transform of the line $y = mx$, for a fixed m.

Picture 3.1

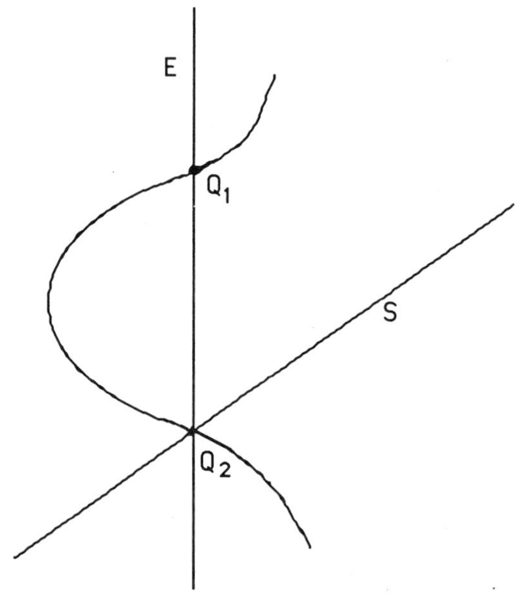

$\pi^{-1}(P) = \{Q_1, Q_2\}$;
$E \equiv$*exceptional divisor;*
$S \equiv$ *strict transform of a line passing through* P.

Set, as usual, $\theta_P^{-1}(P) = \{Q_1, Q_2\}$, and suppose that Q_1 refers to the value $m = 1$ of the angular coefficient, and Q_2 to the value $m = -1$. One has (cfr.previous examples):
(3.3)

$$\begin{array}{cccccc} d(\Sigma) \geq & d(\Sigma - Q_1) \geq & d(\Sigma - 2Q_1) \geq & d(\Sigma - 3Q_1) \geq & d(\Sigma - 4Q_1) \geq & d(\Sigma - 5Q_1) \\ \| & \| & \| & \| & \| & \| \\ 3 & 2 & 1 & 0 & 0 & 0 \end{array}$$

Indeed, imposing that $y = mx$ passes through Q_1 with intersection multiplicity at least 2, (condition for the line be tangent), means to impose the condition $m = 1$, as one can infer from (3.2). The same formula says also that cannot exist any line having a greater multiplicity intersection with the branch corresponding to the point Q_1. In fact that branch does not have any inflection point at P. The same analysis holds for the point Q_2. The Σ-WGS at Q_1 and at Q_2 are hence equal to $< 1, 2, 3 >$. From [G2], formula (4.4), one gets $G(P) =< 1, 2, 3 >$, and $< 0, 2, 6 >$ is its weight sequence.

3.1.b) *Plane quartic having a node P with $< 1, 2, 4 >$ as a WGS.*
Let us consider the plane (affine) quartic:

$$x^4 + x^3 - y^3 + x^2 - y^2 = 0, \tag{3.4}$$

having again a node at the origin P. Now, the transform of the line $y = x$ intersects the curve $\hat{C}$ along 4 points, three of them being coincidents with Q_1 and the fourth with Q_2. The sequence of the Σ-gaps at Q_2 is, as in the previous example, again $< 1, 2, 3 >$, while for Q_1 one has to consider the following chain of inequalities:
(3.5)

$$\begin{array}{cccccc} d(\Sigma) \geq & d(\Sigma - Q_1) \geq & d(\Sigma - 2Q_1) \geq & d(\Sigma - 3Q_1) \geq & d(\Sigma - 4Q_1) \geq & d(\Sigma - 5Q_1) \\ \| & \| & \| & \| & \| & \| \\ 3 & 2 & 1 & 1 & 0 & 0 \end{array}$$

from which $< 1, 2, 4 >$ results to be the Σ-WGS at Q_1. Let us observe, in the meanwhile, that Q_1 is a Weiertraß point for $\hat{C}$ (cfr. Cor. 5.9). By [G2], formula (4.4), $G(P) =< 1, 2, 4 >$, with $\boldsymbol{w}(P) =< 0, 2, 7 >$ as a weight sequences.

3.1.c) *Plane quartic having a node P with $< 1, 2, 5 >$ as a WGS.*
Let us consider:

$$x^4 + x^2 - y^2 = 0 \tag{3.6}$$

where the tangents to the two branches of the curve at the origin are both of inflectional type. The Σ-WGS's at Q_1 and Q_2 are both equal to $< 1, 2, 4 >$. Hence $G(P) =< 1, 2, 5 >$ and $\boldsymbol{w}(P) =< 0, 2, 8 >$.

Remark 3.2. What has been said in Example 3.1, happens in general. Indeed the nodes of plane curves of degree 4 are classified by their weight sequences, which declare how many components of the tangent cone at the node are of inflectional type. See Picture 3.2. Hence on quartic plane curves we can find nodes having 6, 7 or 8 as 3rd weight. Moreover these are the only possible cases.

Picture 3.2

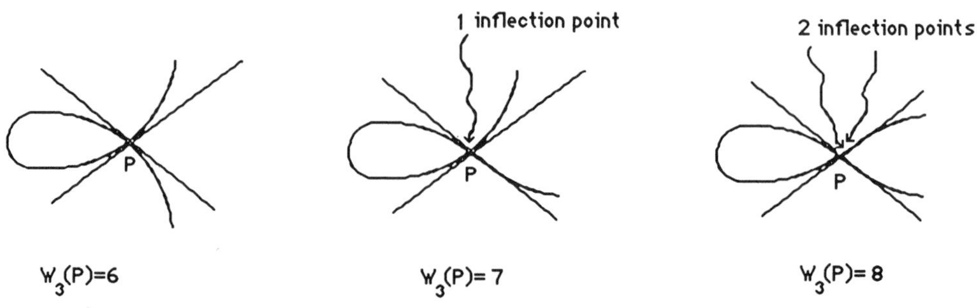

The easy results of the previous examples suggest to study, in a new way, questions posed in [LW2], where the aim was to find all possible distributions of Weierstraß points on a rational nodal Gorenstein curve C of arithmetic genus 3. To this purpose they prove that, for C, the following claims hold:

Proposition 3.3 ([LW2], Prop. 4). *If one node of C has (third) weight 8, then no node of C can have (third) weight 7.*

Proposition 3.4 ([LW2], Prop. 5). *If two nodes of C have weight 8, then the third node also has weight 8.*

Proposition 3.5 ([LW2], Prop. 7). *There is no rational nodal curve of arithmetic genus 3 such that each node has weight 7.*

Our claim is that the above propositions can be proved just by referring to plane models. Indeed, C cannot be embedded in $\mathbb{P}^2$ if and only if its dualizing sheaf is not very ample, i.e. if and only if C is hyperelliptic. Also in the Gorenstein context, we shall say that C is hyperelliptic if and only if there exists a finite morphism $\phi : C \longrightarrow \mathbb{P}^1$ of degree 2. Then the key results is given by a result proved by lax and Widland:

Proposition 3.6. *Suppose that C is a rational curve of arithmetic genus g and suppose that there exists a morphism $\phi : C \longrightarrow \mathbb{P}^1$ of degree two. Then each node on C has (third) Weierstraß weight $g(g-1)$ and there are two non singular Weierstraß points of weight $\frac{g(g-1)}{2}$.*

Proof. [LW2], Prop. 3. □

As a consequence of the previous proposition, it follows that the only possible nodal rational hyperelliptic Gorenstein curve of arithmetic genus 3 must have all its nodes with weight 6. This means that if we want to prove Props. 3.3, 3.4, 3.5 it is sufficient to prove them for plane curves, because, by Prop. 3.6 there will be no hyperelliptic models of such curves. At this point we have:

Proof of Prop. 3.3. Let $[X, Y, Z]$ be the homogeneous coordinates of $\mathbb{P}^2$. Without loss of generality we can suppose that the coordinates of the three nodes are: $[1, 0, 0]$, $[0, 1, 0]$ and $[0, 0, 1]$, so that the generic equation of a rational nodal quartic takes the form:

$$AX^2Y^2 + BX^2YZ + CY^2XZ + DY^2Z^2 + EXYZ^2 + FX^2Z^2 = 0. \tag{3.7}$$

Notice that $D \neq 0$, (for otherwise the polynomial (3.7) would be reducible). Hence, if α and β are suitable complex number such that:

$$DY^2 + EXY + FX^2 = D(Y - \alpha X)(Y - \beta X),$$

the condition ensuring that $[0,0,1]$ is a node of (third) weight 8, can be translated in the following linear system:

$$\text{(3.8)} \qquad \begin{cases} B + C\alpha = 0 \\ B + C\beta = 0 \end{cases}.$$

Now, $\alpha \neq \beta$, because otherwise $[0,0,1]$ would not be an ordinary node, so that system (3.8) is fulfilled if and only if $B = C = 0$. Eqn. 3.6 can hence be rewritten as:

$$\text{(3.9)} \qquad AX^2Y^2 + DY^2Z^2 + EXYZ^2 + FX^2Z^2 = 0.$$

Let us consider now any one of the remaining nodes. By the particular symmetry of the polynomial (3.9) there is no loss of generality in considering the one located in $[0,1,0]$. Having set $AX^2 + DZ^2 = A(X - \gamma)(X + \gamma)$, such a node has weight 7 if (remind the discussion of Ex. 3.1):

$$\begin{cases} E\gamma^2 = 0 \\ E\gamma^2 \neq 0 \end{cases},$$

which is clearly absurd. □

In the same framework of the above proof, by supposing to have one more node with (3rd) weight 8, one easily gets:

Proof of Prop. 3.4. If also $[0,1,0]$ has weight 8, then E must be zero in (3.9). Hence (3.9) takes the final form:

$$\text{(3.10)} \qquad AX^2Y^2 + DY^2Z^2 + FX^2Z^2 = 0,$$

showing that even the point $[1,0,0]$ has weight 8. □

The proof of Prop. 3.5 is completely similar even if it requires some detail more. In fact using the above methods one can quickly prove that the *generic* plane quartic cannot have three nodes each of weight 7. In fact this claim comes from the fact that such a quartic would exist just if a given determinant vanishes. The technical (but not conceptual) complication consists in showing that such a determinant *never* vanishes. Since this means to work out some tedious computation, we shall leave the proof of Prop. 3.5 to the reader's care. Notice that the proof of Prop. 3.4 shows that all plane quartics having three nodes of weight 8 are projectively equivalent to the one represented by (3.10).

Using Props 3.3, 3.4 and 3.5 and some more arguments, Lax and Widland, in [LW2], find all the possible distributions of Weierstraß points on rational nodal curves of arithmetic genus 3 with the only exception of a particular case. Let us look at the following table, taken from [LW2]:

Number of WP's having 3rd weight:

Case	1	2	3	6	7	8	existence
1	6	0	0	3	0	0	Y
2	4	1	0	3	0	0	Y
3	2	2	0	3	0	0	Y
4	0	3	0	3	0	0	Y
5	0	0	2	3	0	0	Y
6	0	0	0	2	1	0	Y
7	3	1	0	2	1	0	Y
8	1	2	0	2	1	0	?
9	4	0	0	2	0	1	Y
10	2	1	0	2	0	1	N
11	0	2	0	2	0	1	Y
12	4	0	0	1	2	0	Y
13	2	1	0	1	2	0	Y
14	0	2	0	1	2	0	Y
15	3	0	0	1	1	1	N
16	1	1	0	1	1	1	N
17	2	0	0	1	0	2	N
18	0	1	0	1	0	2	N
19	3	0	0	0	3	0	N
20	1	1	0	0	3	0	N
21	2	0	0	0	2	1	N
22	0	1	0	0	2	1	N
23	1	0	0	0	1	2	N
24	0	0	0	0	0	3	Y

Table 3.3

The rows of the table 6.3 ending with a "N" represent the distributions of Weierstraß points which are not possible. For instance, the case **15** cannot occur by virtue of Prop. 3.3. The possible distributions are those corresponding to rows ending with a "Y". For instance the distribution of row **24** is realized by the curve having equation (3.10). The unsettled case by Lax and Widland is the one corresponding to row **8**, ending with a "?". But notice that if a *rational nodal curve of arithmetic genus* 3, *having* 1 *normal smooth WP, two smooth WP's of weight* 2, *two nodes of weight* 6 *and one of weight* 7 exists, it must be a plane quartic, by virtue of Prop. 3.6. Using the fact that on a quartic plane curve a WP is a flex (a hyperflex if the (3rd) weight is 2), we shall look for:

(Translation of the Question by Lax and Widland). a rational plane quartic having a simple flex, two hyperflexes, two "simple" nodes (i.e. the node has no inflectional branch (see Pict. 3.2)), and a node having just one inflectional branch.

As a matter of terminology, we recall that an *hyperfex* on a quartic is a flex P whose tangent line intersect the curve with multiplicity 2 at P.

Thanks to this translation, we can fit the gap in table 3.1, by means of the following:

Proposition 3.7. *Let $[X, Y, Z]$ be homogeneous coordinates in $\mathbb{P}^2$ and $[u, v]$ homogeneous coordinates of $\mathbb{P}^1$. The rational curve of parametric equations:*

$$\begin{cases} X = u^4 \\ Y = (u-v)^3(u+iv) \\ Z = v^4 \end{cases} \tag{3.11}$$

where i is the imaginary unit, is an example which realizes situation 8 *of table 3.3.*

Proof. Expression (3.11) has been constructed in such a way to have a hyperflex located at $F_1 \cong [1,1,0]$ with tangent $Z = 0$ and another one located at $F_2 \cong [0, i, 1]$, with tangent $X = 0$. Moreover there is a simple flex in $F_3 \cong [1,0,1]$. In the affine open subset $Z \neq 0$ of $\mathbb{P}^2$, with coordinates $(x = \frac{X}{Z}, y = \frac{Y}{Z})$, (3.11) can be rewritten as:

$$\begin{cases} x = t^4 \\ y = (t-1)^3(t+i) \end{cases}. \tag{3.12}$$

The solutions of the system in the unknown (t_1, t_2):

$$\begin{cases} \dfrac{x(t_1) - x(t_2)}{t_1 - t_2} = 0 \\ \dfrac{y(t_1) - y(t_2)}{t_1 - t_2} = 0 \end{cases},$$

give the localization of the singular points. A straightforward computation shows that each one of the pair of solutions has distinct components, i.e. $t_1 \neq t_2$. Well known facts ensure that (3.11) has just nodes and no cusp or other kinds of unibranch singular points. Moreover, in the affine open set choosen, one has that $F_1 = (x(1), y(1)) = (1,0) = (x(i), y(i))$, which means that F_1 falls into a node. We proceed in looking for other flexes. To this purpose we should notice that for an irreducible quartic plane curve, a basis for the dualizing sheaf is given, locally, by the restriction of the homogeneous coordinates to the curve itself. This means that if P is a smooth point, and s is a local parameter around P, the latter is a Weierstraß point if and only if the *wronskian determinant* (see, e.g. [ACGH], [Gu] or [F]):

$$\begin{vmatrix} X(s) & Y(s) & Z(s) \\ X'(s) & Y'(s) & Z'(s) \\ X"(s) & Y"(s) & Z"(s) \end{vmatrix}, \tag{3.13}$$

vanishes at $s = 0$. In (3.12) the second and third rows are, respectively, the usual first and second derivative of the first row with respect to s. This procedure can be made precise, as in [F], [Gu] or, even, in [G2]. In the affine open set $Z \neq 0$, we can use the local parameter t of formula (3.12).

Hence we have: $(X(t) = x(t) = t^4, Y(t) = y(t) = (t-1)^3(t+i), Z(t) = 1)$. (3.13) is thence:

$$\begin{vmatrix} t^4 & (t-1)^3(t+i) & 1 \\ 4t^3 & (t-1)^2(4t-3i-1) & 0 \\ 12t^2 & 6(t-1)(2t-1-i) & 0 \end{vmatrix}. \tag{3.14}$$

it follows that $t = t_0$ corresponds to a flex of the curve, if and only if it is a solution of the following equation:

$$\begin{vmatrix} 2t^3 & (t-1)^2(4t-3i-1) \\ t^2 & (t-1)(2t-1-i) \end{vmatrix} = 0 \tag{3.15}$$

deduced from the determinat (3.14). Before going on, it is worth to remark that if one thinks about the real picture of the above quartic, determinant (3.15) expresses the well known fact, in elementary differential geometry, that the flexes of plane curves are the points of *vanishing curvature*. Coming back to eqn. 3.15, we observe that it is satisfied for $t = 0$, $t = 1$ and $t = \frac{3i+1}{i+3}$. For $t = 1$ and $t = 0$, one finds again $F_3 \equiv (1,0)$ and $F_2 \equiv (0,i)$ while for $t = \frac{3i+1}{i+3}$ one gets:

$$F_4 = \left(\left(\frac{3i+1}{i+3}\right)^4, -8(i+1)\left(\frac{i-1}{i+3}\right)^4\right).$$

As already noticed, F_2 falls down into a node. As for F_3, it falls down into a node if and only if there exists a $t_0 \neq 0$ such that $P(0) = P(t_0) = (0,i)$. But this would imply that $t_0^4 = 0$, which is impossible. Hence F_3 does not fall into a node. Analogously, F_4 is located at a node if and only if there exists $t_0 \neq \frac{3i+1}{i+3}$ such that $P(t_0) = P(\frac{3i+1}{i+3})$. To check the behaviour we have hence to solve the following system:

$$\begin{cases} t_0^4 = \left(\dfrac{3i+1}{i+3}\right)^4 \\ (t_0-1)^3(t_0-i) + 8(i+1)\left(\dfrac{i-1}{i+3}\right)^4 = 0 \end{cases}. \tag{3.16}$$

The first equation of (3.16) gives the four possible values for t_0:

$$t_0 = \pm\frac{3i+1}{i+3} \quad \text{e} \quad t_0 = \pm i\frac{3i+1}{i+3},$$

where we must exclude $t_0 = \frac{3i+1}{i+3}$. A straightforward check shows that no one of the remaining three values for t_0 satisfies the second equation of system (3.16). hence F_3 does not fall into a node.

Summarizing all what has been said, we conclude that (3.11) is the equation of a plane quartic having three nodes. There must hence be 6 flexes, counted according their own multiplicity. Two of them, F_1 and F_2, are iperflexes and must be counted with multiplicity 2. F_3 falls down into a node, so that F_4 has to be a simple smooth flex. (3.11) is hence an example of Gorenstein curve of arithmetic genus three having 2 WP's of weight 6, a WP of weight 7, two WP's of weight 2 and a WP of weight 1, as prescribed by the 8th row of the table 6.3. □

Remark 3.8. A trivial check shows that the curve:

$$\begin{cases} x = t^4 \\ y = (t-1)^3(t-i) \end{cases}$$

also have the same distribution of WP's prescribed by row **8** of table 3.3. Indeed, it can be proved that all nodal rational plane quartics verifying such a distribution are, in fact, projectively equivalent to one of the curve:

$$\begin{cases} x = t^4 \\ y = (t-1)^3(t \pm i) \end{cases}.$$

By using the manipulation system by Giovini and Niesi, **CoCoA**, one can find the polynomial equations of these two curves in $\mathbb{P}^2$, which are:

$$X^4 + Y^4 - 4X^3Y - 4XY^3 + 2X^2Y^2 - 4X^3Z - (24 \mp 28i)X^2YZ \pm 4iY^3Z +$$
$$\pm 24iXY^2Z - 2Y^2Z^2 + (28 \mp 24i)XYZ^2 \mp 4iYZ^3 \mp XZ^3 + Z^4 = 0.$$

References

[ACGH]. E. Arbarello, M. Cornalba, P.A. Griffiths, J. Harris, *Geometry of Algebraic Curves,* Vol 1, Springer-Verlag, 1984.

[BG1]. E. Ballico, L. Gatto, *On Bounds for the Weierstraß Weight at Singular Points of Gorenstein Curves*, Preprint Politecnico di Torino **27/93** (1993).

[BG2]. E. Ballico, L. Gatto, *Weierstrass Points on Singular Curves*, Preprint Politecnico di Torino (1994).

[F]. O. Forster, *Lectures on Riemann Surfaces*, Springer-Verlag, 1984.

[G1]. L. Gatto, *k-forme wronskiane, successioni di pesi e punti di Weierstraß su curve di Gorenstein*, Tesi di Dottorato-Università di Torino, 1993.

[G2]. L. Gatto, *Weight Sequences versus Gap Sequences at singular Points of Gorenstein Curves*, Geometriae Dedicata, (to appear).

[G3]. L. Gatto, *Weierstrass Loci and Generalizations, I*, in this Volume.

[GN]. A. Giovini & G. Niesi (implemented by), *CoCoA 1.0 (Computational Commutative Algebra)*, Università di Genova, 1990.

[Gu]. R.C. Gunning, *Lectures on Riemann Surfaces*, Princeton Mathematical Notes, 1966.

[H]. R. Hartshorne, *Algebraic Geometry*, Springer-Verlag, 1977.

[L1]. R.F. Lax, *Weierstrass Points on rational Nodal Curves*, preprint.

[LW1]. R.F. Lax and C. Widland, *Gap Sequences at Singularity*, Pacific Journal of Mathematics **150-1** (1991), 111–122.

[LW2]. R.F. Lax and C. Widland, *Weierstrass Points on Rational Nodal Curves of Genus 3*, Canad. Math. Bull. **30** (1987), 286–294.

[W1]. C. Widland, *Weierstrass Points on Gorenstein Curves*, Ph.D Thesis (Louisiana State University) (1984).

[W2]. C. Widland, *Bounds for Weierstrass Weights*, preprint.

[WL]. C. Widland and R.F. Lax, *Weierstrass Points on Gorenstein Curves*, Pacific Journal of Mathematics **142, no 1** (1990), 197–208.

On the Weierstrass Weights at Gorenstein Singularities

FABRIZIO PONZA Dipartimento di Matematica, Politecnico di Torino-Corso Duca degli Abruzzi, 24 - 10129 Torino-Italy

e-mail:
DOTMAT@ITOPOLI.BITNET or
DOTMAT@POLITO.IT.

ABSTRACT. Using, as key tool, the notion of *weight sequence* defined in [G2], some characteristic formulas for extraweights ([W1]) at n-branched singular points on Gorenstein curves are proved. The proofs are essentially of a numerical nature and very formal. An extra geometric information involving the *intersection number* of the branches of P, allows finally a simpler proof of some theorems, contained in [GL2] and [GL3], about expressions for the Weierstrass weights at Gorenstein singularities.

1. Introduction.

In the last few years, several papers ([LW1], [LW2], [W2], [St], [GL1], [LT], [G1], [G2], [BG1], [BG2]) dealt with the theory of *Weierstrass points on Gorenstein curves* introduced by Widland and Lax mainly in [W1] and [WL]. This theory is based on the definition of a suitable *wronskian determinant* ([W1], [WL]) which uses in a relevant way the fact that on a Gorenstein Curve the dualizing sheaf is invertible. Thanks to this theory, Lax and Widland are able to define the *Weierstrass*

Work supported by DOTTORATO DI RICERCA IN MATEMATICA, Università di Torino and, partially, by MURST .

Weight of a singular point, which extends in a natural way the classical one known for smooth curves. Moreover it is possible to show that any Gorenstein singularity P has weight larger than or equal to $\delta_P g(g-1)$, where g is the *arithmetic genus* of the curve and δ_P is a "measure" (see Sect. 2) of the singularity at P.

From a computational point of view, the theory by Lax and Widland has been simplified by the introduction in [G1] and [G2] of the notion of Weierstrass Gap Sequence (WGS in the following) at a singular point. This definition is alternative to the one given in [LW1], which is of a completely different nature, but has relevant technical advantages. In fact, if $\{a_1, \dots, a_g\}$ is the WGS as in [G2], Def. 3.7, and if $w_g(P)$ denotes the Weierstrass weight computed by means of the wronskian defined in [WL], one has:

$$w_g(P) = \delta_P g(g-1) + \sum_{i=1}^{g}(a_i - i), \tag{1.1}$$

so that $\sum_{i=1}^{g}(a_i - i)$ turns out to be what in [W2] has been called *extraweight* and which will be denoted by $E_g(P)$, following [G2]. The nice feature of the previous formula, which reduces to the classical one for smooth points (see [ACGH] or [Gu]), is that, sometimes, it is simpler to compute the WGS at P rather than to write down *explicitly* the wronskian around P itself (see, e.g. ex. 6.8 in [G2]). Indeed a lot of questions arising from the theory by Widland and Lax can be treated by a purely numerical approach based on relation (1.1). A nice example is offered by some formulas found by Garcia and Lax in [GL2] and [GL3] (to which this paper is inspired), where the weight at a singular point P is related with the behaviour of the so-called *semigroup of values* at P, defined at the beginning of Sect. 3.

The main purpose of this short paper is to show how the techniques used to define the WGS at Gorenstein singularities, allow to find the same formulas by Garcia and Lax ([GL2], Thm. 2.4, [GL2] Thm. 3.17 and [GL3] Thm. 6.3) using essentially simple numerical arguments. After some preliminary stuff about notation and basic definitions in Sect. 2, in Sect. 3 the main results of the paper are proved, i.e. Lemma 3.3 and Prop. 3.6. Lemma 3.3, in particular, shows, without knowing anything about the *theory of adjoints* (see [GV] or [Sz]), that if P is a unibranch Gorenstein singularity on a unisingular curve C of arithmetic genus g, then the (affine) dimension of the linear series cut out on C by the dualizing differentials "containing" the *conductor* is exactly $g - \delta_P$.

In Prop. 3.6 we patch together our numerical data to get a formula for the extraweight of any Gorenstein singularity. In that formula some linear systems Σ_i's, living on the partial normalization $\hat{C}$ of the curve C at P, are involved. It turns out that the linear systems V_i's appearing in [GL3], Thm. 6.3 (Prop. 3.7 in this paper), are nothing but $\Sigma_i - \mathcal{I}_i$, where $\mathcal{I}_i$ is the so-called *intersection number* of suitable branches of P, studied in [Ga], [GL2], [GL3].

I am grateful to Prof. S. Greco for encouragement and for having suggested me to work on this subject. Thanks to L. Gatto for discussions. Special thanks are due to Prof. R.F. Lax and Prof. A. Garcia for a couple of important remarks.

2. Preliminaries and Notation.

In this section we shall fix some notation and introduce the basic tools of the main results of next section. To this purpose, in the following C will denote a *projective, integral, Gorenstein Curve* ([S], Chap. 4) of arithmetic genus g over the complex field $\mathbb{C}$. As it is customary to do, at each $P \in C$ we attach a non-negative integer δ_P, defined as:

$$\delta_P = \dim_{\mathbb{C}} \frac{\tilde{O}_{C,P}}{O_{C,P}},$$

where $O_{C,P}$ is the local ring of regular functions at P and $\tilde{O}_{C,P}$ is its *integral closure*. Obviously, $\delta_P \neq 0$ if and only if $O_{C,P}$ is not a DVR, i.e., if and only if the point P is singular. In other words, δ_P measure the "rate" of singularity at P itself. At each point P of C, three fundamental sequences of g integers can be attached (see [G2] for details): the *Weight Sequence* $< w_1(P), \ldots, w_g(P >)$, the *Extraweight Sequence*, $< E_1(P), \ldots, E_g(P) >$ and, finally, the *Weierstraß Gap Sequence* (WGS for short), $< a_1(P), \ldots, a_g(P) >$. The relationships among these three objects are characterized by the following formulas:

$$w_k(P) = \delta_P k(k-1) + E_k(P) \quad \text{for} \quad 1 \leq k \leq g, \tag{2.1}$$

and

$$E_k(P) = \sum_{i=1}^{k} (a_i(P) - i). \tag{2.2}$$

If P is a singular point, $\theta_P : \hat{C} \longrightarrow C$ will denote the *partial normalization* of C at P (see [S], Chap. 4) and, as well known, it turns out that $\hat{g} = g - \delta_P$, where we set $\hat{g}$ =genus of $\hat{C}$. As a matter of notation, we set

$$\Sigma = span(\theta_P^* \omega_1, \ldots, \theta_P^* \omega_g),$$

where $(\omega_1, \ldots, \omega_g)$ is a basis for the global sections of the dualizing bundle $\mathcal{K}_C$. Observe that Σ is a linear subsystem of $H^0(\hat{C}, \theta_P^* \mathcal{K}_C)$, in general not complete unless $\delta_P = 1$. A positive integer b is said to be a Σ-gap at $Q \in \theta_P^{-1}(P)$ if and only if $d(\Sigma - (b-1)Q) > d(\Sigma - bQ)$, where $d(\Sigma - D)$ (D any effective divisor over $\hat{C}$) means the (affine) dimension of the linear system spanned by all elements of Σ containing D. Since $d(\Sigma) = g$ and $d(\Sigma - (2g-1)Q) = 0$ it follows that, at each $Q \in \theta_P^{-1}(P)$, there are exactly g Σ-gaps. Let

$$< b_1(Q), \ldots, b_g(Q) >,$$

be the Σ-Weierstraß Gaps Sequence (Σ-WGS for short) at Q, i.e. the increasing sequence of the Σ-gaps. Define the k-th Σ-*weight* at Q, $w_k^{\Sigma}(Q)$, as:

$$w_k^{\Sigma}(Q) = \sum_{i=1}^{k} (b_i(Q) - i). \tag{2.3}$$

If V is any r-dimensional linear system on $\hat{C}$, define V-gaps and V-weights in the same manner. This will be useful in next section.

It can be shown that ([G2], Prop. 5.5):

$$E_k(P) = \sum_{Q \in \theta_P^{-1}(P)} w_k^{\Sigma}(Q). \tag{2.4}$$

If the Σ-WGS is known for all Q's lying over P, the WGS at P can be computed according to the formula ([G2], Prop. 5.5):

$$a_k(P) = \left(\sum_{Q \in \pi^{-1}(P)} b_k(Q) \right) - (\sharp(\pi^{-1}(P)) - 1) \cdot k \tag{2.5}$$

Before concluding the section let us recall that (see, e.g., [GL2]) a *numerical semigroup* is an additive subsemigroup of the natural numbers having finite complement in $\mathbb{N}$. A non negative integer l will be said a *gap* of the numerical semigroup S if $l \notin S$. If S is a numerical semigroup having δ gaps, say $l_1 < l_2 < \ldots < l_\delta$, the weight, $wt(S)$, of the semigroup S is defined to be $\sum_{i=1}^{\delta}(l_i - i)$. By using such a notion, the following proposition turns out to be another important tool for Sect. 3:

Proposition 2.1. *Let S be a numerical semigroup with δ gaps. Then there are exactly δ elements of S, say $\{n_0 = 0, \ldots, n_{\delta-1}\}$, less than 2δ such that*

$$\sum_{i=0}^{\delta-1}(n_j - j) = \delta(\delta - 1) - wt(S) \tag{2.7}$$

Proof. [GL2], Lemma 2.3. □

3. Expressions for Extraweights.

As is well known (e.g. [Ga], or [St]) to each n-branched singular point of a (Gorenstein) curve a subsemigroup S_P of $\mathbb{N}^n$ can be attached. More precisely, if $Q_1, \ldots, Q_n$ denote the smooth points lying over P in $\tilde{C}$, the *semigroup of values* at P is:

$$S_P = \{(m_1, \ldots, m_n) \subseteq \mathbb{N}^n \backslash m_i = dim_{\mathbb{C}} \frac{O_{\tilde{C},Q_i}}{fO_{\tilde{C},Q_i}}, f \in O_{C,P}\}$$

If $p_i : \mathbb{N}^n \longrightarrow \mathbb{N}$ is the projection onto the i-th factor, it is known that ([GL3]) $S_i = p_i(S_P)$ is a numerical semigroup (i.e. with finite complement in $\mathbb{N}$). Let us denote, following the notation as in Sect. 2, by δ_i the number of gaps of the semigroup S_i (corresponding to the point $Q_i \in \theta_P^{-1}(P)$). By [GL3] it follows that, for each i, $\delta_i \leq \delta_P$. We will need the following

Definition 3.1.

$$\mathfrak{I} = \delta_P - \sum_i \delta_i, \tag{3.1}$$

For the algebraic-geometric meaning of $\mathfrak{I}$ see [GL3], where it is called *intersection degree* of the various branches centered at the Q_i's . We do not need this information here.

The first claim of this section is nothing but a simple translation of the purely numerical Prop. 2.1:

Proposition 3.2. *Let P, $\{Q_i\}$, S_i and δ_i as above. Then:*

$$w^{\Sigma}_{\delta_i}(Q_i) = \delta_i(\delta_i - 1) - wt(S_i). \tag{3.2}$$

Proof. Let $\{1 = b_1, \ldots, b_{\delta_i}\}$ be the first δ_i Σ-gaps at Q_i. By Prop. 5.2 in [G2], it follows that $\{0 = b_1 - 1, \ldots, b_{\delta_i} - 1\}$ are the first δ_i elements of S_i. Hence, by the very definition of k-th Σ-weight $(1 \leq k \leq g)$, and by applying Prop. 2.1, one has:

$$w^{\Sigma}_{\delta_i}(Q_i) = (b_1 - 1) + \ldots + (b_{\delta_i} - i) = \sum_{i=1}^{\delta_i}[(b_i - 1) - (i - 1)] = \delta_i(\delta_i - 1) - wt(S_i).$$

□

Now we proceed in deriving a lemma which is interesting in its own.

Lemma 3.3. *With the same notation as above, we get:*

$$d(\Sigma - 2\delta_i Q_i) = g - \delta_i. \tag{3.3}$$

Proof. By definition of Σ-gaps, one must have:

$$d(\Sigma - b_{\delta_i+1} Q_i) = g - \delta_i - 1.$$

Suppose now that $b_{\delta_i+1} \leq 2\delta_i$. Then, again by Prop. 5.2 in [G2], $b_{\delta_i+1} - 1 \in S_i$, so that we could find at least $\delta_i + 1$ elements of S_i less than $2\delta_i$, against Prop. 2.1. Hence $b_{\delta_i+1} \geq 2\delta_i + 1$, which implies that $d(\Sigma - 2\delta_i Q_i) = g - \delta_i$, as claimed. □

Remark 3.4. Note that the above Lemma agrees with the general theory of the dualizing sheaf of Gorenstein varieties. In fact, in the case of a Gorenstein unibranch singularity, where $\theta_P^{-1}(P) = \{Q\}$ and $\delta = \delta_P$, we get

$$d(\Sigma - 2\delta_P Q) = g - \delta_P, \tag{3.4}$$

a well known result, because $\Sigma - 2\delta_P Q$ is exactly, ([Sz]), the dualizing sheaf of the curve $\hat{C}$, whose global sections form a $(g - \delta_P)$-dimensional $\mathbb{C}$-vector space.

At this point is quite natural, imitating a procedure showed in [BG2], Prop. 4.1, to state the following:

Proposition 3.5. *For each* $Q_i \in \theta_P^{-1}(P)$, *denote by* Σ_i *the linear systems* $\Sigma - 2\delta_i Q_i$. *Then, for each* $1 \leq k \leq g - \delta_i$, c_k *is a* Σ_i*-gap at* Q_i *if and only if* $c_k + 2\delta_i$ *is a* Σ*-gap at* Q_i.

Proof. c_k is a Σ_i-gap at Q_i if and only if

$$d(\Sigma_i - (c_k - 1)Q_i) > d(\Sigma_i - c_k Q_i),$$

i.e., if and only if:

$$d(\Sigma - (2\delta_i + c_k - 1)Q_i) > d(\Sigma - (2\delta_i + c_k)Q_i),$$

which is the same as to say that $c_k + 2\delta_i$ is a Σ-gap at Q_i. □

We can finally state the main result of this paper:

Proposition 3.6. *The* g*-th extraweight at* P *is given by:*

$$E_g(P) = \delta_P(g-1) - \mathfrak{I}(g-1) - \sum_{i=1}^{n} wt(S_i) + \sum_{i=1}^{n} w_{g-\delta_i}^{\Sigma_i}(Q_i) \tag{3.5}$$

Proof. We start by computing the g-th Σ-weight at each Q_i. We have:

$$w_g^{\Sigma}(Q_i) = w_{\delta_i}^{\Sigma}(Q_i) + \sum_{k=1}^{g-\delta_i} [b_{\delta_i+k} - (\delta_i + k)], \tag{3.6}$$

where $\{b_{\delta_i+1}, \ldots, b_g\}$ are the last $g - \delta_i$ Σ-gaps at Q_i. Now, by Prop. 3.5, $b_{\delta_i+k} = c_k + 2\delta_i$, where $\{c_1, \ldots, c_{g-\delta_i}\}$ are the Σ_i-gaps at Q_i. Substituting in (3.6) and using Prop. 3.2, one has:

$$w_g^{\Sigma}(Q_i) = \delta_i(\delta_i - 1) - wt(S_i) + (g - \delta_i)\delta_i + w_{g-\delta_i}^{\Sigma_i}(Q_i),$$

i.e., performing the computations:

$$w_g^{\Sigma}(Q_i) = \delta_i(g-1) - wt(S_i) + w_{g-\delta_i}^{\Sigma_i}(Q_i). \tag{3.7}$$

The g-th extraweight at P is now the sum over $1 \leq i \leq n$ (recall: n=number of branches of P), of the terms looking like (3.7). Hence:

$$E_g(P) = \sum_{i=1}^{n} w_g^{\Sigma}(Q_i) = (g-1)\sum_{i=1}^{n} \delta_i - \sum_{i=1}^{n} wt(S_i) + \sum_{i=1}^{n} w_{g-\delta_i}^{\Sigma_i}(Q_i). \tag{3.8}$$

From (3.1) one has: $\sum_{i=1}^{n} \delta_i = \delta_P - \mathfrak{I}$, so that a simple substitution in (3.8) gives exactly formula (3.5). □

We observe that, if P is a node, one has $\theta_P^{-1}(P) = \{Q_1, Q_2\}$ and, as one can easily check $\delta_1 = \delta_2 = 0$. Hence formula (3.8) gives:

$$E_g(P) = w_g^{\Sigma}(Q_1) + w_g^{\Sigma}(Q_2),$$

a not very significant formula, because it is just the definition of (g-th) extraweight at a node.

As a matter of fact, as a consequence of Prop. 3.6 and a finer analysis of the conductor sheaf one has::

Proposition 3.7 ([GL3], Thm. 6.3). *If P is a n-branched singular point on a Gorenstein curve, one has*

$$(3.9) \qquad w_g(P) = \delta_p(g-1)(g+1) - \mathfrak{I}(g-1) - \sum_{i=1}^{n} wt(S_i) + \sum_{i=1}^{n} w_{g-\delta_i-\mathcal{I}_i}^{V_i}(Q_i)$$

where we have setting $V_i = \Sigma_i - \mathcal{I}_i = \Sigma - 2\delta_i - \mathcal{I}_i$, and $\mathcal{I}_i$ is the intersection number defined in [GL3].

Proof. Using algebraic properties of the local ring at a Gorenstein Singularity, involving the structure of the conductor divisor on $\tilde{C}$, (see remark in [GL3],, page 20, sixth row from the top, or [St], Cor. 2.14), one can prove that:

$$(3.10) \qquad d(\Sigma - (2\delta_i + \mathcal{I}_i)Q_i) = g - \delta_i - \mathcal{I}_i$$

Now, from the chain of inequalities (cfr. Lemma 3.3):

$$\underset{g-\delta_i}{\underset{\|}{d(\Sigma - 2\delta_i)}} \geq \ldots \geq \underset{g-\delta_i-\mathcal{I}_i}{\underset{\|}{d(\Sigma - 2\delta_i - \mathcal{I}_i)}}$$

involving just $\mathcal{I}_i$ terms, it follows that the first $\mathcal{I}_i$ Σ_i-gaps at Q_i are $\{1, 2, \ldots, \mathcal{I}_i\}$. Then the following equality holds:

$$(3.11) \qquad w_{g-\delta_i}^{\Sigma_i}(Q_i) = w_{g-\delta_i-\mathcal{I}_i}^{V_i}(Q_i)$$

Formula (3.5) can be hence rewritten as:

$$(3.12) \qquad E_g(P) = \delta_p(g-1) - \mathfrak{I}(g-1) - \sum_{i=1}^{n} wt(S_i) + \sum_{i=1}^{n} w_{g-\delta_i-\mathcal{I}_i}^{V_i}(Q_i)$$

Our claim hence follows from (3.12) and the relation $w_g(P) = \delta_p g(g-1) + E_g(P)$. □

Remark 3.8. If $n = 1$, i.e. the singularity P is unibranch, one has $\delta_1 = \delta_P$ and $\Sigma_1 = \mathcal{K}_{\hat{C}}$, the dualizing sheaf of $\hat{C}$, the partial normalized curve of C at P. Formula (3.8) hence gives:

$$(3.13) \qquad E_g(P) = \delta_P(g-1) - wt(S) + \hat{w}_{g-\delta_P}(Q)$$

i.e. Cor. 4.5 of [BG2] for $k = g$, where $\hat{w}_{g-\delta_P}(Q)$ is the $g - \delta_P$ Weierstraß weight of $Q = \theta_P^{-1}(P)$. Moreover, recalling that by the theory of Lax and Widland ([W1], [WL]), $w_g(P) = \delta_P g(g-1) + E_g(P)$, one gets, for a unibranch singularity:

$$(3.14) \qquad w_g(P) = \delta_P(g-1)(g+1) - wt(S) + \hat{w}_{g-\delta_P}(Q)$$

i.e. exactly Thm. 2.4 of [GL2].

Remark 3.9. If P is an ordinary node we must put, in (3.12), $\delta_P = 1$, $\mathfrak{I} = \mathcal{I}_1 = \mathcal{I}_2 = 1$. Moreover $V_1 = \Sigma - Q_1 \cong \Sigma - Q_1 - Q_2 \cong V_2$. But $\Sigma - Q_1 - Q_2 = \mathcal{K}_{\hat{C}}$, so that we get, in this particular case:

$$w_g(P) = g(g-1) + \hat{w}_{g-1}(Q_1) + \hat{w}_{g-1}(Q_2),$$

as in [GL2], Cor. 3.18.

References

[ACGH]. E. Arbarello, M. Cornalba, P.A. Griffiths, J. Harris, *Geometry of Algebraic Curves,* Vol 1, Springer-Verlag, 1984.

[BG1]. E. Ballico, L. Gatto, *On the Monodromy of Weierstraß Points on Gorenstein Curves*, preprint, (1993).

[BG2]. E. Ballico, L. Gatto, *On Bounds for the Weierstrass Weight at Singular Points of Gorenstein Curves*, Preprint Politecnico di Torino, **27/93** (1993).

[Ga]. A. Garcia, *Semigroups associated to singular points of plane curves*, J. reine angew. Math. **336** (1982), 165–184.

[G1]. L. Gatto, *k-forme wronskiane, successioni di pesi e punti di Weierstraß su curve di Gorenstein*, Tesi di Dottorato-Università di Torino, 1993.

[G2]. L. Gatto, *Weight Sequences versus Gap Sequences at singular Points of Gorenstein Curves*, Geometriae Dedicata, (to appear).

[GL1]. A. Garcia, R.F. Lax, *Weierstraß points on Gorenstein curves in arbitrary characteristic*, Comm. Algebra (to appear).

[GL2]. A. Garcia, R.F. Lax, *Weierstrass weight of Gorenstein singularities with one or two branches*, Manuscripta Math. (to appear).

[GL3]. A. Garcia, R.F. Lax, *On Canonical Ideals, intersection Numbers and Weierstrass Points on Gorenstein Curves*, preprint , (1993).

[Gu]. R.C. Gunning, *Lectures on Riemann Surfaces*, Princeton Mathematical Notes, 1966.

[GV]. S. Greco, P. Valabrega, *On the Theory of adjoints*, Lecture Notes in Math., **732** (1978).

[H1]. R. Hartshorne, *Algebraic Geometry*, Springer-Verlag, 1977.

[L2]. R.F. Lax, *Weierstrass Points on rational Nodal Curves*, preprint ,.

[LT]. D. Laksov and A. Thorup, *Weierstrass Points and Gap Sequences for Families of Curves*, preprint, (1993).

[LW1]. R.F. Lax and C. Widland, *Gap Sequences at Singularity*, Pac. J. Math. **150-1** (1991), 111–122.

[LW2]. R.F. Lax and C. Widland, *Weierstrass Points on Rational Nodal Curves of Genus 3*, Canad. Math. Bull. **30** (1987), 286–294.

[S]. J.P. Serre, *Groupes Algébriques et Corps de Classes*, Hermann, Paris, 1959.

[St]. K.-0. Stöhr, *On the Poles of Regular Differentials of Singular Curves*, Bol. Soc. Bras. Mat. **24, n.1** (1993), 105–136.

[Sz]. L. Szpiro, *Equations Defining Space Curves*, Springer-Verlag, Tata Institute, 1979.

[W1]. C. Widland, *Weierstrass Points on Gorenstein Curves*, Ph.D Thesis (Louisiana State University) (1984).

[W2]. C. Widland, *Bounds for Weierstrass Weights*, preprint,.

[WL]. C. Widland and R.F. Lax, *Weierstrass Points on Gorenstein Curves*, Pac. J. Math. **142, no 1** (1990), 197–208.

Weierstrass Loci and Generalizations, I

LETTERIO GATTO
Dipartimento di Matematica, Politecnico di Torino,
Corso Duca degli Abruzzi, 24 - 10129 Torino - Italy.
e-mail:
LGATTO@ITOPOLI.BITNET or
LGATTO@POLITO.IT

1. Introduction.

The main purpose of this paper is to survey the main achievements gotten in the last years towards the extension of the notion of *Weierstrass Point* to singular curves. The notion of Weierstrass Point (WP in the following) is one of the most studied in various fields of mathematics, such as complex analysis on Compact Riemann Surfaces, complex analytic geometry and, of course, in algebraic geometry, where the theory has been developped in many interesting directions. From a "philosophycal" point of view a Weierstrass point on a curve (see Sect. 2 for detailed definitions) is a point which asks, with respect to generically chosen points, less conditions than it should for the fulfillment of some property. For instance, on a smooth plane curve of degree 4, the WP's in the classical sense (cfr. [ACGH], [GH], [Gu], [Fo]) are nothing but flexes. Among the flexes of a quartic plane curve one can make distinctions: in fact there are the *simple flexes*, having a contact of order three with the tangent line, and the so called *hyperflexes*, having a contact of order four with the tangent line. But in general, the tangent line to a *generic* point of the quartic, intersects the point at most two times, and no more, otherwise one asks *too much*. Flexes on a quartic plane curve can be hence characterized as the points P for which $(T\cdot C)_P>2$, i.e. points whose intersection multiplicity between the tangent line at P and the curve C itself, is greater than 2. Classical theories say that on a smooth quartic plane curve

---

Work partially supported by GNSAGA-CNR and MURST.

there are at most 24 flexes, i.e. 24 Weierstrass points (Cfr. [G3] for discussion of this example). The example of quartics can be generalized to all plane curves of degree d, considering again flexes, thought as WP's *with respect to the linear system cut out on* C *by the lines of* $\mathbf{P}^2$. It could seem quite natural, hence, to extend the notion of WP's to singular curves adding to the smooth WP's all the singular points. This point of view has been extensively developped in [BG3], where a general theory of Weierstrass Points with respect to any linear system on any projective integral (even singular, not necessarily Gorenstein) curve is constructed.

Since Weierstrass points are, in some sense, "special" points on a curve, they have been very useful to study *moduli problems*. In particular, some "subvarieties" of the moduli space of curves of genus g, M_g (see [ACGH] for a quick account on this subject), formed by isomorphism classes of curves possessing some *non ordinary* WP have been extensively studied, although it still remains much work to do. The *hyperelliptic locus* in M_g, is one of the most common subvarieties of the above type: indeed, given any hyperelliptic curve of genus $g\geq 2$, the WP's are precisely the ramification points (hence, exactly $2g+2$) of the hyperelliptic involution. General questions such as: find the dimension, if non-empty, of the locus of M_g made by curves having a point P with a given Weierstrass Gaps Sequence (WGS) (see Sect. 3 for definitions), establishing if it is irreducible or connected and so on, have been partially answered by many authors, such as Arbarello ([A]), Diaz ([D1]), Lax ([L1]), Patt ([P]), Rauch ([R]). Strictly related with these topics are the approaches showed in [Cu] and [LT1], [LT2] for studyng WP's on families of curves and families of WP's. There are, then, a lot of results aimed to show that even *special curves* try to be *generic* as much as possible have been shown by Coppens in [C1] and [C2]. For instance in [C1], it is shown that *the generic smooth irreducible curve of genus* g *posssessing a Weierstrass points whose first non-gap* (Sect. 2 for definitions) *is equal to* n *for some* $3\leq n\leq g-1$, *thus all the others Weierstrass points of* C *are normal.* Of course it is not possible here to give a full list of references about the mass of literature existing about Weierstrass Points. Nevertheless, many of lacking references can be found in those quoted at the end of this survey.

The aim of this paper is not to speak to an expert reader. The stuff here contained could give some suggestion about general picture of the subject to the non-specialists in the field. We shall not deal neither with moduli spaces nor with deformation theory. We want rather try to explain that behind the notion of Weierstrass point there are some genuine geometrical phenomena which one begins to learn at the undergraduate level. Much more than that, most of the formal language of the theory can be translated in terms of elementary analytic geometry (intersections between lines and curves, points spanning hyperplanes and so on).

In Sect. 2 we shall revisit the classical theory of Weierstrass points for smooth curves, generalizing it to considering WP's with respect to any linear system on the curve. Although in this section the material is not new, we think that rather new is the presentation of it. In fact the theory for smooth curves has been rephrased in such a way to admit a straightforward generalization to Gorestein curve, as shown in Sect. 3, where we get a definition of WP even in terms of a suitably defined (see [G2] or, for positive characteristic, [LT2]) *sheaf of principal part* (cfr. [La]) on the Gorenstein curve itself.

Sect. 3 is hence mainly devoted, as recalled, to explain the generalization gotten in [W1] and [WL] of the theory of WP's to the Gorenstein curves. Again, it is exposed in a non usual way, using extensively tools introduced in [G2] and having in mind the picture drawn in [BG3].

Sect. 4 is concerned with various developments of the theory of Widland and Lax, quoting recent results proved by Lax and Garcia in [GL1], [GL2] (and, in an alternative way, in [Po]) about relations between Weierstrass weight at Gorenstein singularities and the *semigroup of values* (see [De1], [De2], [Ga], [St] for definitions) at the singular point itself.
Leaving out from our discussion some recent results gotten in [BG1] about the monodromy group of the smooth Weierstrass points of Gorenstein curves, the survey ends, in Sect. 4, with some concluding remarks about a property of the the first non-gap (in the sense of [G2], Def. 4.8) for special kind of singular points, such as nodes and tacnodes.

Most of the ideas discussed in this survey have come out by enlighting discussions I had primarily with M. Coppens and S. Greco. Also I owe a lot to the useful e-mail correspondence I had, in the last months, with R.F. Lax, who let me to improve my personal understanding of the subject. I want to thank all of them.

2. Weierstrass Points on Smooth Curves.

For sake of completeness, we start this section by recalling one of the most classical definition of *Weierstrass point* on a smooth projective curve C. From the basic definitions we shall hence try to follow the main ideas which lead to more general formulations studied in the last years. Although many of the results and/or techniques here discussed hold in positive characteristic, we shall assume, from now on, to work over a fixed algebraically closed ground field of characteristic zero, which will be identified with the complex field **C**. In this context our smooth curves are nothing but Compact Riemann Surfaces (CRS).

The most common definition of Weierstrass point is spelled out by means of an analytical language: fixed any point P on a CRS of genus g, we say that P is a Weierstrass point if and only if there exists a meromorphic function f, holomorphic on C-{p}, having at P a pole of order less than or equal to g. This turns out to be a quite strong requirement, because it can be proved that on any CRS there can be at most (g-1)g(g+1) Weierstrass points, or, more significatively, *the generic point on a Compact Riemann Surface is not a Weierstrass point.* As a matter of fact, in a modern language, the existence of a rational function having a pole of order less or equal to g, implies that $h^0(C,O_C(gP))>1$, so that, if K_C is the *canonical bundle* of C, by the Riemann Roch formula one gets the equivalent condition $h^0(C,K_C(-gP))>0$. If the curve is not hyperelliptic think to the canonical embedding. Then last condition means that a Weierstrass Point P does not impose *independent conditions* to the k-osculating hyperplanes ($1\leq k\leq g$) of $\mathbf{P}^{g-1}$ to the curve C at the point P, which is clearly a non generic situation, since for g *generic* points in $\mathbf{P}^{g-1}$ (even *infinitely near*) there passes one and only one hyperplane. Notice that $h^0(C,O_C(P))=1$ unless the curve is rational. Moreover if g=1, the above definition implies that on a elliptic curve there cannot be any Weierstrass Point. From now on, then, we shall suppose that $g\geq 2$. Hence, the condition $h^0(C,O_C(gP))>1$ is equivalent to claim that there exists $1<n_2\leq g$ such that $h^0(C,O_C(n_2P))=2$ and $h^0(C,O_C((n_2-1)P)=1$. In general, set :

$$G_P=\{n \,/\, \text{there does not exist any holomorphic function on } C-\{P\} \text{ such that } (f)_\infty=nP\}.$$

It is a straightforward check to show that such a definition is equivalent to the other one:

$$(2.1) \qquad G_P=\{n \,/ h^0(C,O_C((n-1)P)=h^0(C,O_C(nP)).$$

The elements of G_P are said to be the *Weierstrass gaps at the point* P. They are finitely many, since by the Riemann-Roch formula $h^0(C,O_C((2g-1)P)=g$, so that *at any point* P there are exactly g gaps, forming the so-called Weierstrass Gap Sequence (WGS for short) at P:

$$(2.2) \qquad G_P=<1=n_1,n_2,\ldots,n_g>.$$

Hence a point P is a Weierstrass point if and only if its WGS is not <1, 2,...,g>. Notice that by the very defintion of G_P it follows that $H_P=\mathbf{N}-G_P$ is an additive subsemigroup of $\mathbf{N}$. If P is a WP having <1, 2, ...,g-1, g+1> as WGS it is said to be *normal.* and *special* otherwise. It is well known that the generic CRS has only *normal* WP's (see [GH] for references).

Even in the smooth context there are many way to reformulate the definition of Weierstrass point on a curve C or to generalize it. To start with, we should begin in noticing that the condition $h^0(C, K_C(-gP))>0$ defining a WP is much more significant than $h^0(C,O_C(gP))>1$. In fact, let :

(2.3) $$\omega=(\omega_1,\omega_2,\ldots,\omega_g)$$

be a basis over **C** of $H^0(C, K_C)$. P is a WP of C if and only if one is able to find a *positive dimensional* vector **C**-subspace made by global holomorphic differential vanishing at P with order at least g. This request imposes a compatibility condition of a system of linear equations (see, e.g. [Fo], [Gu], [ACGH]), which can be expressed as the vanishing, at z=0, of the following *wronskian* determinant:

(2.4) $$\begin{vmatrix} f_1(z) & f_2(z) & \ldots & f_g(z) \\ f'_1(z) & f'_2(z) & \ldots & f'_g(z) \\ \ldots & \ldots & \ldots & \ldots \\ f_1^{(g-1)}(z) & f_2^{(g-1)}(z) & \ldots & f_g^{(g-1)}(z) \end{vmatrix},$$

at z=0, where z is a local coordinate defined on a neighbourhood U around P, such that z(P)=0 and ω_i is expressed locally as:

(2.5) $$\omega_{i|U}=f_i(z)dz.$$

The derivatives in formula (2.4) are intended to be taken, of course, with respect to the local parameter z.

As a matter of fact one can show ([Fo], [ACGH], [Gu]) that the vanishing of expression (2.4) depends neither by the choice of the local parameter around P nor by the local representatives (2.5) chosen for the basis $\omega=(\omega_1,\omega_2,\ldots,\omega_g)$ of $H^0(C, K_C)$.

For this reason, as a matter of notation, we shall use, instead of determinant (2.4), the shorter *global* expression:

(2.6) $$\omega\wedge D\omega\wedge\ldots\wedge D^{g-1}\omega,$$

naming it *wronskian* g-form. *Locally*, a*wronskian* g-form can be represented as the rational function (2.4) regular at U, while *globally*, ([ACGH]) it represents an holomorphic section of the line bundle, over C, $(K_C)^{\otimes\frac{g(g+1)}{2}}$. It must be remarked that the vanishing locus of (2.6) is also independent by the choice of the basis ω, so that all the objects we are dealing with are surely well defined.

Define the *Weierstrass weight* $w_g(P)$ of P as the order of vanishing of

$\omega\wedge D\omega\wedge\ldots\wedge D^{g-1}\omega$ at P. Then *P is a Weierstrass Point if and only if* $w_g(P)>0$. Moreover the total weight of the Weierstrass points is:

$$\sum_{P\in C} w_g(P) = \deg\left((K_C)^{\otimes\frac{g(g+1)}{2}} \right) = (2g-2)\frac{g(g+1)}{2} = (g-1)g(g+1), \tag{2.7}$$

as previously claimed.

Furtherly one can prove that (see [Gu]) if $G_P = <1=n_1, n_2, \ldots, n_g>$ is the WGS at P, then: $w_g(P)=(n_1-1)+(n_2-2)+\ldots+(n_g-g)$, so that a WP is *normal* (see def. above) if and only if $w_g(P)=1$. If one assumes as definition of Weierstrass Point the one saying that P must satisfy the inequality $h^0(C,K_C(-gP))>0$, one understands that P is a Weierstrass Points if and only if is an *inflection point* (cfr. [ACGH]) for the linear system $H^0(C, K_C)$. This suggests to generalize the definition of Weierstrass point with respect to any linear system (cfr. [Cu], [EH1]) $(\mathcal{L}, V)$ on C, where $\mathcal{L}$ is a line bundle over C of degree d and V is an r-dimensional vector **C**-subspace of $H^0(C, \mathcal{L})$ (in other words, we are dealing with a g^r_d over C).

Let us denote by $H^0(C, \mathcal{L}/\mathcal{L}(-rP))$ the vector space of all global sections of $\mathcal{L}$ vanishing at P with order at least r. Then P is an *inflection point* of the linear system $(\mathcal{L}, V)$ if and only if the kernel of the natural map:

$$V \longrightarrow H^0(C, \mathcal{L}/\mathcal{L}(-rP)), \tag{2.8}$$

is not zero, or, equivalently, that:

$$\mathrm{rk}(V \longrightarrow H^0(C, \mathcal{L}/\mathcal{L}(-rP))) < r. \tag{2.9}$$

As a matter of notation let us denote by $D^r_{P|V}$ the map (2.8), whose meaning is going to be explained below. To start with, condition (2.9) suggests to introduce a new natural object. In a more general setting, fix a positive integer k and consider, then, the family of vector spaces, parametrized by C, $H^0(C, \mathcal{L}/\mathcal{L}(-kP))$. Such a family defines (see [ACGH], [La]) a rank k vector bundle or, which is the same, a rank k locally free sheaf of O_C-modules said the *sheaf of k-principal parts*, whose general construction for arbitrary schemes is shown in [Gr]. Here we sketch the construction just for curves, as in [ACGH], pag. 38. To this purpose, let C×C be the fiber product (over Spec(**C**)) of two copies of C, and let p and q be the projections of C×C respectively onto the first and the second factor. Let Δ be the diagonal of C×C and let I_Δ be the ideal sheaf defining Δ inside C×C. Then, the sheaf of k-principal parts associated to $\mathcal{L}$ is:

(2.10) $$P^k(\mathcal{L})=p_*\left(q^*\mathcal{L}\otimes\frac{O_{C\times C}}{I_\Delta^{k+1}} \right).$$

and it can be shown that the fiber at the point P of the associated rank k vector bundle is exactly $H^0(C, \mathcal{L}/\mathcal{L}(-kP))$.

Hence, to say that P is an *inflection point* for V, and hence, from now on and by definition, a*Weiertsrass point* with respect to the linear system $(\mathcal{L},V)$, may be rephrased by saying that:

(2.11) $$rk\,(D^r_{P|V} :V\longrightarrow P^r(\mathcal{L})_P)<r$$

or that the Weierstrass points with respect $(\mathcal{L},V)$, are those points for which the map:

(2.12) $$D^r_{|V} : C\times V\longrightarrow P^r(\mathcal{L}),$$

decreases in rank. As a matter of terminology recall that a WP with respect the complete linear system $H^0(C, K_C)$ will be simply said Weierstrass point.

When working over **C** there is a very simple analytical characterization of the sheaf of k-principal parts $P^k(\mathcal{L})$. In fact, you can cover the curve C by means of open sets $\{U_\alpha,\ \alpha\in\mathcal{A}\}$ domains of a local coordinate (z_α) (recall we are working on a smooth curve). If $\lambda\in H^0(C,\mathcal{L})$, then:

(2.13) $$\lambda_{|U_\alpha}=\lambda_\alpha(z_\alpha)\cdot\psi_\alpha,$$

where $\lambda_\alpha(z_\alpha)$ is a regular function over the open set U_α, and ψ_α generates $H^0(U_\alpha, \mathcal{L})$ over the regular functions on U_α.

The local data $\{U_\alpha\,;\, \lambda_\alpha(z_\alpha), \lambda'_\alpha(z_\alpha), ..., \lambda_\alpha^{k-1}(z_\alpha)\}$, of open sets and C^k-valued regular functions, patch together according to well defined functions:

(2.14) $$j^k\ell_{\alpha\beta} : U_\alpha\cap U_\alpha \longrightarrow Gl(k,\mathbf{C}),$$

satisfying the cocycle conditions $j^k\ell_{\alpha\alpha}=\mathbf{1}$ and $j^k\ell_{\alpha\beta}\, j^k\ell_{\beta\gamma}=j^k\ell_{\alpha\gamma}$.

Consider then the disjoint union $\cup_{\alpha\in\mathcal{A}}\{\alpha\}\times U_\alpha\times \mathbf{C}^k$ and takes the quotient with respect to the equivalence relations induced by the *glueing maps* $j^k\ell_{\alpha\beta}$, in such a way to get a rank k vector bundle over C. Such a vector bundle is often denoted by $J^k\mathcal{L}$, which is nothing but the analytical description of the sheaf of the k-principal parts associated to the line bundle $\mathcal{L}$. Indeed, the map (2.8) is the particular case for k=r of the map $D^k_{p|V}$ which can be explicitly described as follows:

(2.15)
$$\begin{cases} D^k_p : H^0(C,\mathcal{L}) \longrightarrow J^k_p\mathcal{L} \\ \lambda \longmapsto (\lambda_\alpha(z_\alpha(P)), \lambda'_\alpha(z_\alpha(P)), \ldots, \lambda^{(k-1)}_\alpha(z_\alpha(P))) \end{cases},$$

where it has been supposed that $P \in U_\alpha$, while the meaning of the $\lambda^{(m)}_\alpha$'s (the m-*th derivatives* taken with respect to the local coordinate z_α, $0 \le m \le k-1$) is clear after (2.13). In other words, D^k_p associates to any section $\lambda \in H^0(C, \mathcal{L})$ its *Taylor expansion* up to the order k around the point P. Even if it should be clear from the context, it turns out that:

(2.16)
$$J^k_p\mathcal{L} \cong P^r(\mathcal{L})_P \cong H^0(C, \mathcal{L}/\mathcal{L}(-rP)).$$

The time has come to go back to the definition of Weierstrass point, with respect to the r-dimensional linear system $(\mathcal{L},V)$. From the above discussion it turns out that $P \in C$ is a Weierstrass point with respect to $(\mathcal{L},V)$, if and only if $rk(D^r_{p|V})<r$. After the explicit expression (2.15), it is a straightforward exercise to show that, if $\mathbf{v}=(v_1,...v_r)$ is a basis of V over $\mathbf{C}$, and $P \in U_\alpha$, then:

(2.17)
$$rk(D^k_{p|V})<r \Leftrightarrow \begin{vmatrix} v_{1,\alpha}(z_\alpha(P)) & v_{2,\alpha}(z_\alpha(P)) & \ldots & v_{r,\alpha}(z_\alpha(P)) \\ v'_{1,\alpha}(z_\alpha(P)) & v'_{2,\alpha}(z_\alpha(P)) & \ldots & v'_{r,\alpha}(z_\alpha(P)) \\ \ldots & \ldots & \ldots & \ldots \\ v^{(r-1)}_{1,\alpha}(z_\alpha(P)) & v^{(r-1)}_{2,\alpha}(z_\alpha(P)) & \ldots & v^{(r-1)}_{r,\alpha}(z_\alpha(P)) \end{vmatrix} = 0.$$

Of course (see e.g. [G2] or [BG3]), the vanishing of the *wronskian* determinant (2.17) does not depend by the local representation chosen (as it must be, since $rk(D^r_{p|V})<r$ has to be an intrinsic expression), and similarly to what seen for holomorphic differentials (cfr. (2.6)), we shall refer to the determinant (2.17) by means of the shorter expression:

(2.18)
$$\mathbf{v} \wedge D\mathbf{v} \wedge \ldots \wedge D^{r-1}\mathbf{v},$$

which is easily seen to be a global section of the bundle $\mathcal{L}^{\otimes r}\otimes(K_C)^{\otimes\frac{r(r-1)}{2}}$, since it is locally represented by the *wronskian* (2.17). As a matter of notation, (see [G2] for rigorous meaning) we set, of course:

$$D^k\mathbf{v}=(D^k v_1, D^k v_2,\ldots, D^k v_r) \quad \text{for} \quad 1\leq k\leq r,$$

agreeing in using the abbreviated form D instead of D^1.

To the basis **v** of V we have hence associated the *wronskian* r-*form* $\mathbf{v}\wedge D\mathbf{v}\wedge\ldots\wedge D^{r-1}\mathbf{v}$. If $w_r^V(P)$ denotes the order of vanishing of the section $\mathbf{v}\wedge D\mathbf{v}\wedge\ldots\wedge D^{r-1}\mathbf{v}$ at the point P, the so-called r-th V-*weight* of P, from what has been said it follows that *the total* r-th *weight* of the $(\mathcal{L},V)$-Weierstrass points on C is given by:

$$(2.19)\quad \sum_{P\in C} w_r^V(P)=\#\{P\in C: \mathrm{rk}((D^k_{P|V})<r\}=\deg\left(\mathcal{L}^{\otimes r}\otimes(K_C)^{\otimes\frac{r(r-1)}{2}}\right)=rd+(g-1)r(r+1),$$

well known as the *Brill-Segre* formula. In the above discussion, note that if we choose the pair $(\mathcal{L},V)$ as $(K_C, H^0(C, K_C))$ then we get the *classical* theory of Weierstrass Points with respect to the canonical sheaf, so that, for instance, the *Brill-Segre* formula gives (because of d=2g-2 and r=g) exactly formula (2.7).

As shown in [G1], [G2] and [BG3], it can be useful to introduce what has been called in [G2] the *weight sequence* of a linear system $(\mathcal{L},V)$ at a point P of the curve. The idea consists in considering all the wronskian k-forms $(1\leq k\leq r)$ $\mathbf{v}\wedge D\mathbf{v}\wedge\ldots\wedge D^{k-1}\mathbf{v}$. They are, in fact, global holomorphic sections of the rank $\binom{r}{k}$-vector bundle:

$$\mathcal{L}^{\otimes k}\otimes(K_C)^{\otimes\frac{k(k-1)}{2}}\otimes\wedge^k V,$$

whose components are represented by the k×k minors of the k×r wronskian matrix:

$$(2.20)\qquad \begin{pmatrix} v_1 & v_2 & \ldots & v_r \\ Dv_1 & Dv_2 & \ldots & Dv_r \\ \ldots & \ldots & \ldots & \ldots \\ D^{k-1}v_1 & D^{k-1}v_2 & \ldots & D^{k-1}v_r \end{pmatrix},$$

i.e. global sections of the line bundle $\mathcal{L}^{\otimes k}\otimes(K_C)^{\otimes\frac{k(k-1)}{2}}$. Notice that the ideal of the local ring $O_{C,P}$ of the regular function at P, defined by the k×k minors of (2.20), $I_k O_{C,P}$, is principal (since P is smooth).

Set, for each $1\leq k\leq g$:

(2.21) $$w_k^V(P)=\mathrm{ord}_P\, \mathbf{v}\wedge D\mathbf{v}\wedge\ldots\wedge D^{k-1}\mathbf{v} = \dim_C\left(\frac{O_{C,P}}{I_k O_{C,P}}\right),$$

and name the non-decreasing sequence $< w_1^V(P), w_2^V(P),\ldots, w_r^V(P)>$ the V-*weight sequence* at the point P. We say that $G_p^V=<n_1,n_2,\ldots,n_r>$ is the V-*Weierstrass Gaps Sequence* (V-WGS for short) if and only if the following relation holds:

(2.22) $$w_k^V(P)=\sum_{i=1}^{k}(n_i-i).$$

It can be easily proved that n_i is a V-gap if and only if there exists a section $v\in V$, which vanish at least at the order n_i-1, or, rephrasing by means of the operator D, if and only if $\mathrm{rk}(D_p^{n_i-1})<i$.

Before concluding the section, it can be useful to explicit in another (of course equivalent) way the meaning of the V-WGS, holding because of a suitable (easy) readapting of the proof of [Gu], Thm. 15. To this purpose, pick a point $P\in C$. Let us denote by V-kP the vector subspace of V containing all sections of V vanishing at P with order at least k. Then, keeping in account that V has degree d, we can consider the following chain of inequalities:

(2.23) $$r=\dim(V)\geq\dim(V-P)\geq\dim(V-2P)\geq\ldots\geq\dim(V-dP)\geq\dim(V-(d+1)P)=0,$$

whose "extremal" data ($\dim(V)=r$ and $\dim(V-(d+1)P)=0$) say that there must be exactly r jumps of dimension, corresponding to positive integers $<m_1, m_2,\ldots, m_r>$, such that:

(2.24) $$\dim(V-(m_i-1)P) > \dim(V-m_iP).$$

The remarkable result, which shall be used in the following, is that, actually, the increasing sequence of integers $<m_1, m_2,\ldots, m_r>$ is nothing but the V-WGS at P. As a matter of terminology, recall that $<n_1-1, n_2-1,\ldots,n_r-1>$ is said to be the *ramification sequence* of V at P ([ACGH]). The ramification sequence starts with a zero if P is not a base point of V.

In the next section we shall discuss how, in a sense, quite surprisingly, almost the same framework can be imitated to build up a theory of Weierstrass point for Gorenstein Curves.

3. Weierstrass Points on Gorenstein Curves.

In a series of papers ([W1], [WL], [W2], [LW1], [LW2], [L2], [L3]), C. Widland and R. F. Lax founded and developped a theory of Weierstrass points on Gorenstein curves. Indeed, Gorenstein curves are the most natural framework, among the singular ones, to generalize almost straightforwardly the theory of Weierstrass points known for smooth curves. First of all we should notice that a Weierstrass Point theory on singular curves is mainly motivated by the try to find tools suitable for studying degeneration problems. For instance, it was fairly known that certain kinds of singular points may be viewed as limit of smooth Weierstrass points. One of the simplest result in this sense is due to Diaz, who shows, in [D2], that the generic node on a generic uninodal curve of arithmetic genus g is the limit of g(g-1) smooth WP's lying on "neighbouring" smooth curves. Other deeper results for so called "limit linear series" have been found in a series of papers by Eisembud and Harris (e.g. [EH1]. We refer to [BG1] for further references). From this point of view, the theory by Lax and Widland has the merit to define precisely what one should mean by Weierstrass point on a Gorenstein curve, without referring to any deformation setting.

To start with, let C be, from now on, a projective integral Gorenstein curve of arithmetic genus g (see, e.g., [S], Chap. 4). The main ingredients of the theory by Lax and Widland are the two possible definition of Gorenstein curve. The first one, sufficient *to set up* all the key definitions (as we shall show below) is to say that C is Gorenstein if and only if its *dualizing sheaf* (existing by [H1]-III-7), which by analogy with the previous section we shall denote again by K_C, is an *inverible sheaf*. As is shown in [S] the dualizing sheaf is nothing but the sheaf of the *regular differentials*, which is a *line bundle* in the Gorenstein case ([S], [W1]). The second definition needed in [W1] or [WL] to *develop* the theory is the following one. Let $O_{C,P}$ be the local ring of rational functions on C regular at P, $\tilde{O}_{C,P}$ its *integral closure*, and $\mathbf{c}_P$ the *conductor ideal* of $O_{C,P}$ in $\tilde{O}_{C,P}$, i.e. the largest ideal of $\tilde{O}_{C,P}$ contained in $O_{C,P}$ (for instance, if $\tilde{O}_{C,P} = O_{C,P}$, then $\mathbf{c}_P=\tilde{O}_{C,P}$). Define:

$$(3.1) \qquad \delta_P = \dim_{\mathbf{C}} \frac{\tilde{O}_{C,P}}{O_{C,P}} \quad \text{and} \quad n_P=\dim_{\mathbf{C}} \frac{\tilde{O}_{C,P}}{\mathbf{c}_P} .$$

Notice that P is smooth if and only if $\delta_P = n_P=0$. Then the curve C is said to be Gorestein if and only if for each $P\in C$, one has:

$$(3.2) \qquad n_P=2\delta_P .$$

It is not very difficult to meet Gorenstein Curves in algebraic geometry. Trivially, all smooth curves are. By (3.2) it follows quite easily that all nodal or cuspidal curves are Gorenstein, as well as curves which are complete intersections. In particular, all plane curves are Gorenstein. On the other hand it is not even difficult to find examples of non Gorenstein singularities. For instance $\mathrm{Spec}(\mathbf{C}[T^3,T^4,T^5]_{(T^3,T^4,T^5)})$ is not a Gorenstein singularity ($\delta_P=2$, $n_P=3$).

The idea of Lax and Widland to extend the notion of Weierstrass point on Gorenstein curves is based on the following key remark. Let

(3.3) $$d: O_C \longrightarrow \Omega^1_{C/\mathbf{C}},$$

the sheaf homomorphism given by the exterior differentiation. As is well known, if the curve C is not smooth, the coherent sheaf of differentials, $\Omega^1_{C/\mathbf{C}}$, is not locally free, so that, in principle, it should not be possible to construct something analogous to the principal parts. Nevertheless, Widland in [W1] notices that the differential of a regular function at P is a regular differential at P. In other words:

$$\mathrm{Im}(d_P: O_{C,P} \longrightarrow \Omega^1_{C/\mathbf{C},P}) \subseteq K_{C,P},$$

and, since $K_{C,P}$ (the stalk at P of the sheaf K_C) is free of rank 1 (by the Gorenstein hypothesis), there is a generator σ of $K_{C,P}$ over $O_{C,P}$, so that it is possible to define, for each $f \in K_{C,P}$, a new regular function at P, f', by means of the obvious relation:

(3.4) $$d_P f = f'\sigma.$$

Moreover one can set :

(3.5) $$d_P(f^{(n-1)}) = f^{(n)}\sigma,$$

which corresponds to define the "derivatives" of any order of f at P with respect to the local generator σ of $K_{C,P}$ over $O_{C,P}$. Notice that if P is a smooth point, and z is a local parameter at P, one can choose $\sigma=dz$. At this point, given any linear system $(\mathcal{L}, V)$ of degree d and dimension r, on the curve C, it is possible to mimic, thanks to the invertibility of the dualizing sheaf , the same construction as in Sect. 2. We sketch briefly the construction whose details can be found in [G2]. As usual, $\{U_\alpha, \alpha \in \mathcal{A}\}$ will be an open cover of C,

enjoying the property that, for any $\alpha\in\mathcal{A}$, there exists a generator σ_α of $H^0(U_\alpha,K_C)$ over $H^0(U_\alpha,O_C)$ ("most" of such σ_α will be of the form dz_α).
If $\lambda\in H^0(C,\mathcal{L})$, then, as in (2.13):

$$\lambda_{|U_\alpha}=\lambda_\alpha\cdot\psi_\alpha, \tag{3.6}$$

where λ_α is a regular function over the open set U_α, and ψ_α generates $H^0(U_\alpha,\mathcal{L})$ over $H^0(U_\alpha,O_C)$. The local data $\{U_\alpha\,;\,\lambda_\alpha, \lambda'_\alpha, ..., \lambda_\alpha^{k-1}\}$, where $\lambda^{(m)}$ is defined according to relations (3.4) and (3.5), of open sets and $\mathbf{C}^k$-valued regular functions patch together again by means of functions:

$$j^k\ell_{\alpha\beta} : U_\alpha\cap U_\alpha \longrightarrow Gl(k,\mathbf{C}), \tag{3.7}$$

satisfying the cocycle conditions $j^k\ell_{\alpha\alpha}=\mathbf{1}$ and $j^k\ell_{\alpha\beta}\, j^k\ell_{\beta\gamma}=j^k\ell_{\alpha\gamma}$.

As in Sect. 2 define $J^k\mathcal{L}$ as the rank k vector bundle having the (3.7) as transition functions with respect to the cover $\{U_\alpha\}$, and notice that if the curve is smooth this is exactly the construction worked out in Sect. 2. Setting $J^0\mathcal{L}=\mathcal{L}$, for each $k\geq 0$, we have hence a map:

$$D^k: H^0(C,\mathcal{L}) \longrightarrow H^0(C,J^k\mathcal{L})$$

such that D^0=identity and:

$$\left\{\begin{array}{l} D^k_p : H^0(C,\mathcal{L}) \longrightarrow J^k_p\mathcal{L} \\ \lambda \longmapsto (\lambda_\alpha(P), \lambda'_\alpha(P), ..., \lambda_\alpha^{(k-1)}(P)) \end{array}\right. \tag{3.9}$$

which allows once more to define the ($\mathcal{L}$, V)-Weierstrass points as the points $P\in C$ such that $rk(D^r_{p|V})<r$. In a totally analogous way, this condition implies that the *wronskian* r-*form* $\mathbf{v}\wedge D\mathbf{v}\wedge...\wedge D^{r-1}\mathbf{v}$ (cfr. Sect. 2) vanish at P, and the r-th V- weight of P, $w_r^V(P)$ is defined to be $\dim_C\left(\frac{O_{C,P}}{I_rO_{C,P}}\right)$. Let us remark that $\mathbf{v}\wedge D\mathbf{v}\wedge...\wedge D^{r-1}\mathbf{v}$ is just a shortened expression of a *wronskian determinant*, which, as a matter of fact, has been defined (for linear system ($\mathcal{L}$, $H^0(C,\mathcal{L})$) and, in particular, for (K_C, $H^0(C,K_C)$)) by Widland and Lax in [W1] and [WL]. Of course, $\mathbf{v}\wedge D\mathbf{v}\wedge...\wedge D^{r-1}\mathbf{v}$ is now a global section of $\mathcal{L}^{\otimes r}\otimes(K_C)^{\otimes\frac{r(r-1)}{2}}$ (where K_C is now the *dualizing sheaf*) whose degree is hence again given by formula (2.19). The

reason for which, in general, the wedge product $v\wedge Dv\wedge\ldots\wedge D^{k-1}v$ lies in $H^0(C, \mathcal{L}^{\otimes k}\otimes(K_C)^{\otimes\frac{k(k-1)}{2}}\otimes\wedge^k V)$ is due to the fact, as is implicitly showed in [G2], that the following exact sequence holds:

$$0\longrightarrow\mathcal{L}\otimes K_C^{\otimes k}\longrightarrow J^k\mathcal{L}\longrightarrow J^{k-1}\mathcal{L}\longrightarrow 0,$$

which was well known for smooth curves (see [ACGH]). We remark here that an extension of principal parts for Gorenstein curves in characteristic p has been gotten by Laksov and Thorup in [LT2].

Using now the Gorenstein condition on local rings, in [W1] and [L4] is shown that if P is a singular point, then:

(3.10) $$w_r^V(P)\geq\delta_P r(r-1),$$

which means that if the dimension of the linear system is greater than 1, then *all singular points turns out to be* $(\mathcal{L},V)$-Weierstrass points. At this point we must remark that if we let $\mathcal{L}$ to be the dualizing sheaf, and $V=H^0(C,K_C)$, then (3.10) sounds as $w_g^V(P)\geq\delta_P g(g-1)$, and if P is a node one has $\delta_P=1$, so that:

(3.11) $$w_g^V(P)\geq g(g-1)$$

for an ordinary node. If one does not find any difficulty to believe that *generically* the equality holds in (3.11), then one can view expression (3.10) as a generalization of the phenomena described by Diaz in [D2], which, in a sense, have been studied by Lax in [L3].

Coming back to our story, to complete the formula (3.10) we have to introduce the concept of r-th V-*extraweight* $E_r^V(P)$, which is the lacking part of $\delta_P r(r-1)$ to join $w_r^V(P)$. In other words we shall write:

(3.12) $$w_r^V(P)=\delta_P r(r-1)+E_r^V(P).$$

Remark that if P is smooth, to say (r-th)-weight is the same as saying (r-th)-extraweight ($\delta_P=0$)

Again, the reason for writing a subscript ("r" in formula (3.12)) to weights and extraweights is motivated from the fact that we wish to consider the "full" weight sequence. At this point, keeping in account that we are dealing with singular curves, various definitions are possible and, of course, the choice has fallen on the one which is the most useful for explicit computations. Referring to [G1] or [G2] for details, we shall define, for each $1 \leq k \leq r = \dim V$ the k-th weight at P as:

$$w_k^V(P) = \dim_C \left(\frac{\tilde{O}_{C,P}}{I_k \tilde{O}_{C,P}} \right), \tag{3.13}$$

where $I_k \tilde{O}_{C,P}$ is the extension in $\tilde{O}_{C,P}$ (the integral closure of $O_{C,P}$) of the ideal $I_k O_{C,P}$ in $O_{C,P}$, generated by all k×k minors of the k×r wronskian matrix defined as in (2.20) where the operator D is now the one defined through relations (3.4) and (3.5), holding for Gorenstein curves. Notice that since $I_r O_{C,P}$ is principal in $O_{C,P}$, definition (3.13), for k=r, agree with the one previously given for the r-th weight. In [G2] it is proved that, as a matter of fact, inequality (3.10) generalizes to the k-th weights, giving, for $1 \leq k \leq r$:

$$w_k^V(P) \geq \delta_P k(k-1),$$

so that we can write:

$$w_k^V(P) = \delta_P k(k-1) + E_k^V(P). \tag{3.14}$$

The non decreasing sequence (cfr. [G2])of r integers $\mathbf{E}^V(P) = \langle E_1^V(P), E_2^V(P), \ldots, E_r^V(P) \rangle$,

will be said the V-*extraweight sequence*. The idea is now: use the extraweight sequence to define the notion of WGS at a (possibly singular) point P of a Gorenstein curve, analogously to what has been done for smooth curves. We *could* say hence that $G_P = \langle n_1, n_2, \ldots, n_r \rangle$ is the V-WGS at P if and only if the k-th extraweight can be computed according to the formula:

$$E_k^V(P) = \sum_{i=1}^{k} (n_i - i). \tag{3.15}$$

This is, in fact, the way followed in [G2] which has revealed very useful to compute Weierstrass weights at singular points P with respect to certain linear systems for which it is not so easy to compute explicitly the wronskians. Examples in this sense are showed in [G3] using as linear system the dualizing sheaf together with its g-dimensional vector space of global sections, $H^0(C, K_C)$.

In this case we have a WGS of g integers, $G_P = \langle 1 = n_1, n_2, \ldots, n_g \rangle$ which, of course, coincides with the classical one for smooth point. At this point we need a digression. In fact, as seen in Sect. 2, to say that n is a Weierstrass gap at P on a smooth curve, with respect to the canonical bundle, means that there does not exist a rational function regular on C-{p} such that $(f)_\infty=nP$. In [LW1], Lax and Widland try to generalize the definition of WGS (with respect to the dualizing sheaf) aiming to mantain something of the above property. The problem is that at a singular point it is not possible to speak about divisors supported on it, but one should rather consider all the zero-dimensional closed subschemes supported on it of given degree ([LW1] for details). As a consequence the two Authors are lead to consider the so-called *Weierstrass chains* of ideals, which unfortunately are, in general, more than one, producing in this way different sequences of g integers , none having any relation with the Weierstrass weight of the point P, computed by means of the wronskian. By a Weierstrass chain in the local ring $O_{C,P}$ of regular functions at P, is meant, in [LW1], a chain $\mathcal{C}$ of ideals:

$$\mathcal{C}\colon \mathbf{m}_P=J_1\supset J_2\supset\ldots\supset J_{2g-1}, \tag{3.16}$$

such that $col(J_k)=\dim_C\frac{O_{C,P}}{J_k}$, the *colenght of* J_k, is exactly k, for $1\leq k\leq 2g-1$.

The idea is now to use the following ingredients: first, the sheaf of ideals $\mathcal{J}_Z$ of a zero-dimensional subscheme Z supported on P, which can be defined stalk by stalk by setting $\mathcal{J}_{Z,Q}=O_{C,Q}$, for Q≠P, and $\mathcal{J}_{Z,P}$ to be some proper ideal contained in $O_{C,P}$ of given colenght. Secondly a Riemann-Roch formula for zero-dimensional closed subschemes of C which sounds as:

$$\ell(Z) - i(Z)=1- g + \deg Z, \tag{3.17}$$

where we set:

$\ell(Z)=\dim_C \mathrm{Hom}_{O_C}(\mathcal{J}_Z, O_C)$, $i(Z)=H^0(C, \mathcal{J}_Z\otimes K_C)$ and, finally, $\deg(Z)=\sum_{Q\in supp(Z)} col(\mathcal{J}_{Z,Q})$.

Then Widland and Lax prove what they call the *Weierstrass Gap Theorem for Gorenstein curves:* given any Weierstrass chain:

$$\mathcal{C}\colon \mathbf{m}_P=J_1\supset J_2\supset\ldots\supset J_{2g-1},$$

in $O_{C,P}$, there exist precisely g integers:

$$1=\gamma_1<\gamma_2<\gamma_3<\ldots<\gamma_g\leq 2g-1,$$

such that $h(J_{\gamma_k})=h(J_{\gamma_{k-1}})$ for k=2, 3, ...,g, h(J) being defined as $\dim_{\mathbb{C}}\mathrm{Hom}_{O_C}(\mathcal{J}_Z, O_C)$, where $\mathcal{J}_Z$ is the ideal sheaf supported on P defined by $\mathcal{J}_{Z,P}=J$.

The weight of a Weierstrass chain is defined to be :

$$w(\mathcal{C})=\sum_{k=1}^{g}(\gamma_k - k).$$

It can be proved that ([LW1], Thm. 1.9) if g>1 and P is a singular point of a Gorenstein curve C, then there exists a Weierstrass chain in $O_{C,P}$ of positive weight. Of course the definition of WGS for singular points relative to the dualizing sheaf given in [LW1] coincides with the classical one for smooth points. Nevertheless the two Authors wonder about the "relationship between the Weierstrass weight of a singular point, defined by means of the wronskian, and the weights of the various Weierstrass chains in the local ring of that singularity" ([LW1]).

As a matter of fact, at the singular point (0,0), having "wronskian weight" equal to 22 (see [LW1], Ex. 2.1, or [G3], Ex. 2.1) of the rational quartic having affine equations:

$$\begin{cases} x=t^3 \\ y=t^4 \end{cases},$$

it is possible to find "14 Weierstrass chains of weight 0, five Weierestrass chains of weight 1 each, and one Weierstrass chain of weight 2". This claim ([LW1]) , as showed in [G3], Remark 2.4, is sufficient to conclude there cannot be any relationship between the definition of WGS given by Lax and Widland and the one expressed by formula (3.15) which, on the contrary, and by its very construction, is strictly related to the *wronskian* Weierstrass weight and *works very well* in computing weights on interesting examples of, for instance, non-rational curves (See [G3], Ex. 2.10).

The last claim forces us to go back to our first definition of V-WGS (which has nothing to do with the definition by Lax and Widland, for the simple reason that they describe different kind of phenomena), to show and explain how it is, in principle, very easy to compute the WGS at a singular point. To this purpose, we need to analyze the structure of the extraweight in a deeper way. Explicit expression for extraweights are given in [W1], [WL] and [G1], [G2]. But surely, the most intrinsic way to define it (see [BG3]) is the following one.

Let $(\mathcal{L},V)$ be an r-dimensional linear system, as always, and let us allow to come into play the normalization of the curve C, $\pi : \tilde{C} \longrightarrow C$. Let P be a singular point of C, and consider a basis $\mathbf{v}=(v_1,...,v_r)$ of V. On $\tilde{C}$ consider then the following linear system $(\pi^*\mathcal{L},\tilde{V})$,

such that $h(J_{\gamma_k})=h(J_{\gamma_{k-1}})$ for $k=2, 3, ..., g$, $h(J)$ being defined as $\dim_{\mathbb{C}} Hom_{O_C}(\mathcal{J}_Z, O_C)$, where $\mathcal{J}_Z$ is the ideal sheaf supported on P defined by $\mathcal{J}_{Z,P}=J$.
The weight of a Weierstrass chain is defined to be :

$$w(C)=\sum_{k=1}^{g}(\gamma_k-k).$$

It can be proved that ([LW1], Thm. 1.9) if $g>1$ and P is a singular point of a Gorenstein curve C, then there exists a Weierstrass chain in $O_{C,P}$ of positive weight. Of course the definition of WGS for singular points relative to the dualizing sheaf given in [LW1] coincides with the classical one for smooth points. Nevertheless the two Authors wonder about the "relationship between the Weierstrass weight of a singular point, defined by means of the wronskian, and the weights of the various Weierstrass chains in the local ring of that singularity" ([LW1]).
As a matter of fact, at the singular point (0,0), having "wronskian weight" equal to 22 (see [LW1], Ex. 2.1, or [G3], Ex. 2.1) of the rational quartic having affine equations:

$$\begin{cases} x=t^3 \\ y=t^4 \end{cases},$$

it is possible to find "14 Weierstrass chains of weight 0, five Weierestrass chains of weight 1 each, and one Weierstrass chain of weight 2". This claim ([LW1]) , as showed in [G3], Remark 2.4, is sufficient to conclude there cannot be any relationship between the definition of WGS given by Lax and Widland and the one expressed by formula (3.15) which, on the contrary, and by its very construction, is strictly related to the *wronskian* Weierstrass weight and *works very well* in computing weights on interesting examples of, for instance, non-rational curves (See [G3], Ex. 2.10).

The last claim forces us to go back to our first definition of V-WGS (which has nothing to do with the definition by Lax and Widland, for the simple reason that they describe different kind of phenomena), to show and explain how it is, in principle, very easy to compute the WGS at a singular point. To this purpose, we need to analyze the structure of the extraweight in a deeper way. Explicit expression for extraweights are given in [W1], [WL] and [G1], [G2]. But surely, the most intrinsic way to define it (see [BG3]) is the following one.
Let $(\mathcal{L},V)$ be an r-dimensional linear system, as always, and let us allow to come into play the normalization of the curve C, $\pi : \tilde{C} \longrightarrow C$. Let P be a singular point of C, and consider a basis $v=(v_1,...,v_r)$ of V. On $\tilde{C}$ consider then the following linear system $(\pi^*\mathcal{L},\tilde{V})$,

i) *if* P *is an ordinary node* (δ_P=1) *on a Gorenstein curve C, and if* Q_1 *and* Q_2 *lies over* P *in the partial normalization of* C, *then:*

(3.20) $$w_g(P)=g(g-1)+\hat{w}_{g-1}(Q_1)+\hat{w}_{g-1}(Q_2);$$

and

ii) *if* P *is a simple cusp* (δ_P=1) *on a Gorenstein curve C and if* Q *lies over* P *in the partial normalization of* C, *then:*

(3.21) $$w_g(P)=(g-1)(g+1)+\hat{w}_{g-1}(Q),$$

where $w_g(P)$ is the (g-th)-Weierstrass weight of P with respect to the dualizing sheaf of C, while $\hat{w}_{g-1}(R)$ represents the ((g-1)-th)-Weierstrass weight of a point R with respect to the *dualizing sheaf of the partial normalized curve* $\hat{C}$.

Notice that in the two above propositions we are using the fact that the dualizing sheaf of the partial normalized curve is still invertible. The above results, especially the one for cusps, is deeper than it seems to be. In fact it is the particular case of a more general proposition which uses in a striking way the tool of the V-WGS in the sense of formula (3.15), which holds in a much more general context (see [BG3] for detailed proofs and explanations). We shall limit here to the Gorenstein context (although it is not strictly necessary) but coming back to a more general situation. Let us start, hence, in considering a g_d^{r-1} ($\mathcal{L}$, V) on a integral projective Gorenstein curve C. Notation as above. Suppose that P is a n-branched singular point having branches centered at $Q_1,\ldots,Q_n$ in the partial normalization $\hat{C}$ (nothing changes if one consider the "full" normalization, but for our purposes it is useful to focus attention on the point P only), each one counted with its own multiplicity $m_1, \ldots, m_n$ (for instance, for the ordinary node: n=2, $m_1=m_2=1$, for the simple cusp: n=1, m=2). Let us denote by $\hat{V}$ the analogous of the $\tilde{V}$ defined above, i.e. set $\hat{V}$= span (θ_P^*v) in $H^0(\hat{C}, \theta_P^*\mathcal{L})$, $\hat{V}'$ the linear system of sections of $\hat{V}$ vanishing at least once at any of the point $Q_i \in \theta_P^{-1}(P)$. Then, by hypotheses on the multiplicities of the various Q_i's, it follows that if $\hat{V}$ vanishes at any Q_i at least once, then it vanishes at the divisor $m_1Q_1+\ldots+m_nQ_n$. This means that: *if no* Q_i *is a base point for* $\hat{V}$ (and we suppose to be in this situation, though the more general one does not involve any difficulty), *then the dimension of the sections,* $\hat{V}'$, *vanishing at each* Q_i *at least* m_i *times is equal to* r -1. Formally:

(3.22) $$\dim(\hat{V}') = \dim(\hat{V} - m_1Q_1 + \ldots + m_nQ_n) = r - 1.$$

Hence, at each Q_i we can consider the respective $\hat{V}'$-WGS, i.e. the r -1 integers:

$$<b_{i1}, b_{i2}, \ldots, b_{i,r-1}>$$

such that (cfr. end of Sect. 2):

$$\dim(\hat{V}' - b_{ik}Q_i) > \dim(\hat{V}' - (b_{ik}-1)Q_i).$$

From the hypotheses and by construction, the following easy proposition holds: $<1=a_{i1}, a_{i2}, \ldots, a_{i,r}>$ *is the* $\hat{V}$-WGS *at* Q_i, *if and only if, for each* $2 \leq k \leq r$:

(3.23) $$a_{ik} = b_{ik} + m_i.$$

The reader can perform by itself the needing computation for linking the V-WGS to the various $\hat{V}'$-WGS's at $Q \in \pi^{-1}(P)$. For instance, in the case of ordinary nodes for the dualizing linear system $(K_C, H^0(C, K_C))$, one has that (see [Sz], pp. 25-26):

$$\hat{V}' = \hat{V} - Q_1 - Q_2 = K_{\hat{C}},$$

while for simple cusps:

$$\hat{V}' = \hat{V} - 2Q = K_{\hat{C}},$$

so that one can prove that ([G2], Props. 6.1 and 6.3):

a) *if* P *is a simple cusp, the* WGS *at* P *is given by:*

$$G_P = < 1, 2 + b_{k-1} : 2 \leq k \leq g >,$$

where $<b_1, b_2, \ldots, b_{g-1}>$ *is the* WGS *at the smooth point* $Q \in \theta_P^{-1}(P)$, *with respect to the dualizing sheaf of* $\hat{C}$.

b) *if* P *is an ordinary node, the* WGS *at* P *is given by:*

$$G_P = < 1, b_{1,k-1} + b_{1,k-1} - (k - 2) : 2 \leq k \leq g >,$$

(3.22) $$\dim(\hat{V}') = \dim(\hat{V} - m_1Q_1 + \ldots + m_nQ_n) = r - 1.$$

Hence, at each Q_i we can consider the respective $\hat{V}'$-WGS, i.e. the r -1 integers:

$$<b_{i1}, b_{i2},\ldots,b_{i,r-1}>$$

such that (cfr. end of Sect. 2):

$$\dim(\hat{V}' - b_{ik}Q_i) > \dim(\hat{V}' - (b_{ik}-1)Q_i).$$

From the hypotheses and by construction, the following easy proposition holds: $<1=a_{i1}, a_{i2},\ldots,a_{i,r}>$ *is the* $\hat{V}$-WGS *at* Q_i, *if and only if, for each* $2\leq k\leq r$:

(3.23) $$a_{ik} = b_{ik} + m_i.$$

The reader can perform by itself the needing computation for linking the V-WGS to the various $\hat{V}'$-WGS's at $Q\in\pi^{-1}(P)$. For instance, in the case of ordinary nodes for the dualizing linear system $(K_C, H^0(C, K_C))$, one has that (see [Sz], pp. 25-26):

$$\hat{V}' = \hat{V} - Q_1 - Q_2 = K_{\hat{C}},$$

while for simple cusps:

$$\hat{V}' = \hat{V} - 2Q = K_{\hat{C}},$$

so that one can prove that ([G2], Props. 6.1 and 6.3):
a) *if* P *is a simple cusp, the* WGS *at* P *is given by:*

$$G_P = < 1, 2 + b_{k-1} : 2\leq k\leq g >,$$

where $<b_1, b_2,\ldots,b_{g-1}>$ *is the* WGS *at the smooth point* $Q\in\theta_P^{-1}(P)$, *with respect to the dualizing sheaf of* $\hat{C}$.

b) *if* P *is an ordinary node, the* WGS *at* P *is given by:*

$$G_P = < 1, b_{1,k-1} + b_{1,k-1} - (k - 2): 2\leq k\leq g >,$$

where $<b_{i1}, b_{i2},\dots,b_{i,g-1}>$ *is the* WGS *at the smooth point* Q_i *with respect to the dualizing sheaf of* $\hat{C}$.

An easy check shows that (3.20) and (3.21) are corollaries of the Props. a) and b) above. Anyway we want to go back again to our general situation, letting V to be part of an arbitrary g_d^{r-1} . Hence, formula (3.23) easily implies (see again [BG3] for details or even [BG2] for suggestions):

$$w_k^{\hat{V}}(Q_i)=w_{k-1}^{\hat{V}'}(Q_i)+(m_i-1)(k-1) \qquad \text{for } 2\le k\le r, \tag{3.24}$$

the same formula holding if replacing $\hat{V}$ by $\tilde{V}$. Hence, it turns out that (3.24), together with (3.18) and (3.14), gives rise to the most natural and significant (and we shall see why in a minute) generalization of (3.20) and (3.21), i.e.:

$$w_r^V(P)=\delta_P r(r-1)+(M-n)(r-1)+\sum_{i=1}^{n} w_{r-1}^{\hat{V}'}(Q_i), \tag{3.25}$$

where we set $M=\sum_{i=1}^{n} m_i$. Expression (3.5) says also that the r-th V-extraweight is always greater (or equal) than (M-n)(r-1). If $V=H^0(C, K_C)$, and P is a cusp, one has $\delta_P=1$, M=2, n=1, r=g, and one gets again (3.21). At this point it is not possible to avoid an example showing that formula (3.25) is very geometrical. The example consists in considering an irreducible plane curve of degree d having ordinary nodes and simple cusps only. Let α be the number of nodes and β the number of cusps. On C consider as V the complete linear system g_d^2 cut out by the net of lines of $\mathbf{P}^2$. By (3.25), the third V-weight at a node is $w_3^V(P)=6+w_2^{\hat{V}'}(Q_1)+w_2^{\hat{V}'}(Q_2)$, because r=3, M=n=2. on the other hand, for a cusp M=2 and n=1, so that, $w_3^V(P)=6+2+w_2^{\hat{V}'}(Q)$. Moreover by the *adjunction formula* for plane curves (or, if you prefer, by the *genus formula*), we have 2g-2=d(d-3) (see also [GV] for general facts on adjoint curves). If $N_1,\dots,N_\alpha$ and $K_1,\dots, K_\beta$ are, respectively, the nodes and the cusps of C, keeping in account the Brill-Segre formula (2.19) (which, as previously remarked, holds for Gorenstein curves, too), the tautologous statement:

Total Weight of Smooth WP's = Total Weight of all WP's - Total Weight of singular WP's

gives:

$$\textit{Total Weight of Smooth WP's} = 3d + 3d(d-3) - \sum_{i=1}^{\alpha} w_3^{V}(N_i) - \sum_{i=1}^{\beta} w_3^{V}(K_i),$$

i.e., for what have been said:

$$\textit{T. Weight of Smooth WP's} = 3d(d-2) - 6\alpha - 8\beta - \sum_{i=1}^{\alpha} (w_2^{\hat{V}'}(Q_{i1}) + w_2^{\hat{V}'}(Q_{i2})) - \sum_{i=1}^{\beta} w_2^{\hat{V}'}(Q_i).$$

It is well known, now, that the *smooth Weierstrass points* with respect the net of lines are nothing but the flexes falling at smooth points. Moreover it can be easily checked (cfr. also [G3], Sect. 3) that the term $\sum_{i=1}^{\alpha} (w_2^{\hat{V}'}(Q_{i1}) + w_2^{\hat{V}'}(Q_{i2}))$ represents the total weight of flexes falling into one or the other (or both!) branches of each node, and $\sum_{i=1}^{\beta} w_2^{\hat{V}'}(Q_i)$ represents the total weight of flexes located at the center of the branches of the various cusps. We have hence gotten:

(3.26) $$\#(\text{flexes}) = 3d(d-2) - 6\alpha - 8\beta.$$

i.e. the very classical *Plücker formula* for the number of flexes (counted according their multiplicities) on a plane curve having nodes and/or cusps only. The reader can easily mimic the same procedure for a g_2^1 on a plane curve of degree d (for instance, the pencil cut out by lines passing through a point not belonging to the curve itself) having nodes and cusps only, to get the *Plücker formula* for the *class*, i.e. the *"number" of tangents of the curve which can be drawn by a generic point of the plane:*

(3.27) $$\textit{(class of the curve)} = d(d-1) - 2\alpha - 3\beta.$$

Before concluding the section, notice that we have not just gotten old formulas in a new way, but we give an explicit interpretation of the difference between the coefficients relatives to nodes and cusps respectively. Moreover this method is easily generalizable for other kind of singularities, for any linear system (e.g. what is the *Plücker Formula* for the linear system cut out on C by conics ?), and, indeed for any curve, plane or not, and even non Gorenstein, as it is shown in [BG3].

4. Weierstrass Weights and WGS at Gorenstein Singularities: Concluding Remarks.

In this section we shall concentrate once more on the theory of Weierstrass points with respect to the dualizing sheaf of a Gorenstein curve, which is formally the same as for smooth curves. This particular situation has been extensively studied again by Widland ([W2]), Lax and Garcia ([GL1], [GL2], [GL3]). In [GL3] is contained a generalization of the theory by Widland and Lax in caracteristic p. Other approaches in this sense are the ones by O. Stöhr ([St]) and [LT2]. In the latter the approach consists in studying gaps sequences for families of curves. It is not our aim to deal with generalization in positive characteristic, because in this survey we are much more interested to explain the basic underlying ideas of Weierstrass points' theories for singular curves.

Recalling that, from now on, the theory will be considered with respect to the complete linear system $H^0(C, K_C)$, we shall rather concentrate about the following quite relevant question. In general, it is not difficult to believe (even if we should make this statement more precise) that at the generic singularity the extraweight is precisely (M-n)(g-1) as explained in Sect. 2. It is hence natural, as done by Widland in [W2], to ask how much can the weight increase. In other words we are looking for an upper bound (the lower being already found in Sect. 3) for the extraweight. From the point of view of WGS's, this turns out to be a very interesting question. In fact, for smooth points it is well known that the last gap cannot be greater than 2g-1. Does this result hold true for singular points? In [BG2] it is shown that this is always true for unibranch singularities. As a matter of fact, if n is the multiplicity of a unibranch singularity, the last gap in the sense of [BG2], Def. 4.7 (cfr. formula 3.15), cannot exceed 2g-n. The situation is a little bit more complicated for multibranched singularities: in [BG2] some sufficient conditions are found, but it is also remarked that the purely numerical approach cannot give any information if not supported by a deeper geometrical analysis.

As an example of this situation we want quote an upper bound found by Widland for n-branched Gorenstein singularities. He is able to prove that, in this situation, the (g-th) extraweight cannot be higher than $\frac{n}{2}(g-1)(g-n)$. On the other hand, in [BG2], Prop. 3.4, it is proved that, in the same situation, the following bound holds:

$$E_g(P) \leq M-n + (g-2)[n(g-1) - M(n-1)], \tag{4.1}$$

where, as in Sect. 3, M is the sum of the multiplicities at the various branches.

The story is very curious, because none of the two bounds, neither the Widland one nor bound (4.1) can be said better than the other. For instance, if P is an ordinary node, Widland bound gives:

(4.2) $$E_g(P) \leq (g-1)(g-2),$$

and this bound is sharp, because of the formula (3.20) in case both Q_i's are hyperelliptic. On the other hand, if P is a node, formula (4.1) gives:

(4.3) $$E_g(P) \leq 2(g-2)^2,$$

which of course is much worse of the bound (4.2).
Nevertheless, let us consider a n-branched g-fold point. For instance consider the origin of the affine plane which is a 3-branched triple point of the rational quartic (see also [G3], Sect. 2):

$$x^4-y(y-x)(y+x)=0.$$

For n-branched g-fold points, estimate (4.1) gives:

$$E_g(P) \leq (g-1)(g-n),$$

which is of course better than Widland's one. Notice that since $E_g(P) \geq (g-1)(g-n)$, (cfr. Sect. 3), we have that $E_g(P)=(g-1)(g-n)$. In general ([BG2]) the equality hold for all k-extraweights, i.e. $E_k(P)=(k-1)(g-n)$, so that, for this special kind of singularities, the WGS has necessarily to be:

$$<1,\ g-n+2, \ldots,\ 2g-n>.$$

We remark that, with the only exception of plane quartics, such singularities, by [M], Corollary 13.8 cannot be of *plane* type.

It is worth to notice that (4.1) is a *numerical sharp* bound in the sense that any sequence of g integers candidates to be a WGS at a point P (see [BG2] for a more precise statement), cannot have weight exceeding the right hand side of (4.1). But Widland estimate shows that from a geometrical point of view, some situations occur in such a way to show that bound (4.1) is not the best. It is thence necessary, in further studies, to find some geoemetrical properties (such as the semigroup property of the complement in the case of smooth curve, see Sect. 2) able to exclude sequences which cannot occur as WGS of a singular point. See also discussion in [G3], Sect. 2. Note, however, that in the case of unibranch singular points, formula (4.1) specializes into:

(4.4) $$E_g(P) \leq n-1 + (g-2)(g-1) \leq g-1 + g-2)(g-1) = (g-1)^2,$$

where the very right side is the estimate found by Widland in [W1] for unibranch singularities.

To come back to the general framework, in [W2] some considerations are done about estimates and the least number of distinct Weierstrass points on a curve. For instance, Widland proves that a curve having only unibranch or two-branched singularities have no less than $2\gamma-2$ distinct WP's, γ being the geometric genus of the curve C. The same result can be proved, for unibranch singularities only, by using the following fact ([BG2], Prop. 4.1):

let P *be a* m*-fold unibranch singular point on* C. *Let* $Q \in \theta_P^{-1}(P)$ *be the smooth point lying over* P *in the partial normalization* $\hat{C}$ *of* C *at* P. If $<c_1, c_2, \ldots, c_{g-\delta_P}>$ *is the* WGS *at* Q *with respect to* $K_{\hat{C}}$, *and* $<1, m+1, a_3, \ldots, a_g>$ *is the* WGS *at* P, *then* $a_{\delta_P+i} = c_i + 2\delta_P$, for $1 \leq i \leq g - \delta_P$,

which generalizes Prop. 6.1 in [G2] (cfr. Sect. 3).

As a matter of fact, the previous proposition allows to find the following bounds for extraweights ([BG2], Prop. 4.2):

(4.5) $$E_k(P) \leq (k-1)\delta_P \quad \text{for} \quad 1 \leq k \leq \delta_P,$$

and

(4.6) $$E_k(P) \leq (k-1)\delta_P + \hat{w}_{k-\delta_P}(Q) \quad \text{for} \quad \delta_P + 1 \leq k \leq g,$$

where $\hat{w}_{k-\delta_P}(Q)$ represents the $(k-\delta_P)$-th Weierstrass weight at the point $Q = \theta_P^{-1}(P)$ with respect to $K_{\hat{C}}$. A better version of bound (4.6), for k=g, was already known to Garcia and Lax, which, in [GL1], prove that at a unibranch singular point one has:

(4.7) $$w_g(P) = \delta_P(g-1)(g+1) - wt(S_P) + \hat{w}_{g-\delta_P}(Q),$$

where $S_P = \left\{ n = \dim_{\mathbb{C}} \frac{O_{C,P}}{fO_{C,P}} \ : \ f \in O_{C,P} \right\}$ is the so-called *semigroup of values* at the unibranch point P. It is in fact obvious that S_P is a semigroup. Moreover it is a *numerical semigroup* , i.e. it has finite complement in **N**. Let $<g_1,...g_\delta>$ be the *gaps* of an arbitrary numerical semigroup, i.e. the δ positive integers not lying in **N**. Define the weight of the semigroup, wt(S) as $(g_1-1)+...+(g_\delta-\delta)$. Then formula (4.7) found by Lax and Garcia is full explained.

That formula has been generalized for Gorenstein singularities having an arbitrary number of branches in [GL2]. There the semigroup of values at P is defined as:

$$S_P=\left\{ (n_1,\ldots,n_k)\subseteq \mathbf{N}^k / \dim_{\mathbf{C}}\frac{O_{\tilde{C},Q_i}}{fO_{\tilde{C},Q_i}} \ :\ f\in O_{C,P} \right\} ,$$

(see [De1], [De2], [St], [Ga] for detailed definitions and properties), where, as usual, π: $\tilde{C}\longrightarrow C$ is the normalization of C and $Q_i\in\pi^{-1}(P)$ are the centers of the branches of P in the normalization. By using the same language introduced in Sect. 3 of this survey (i.e. the same language introduced in [G2]), one can provide a simple proof of the Thm. 6.3, in [GL2], generalizing formula (4.7). The idea is to separate the purely numerical informations from the more geometrical one. We refer to [Po], for all details and discussions.

We began the first part of this survey by speaking of Weierstrass points by means of the WGS and analyzing their meaning in terms of the existence of certain meromorphic functions, having a prescribed order at its only pole. We shall conclude in almost the same manner, by quoting one more property of the notion of WGS's introduced in [G2], which refines, at special kind of singularities, a theorem by Lax and Widland proved in [WL]. Referring to the Riemann-Roch formula for Gorenstein curves quoted in Sect. 3, we remind that, following Kleiman ([K]), a closed zero-dimensional subscheme Z of C is said to be r-*special* if and only if $\ell(Z) > r$. Then in [WL] it is shown that the following theorem holds: P *is a Weierstrass point of* C, *if and only if there is a* 1-*special closed subschemes of* C *with support* P *and degree* g (see Sect. 3 for basic definitions).

Notice that we already know the proposition to be true for WP's on smooth curve, since (see Sect. 2) $P\in C$ is a WP if and only if $\ell(gP)=h^0(C, O_C(gP))>1$, and the divisor gP corresponds to the *unique* zero-dimensional subscheme having support P and degree g. Moreover, for smooth curves some stronger result holds. In fact, let P be a WP having n as *first* non-gap. Hence, by the very definition of non-gap, it follows that $\ell(nP)=2$, so that, in the Kleiman terminology, nP is a (in fact the unique!) *1-special zero-dimensional subscheme supported at P and degree n* . Then the following question naturally arises: does the last italicized sentence generalizes to Gorenstein curves, where it is unmeaningful to talk about divisors at singular points, while one must rather consider all the closed subschemes supported at P?

We have no complete answer to this question. Nevertheless the following fact can be proved ([G2], Thm. 7.10):

suppose that P *is a singular Weierstrass point on a Gorenstein curve* C, *having* k *as first non gap* (in the sense of [G2], Def. 4.8). *If* P *is a* (i) *an ordinary node,* (ii) *a tacnode* (see [H1], page 36) *or* (iii) *an ordinary triple point, then there exists a 1-special subscheme having support* P *and degree* k.

Nothing is actually known for other kinds of singularities. It may be possible that, either to prove the proposition in the general case, or to find a counterexample, one has to change the strategy for demonstrate it, which in the above cases is based on purely numerical constraints, which luckily occur for nodes, tacnodes or ordinary triple points.

References

[A]. E. Arbarello, *Weierstrass Points and Moduli of Curves,* Compositio Math., **29**-3 (1974), 325-342.

[ACGH]. E. Arbarello, M. Cornalba, P.A. Griffiths, J. Harris, *Geometry of Algebraic Curves* , Vol 1, Springer-Verlag, 1984.

[BG1]. E. Ballico, L. Gatto, *On the Monodromy of Weierstrass Points on Gorenstein Curves,* preprint, 1993

[BG2]. E. Ballico, L. Gatto, *On Bounds for the Weierstrass Weight at Singular Points of Gorenstein Curves,* Preprint Politecnico di Torino, **27/93,** 1993.

[BG3]. E. Ballico, L. Gatto, *Weierstrass Points on Singular Curves,* Preprint Politecnico di Torino, 1994.

[C1]. M. Coppens, *The Number of Weierstrass Points on some Special Curves I,* Arch. Math., **46**, (1986), 453-465.

[C2]. M. Coppens, *Weierstrass Points with two Prescribed non-Gaps,* Pac. J. Math., **131**-1, (1988), 71-104.

[Cu]. F.M. Cukierman, *Families of Weierstrass Points*, Ph.D Thesis, Brown University, 1987.

[D1]. S. Diaz,*Tangent spaces in Moduli via Deformations with applications to Weierstrass Points*, Duke Math. J., **51**, (1984), 905-922.

[D2] S. Diaz, *Exceptional Weierstrass points and the divisor on moduli space that they define*, Memoirs of the American Mathematical Society, **56**, number 327, 1985.

[De1]. F. Delgado,*The semigroup of values of a curve singularity with several branches*, Manuscripta Math. **59**, (1987), 347-374.

[De1]. F. Delgado, *Gorenstein curves and symmetry of the semigroup of values*, Manuscripta Math. **61**, (1988), 285-296.

[EH1]. D. Eisenbud, J. Harris, *Limit Linear Series: Basic Theory,* Invent. Math. **85**, (1986), 337-371.

[EH2]. D. Eisenbud, J. Harris, *The Monodromy of Weierstrass Points,* Invent. Math. **87**, (1987), 495-515.

[Fo]. O. Forster, *Lectures on Riemann Surfaces,* Springer-Verlag, GTM, 1984.

[Ga]. A. Garcia, Semigroups associated to singular points of plane curves, J. reine angew. Math., **336**, (1982), 165-184.

[G1]. L. Gatto, k-*forme wronskiane, successioni di pesi e punti di Weierstra{\ss} su curve di Gorenstein* , Tesi di Dottorato-Università di Torino 1993 (umpublished).

[G2]. L. Gatto, *Weight Sequences versus Gap Sequences at singular Points of Gorenstein Curves* , Geometriae Dedicata, (to appear).

[G3]. L. Gatto, *Computing Gaps Sequences at Gorenstein Singularities,* in this Volume.

[GH]. Ph. A. Griffiths, J. Harris, *Principles of Algebraic Geometry,* Wiley & Sons, 1984.

[Gr]. A. Grothendieck (with J. Dieudonné), *Eléments de Géométrie Algebrique,* Chap. IV, 4ème partie, Inst. Hautes Études Sci., Publ. Math. **32**, 1967.

[GL1]. A. Garcia, R. F. Lax, *Weierstrass Weight of Gorenstein Singularities with one or two Branches*, Manuscripta Math., (to appear).

[GL2]. A. Garcia, R.F. Lax, *On Canonical Ideals, Intersection Numbers and Weierstrass Points on Gorenstein Curves* , (1993).

[GL1]. A. Garcia, R.F. Lax,*Weierstrass Points on Gorenstein curves in arbitrary characteristic*, Comm. Algebra, (to appear).

[Gu]. R.C. Gunning, *Lectures on Riemann Surfaces* , Princeton Mathematical Notes, 1966.

[GV]. S. Greco, P. Valabrega, *On the Theory of adjoints* , Lecture Notes in Math., **732**, Springer-Verlag, 1978.

[H1]. R. Hartshorne, *Algebraic Geometry* , Springer-Verlag, GTM **52**, 1977.

[H2]. R. Hartshorne, *Generalized divisors on Gorenstein curves and a theorem of Noether* , J. Math. Kyoto Univ., **26**-3, (1986), 375-386.

[K]. S. L. Kleiman, r-*Special Subschemes and an Argument of Severi's,* Advances in Math, **22** (1976), 1-23.

[La]. D. Laksov, *Weierstrass Points on Curves,* Astérisque, vols. **87-88**, (1981), 221-247.

[L1]. R.F. Lax, *Weierstrass Points of the Universal Curve,* Math. Ann., **42**, (1975), 35-42.

[L2]. R.F. Lax, *Weierstrass Points on rational Nodal Curves*, preprint.

[L3]. R.F. Lax, *Weierstrass Weights and Degeneration* , Proc. Amer. Math. Soc., **101**, (1987), 8-10.

[L4]. R.F. Lax, *On the Distribution of Weierstrass points on singular curves,* Israel J. Math., **57** (1987), 107-115.

[LT1]. D. Laksov, A. Thorup, *The Brill-Segre Formula for Families of Curves*, in: Enumerative Algebraic Geometry (Copenaghen, 1989), Contemp. Math. **123**, AMS, (1991), 131-148.

[LT2]. D. Laksov, A. Thorup, *Weierstrass Points and Gap Sequences for Families of Curves,* preprint, (1993).

[LW1]. R.F. Lax and C. Widland, *Gap Sequences at Singularity* , Pac. J. Math. **150**-1, (1991), 111-122.

[LW2]. R.F. Lax, C. Widland,*Weierstrass Points on Rational Nodal Curves of Genus 3*, Canad. Math. Bull., **30**, (1987), 286-294.

[M]. E. Matlis, 1-*Dimensional Cohen Macaulay Rings,* Lecture Notes in Math., **327**, Springer-Verlag, 1973.

[P]. C. Patt, *Variations of Teichmueller and Torelli Surfaces,* J. Analyse Math., **11**, (1963), 221-247.

[Po]. F. Ponza, *On the Weierstrass Weights at Gorenstein Singularities,* in this Volume.

[R]. H.E. Rauch, *Weierstrass Points, Branch Points, and Moduli of Riemann Surfaces,* Comm. in Pure and Applied Mathematics, **XII**, (1959), 543-560.

[S]. J.P. Serre, *Groupes Algébriques et Corps de Classes*, Hermann, Paris, 1959.

[St]. K.-0. Stöhr, *On the Poles of Regular Differentials of Singular Curves* , Bol. Soc. Bras. Mat., **24**, n.1, (1993), 105-136.

[Sz]. L. Szpiro, *Equations Defining Space Curves*, Springer-Verlag, Tata Institute, 1979.

[W1]. C. Widland, *Weierstrass Points on Gorenstein Curves,* Ph.D Thesis (Louisiana State University), 1984.

[W2]. C. Widland, *Bounds for Weierstrass Weights* , preprint.

[WL]. C. Widland, R.F. Lax, *Weierstrass Points on Gorenstein Curves*, Pac. J. Math., **142**, n. 1, (1990), 197-208.

Weierstrass Loci and Generalizations, II

E. Ballico
Dept. of Mathematics, University of Trento, 38050 Povo (TN), Italy
e-mail: (bitnet) ballico@itncisca or ballico@itnvax.science.unitn.it
fax: italy + 461881624

This is the second part (see [Ga] in this volume for the first one) of a survey on Weierstrass points, ramification loci, their generalizations and their use in Projective Geometry. However, it is completely independent from the first part [Ga].

We list two broad cases in which the notion of Weierstrass point is generalized to the one of Weierstrass locus.

(A) From the classical point of view of Weierstrass points on smooth curves one possible generalization is the notion of Weierstrass pairs (see [ACGH], p. 365). Several variations are possible. Here is one. Fix a smooth complete curve C of genus g, a linear system, V, on C and integers t, s, and $a_i, 1 \le i \le s$, with $t>0$, $s>1$, $a_i>0$; the Weierstrass locus of V with respect to $(t,s,\{a_i\})$ is the ordered subset of C^s formed by points $(P_1,...,P_s)$ such that the divisor $\Sigma_{1\le i\le s} a_i P_i$ imposes at most $\Sigma_{1\le i\le s} a_i - t$ conditions to V.

(B) Very important situations can be described by the key words "versal deformations for multigerms". Some of these situations arise in a natural way in the projective geometry of embedded varieties. See §1 and §2 for two such situations.

In §3 (following [Pe]) we will consider a good theory for vector bundles on a smooth projective curve. In §4 we will consider Weierstrass loci for varieties of dimension >1.

The author was partially supported by MURST and GNSAGA of CNR (Italy).

This paper is in final form and no part of it will be published elsewhere.

§1. In [MSV] the authors analyzed from the point of view of Singularity Theory the geometrical properties of the Gauss map of a general hypersurface of $\mathbf{CP}^4$. The main result of [MSV] states that the Gauss map ψ: $X \to \mathbf{CP}^{4*}$ of a general hypersurface $X \subset \mathbf{CP}^4$ with d:= deg(X)≥2 is stable. The authors gave two proofs of this theorem. The first proof uses versal deformations of singularities of projective varieties. The second proof uses unfoldings of singularities of maps and jet transversality. They classify the possible analytic type of the singularities of a Gauss-stable hypersurface. Furthermore, they study the geometry and the enumerative formulas of the loci of X which are related to the Gauss map ψ: the parabolic surface, the singular locus of the parabolic surface, the cuspidal curve, the higher order cusps. The parabolic surface Π of X is the scheme on which $d\psi$ has rank ≤ 2. The parabolic surface is the intersection of $X = \{F = 0\}$ with its Hessian hypersurface $Hess(X) := \{Hess(F) = 0\}$ ([MSV], Prop. 1.8). Let $\Pi' := \{x \in X: d\psi(x) \leq 1\}$; $\Pi \backslash \Pi'$ is smooth ; if ψ is stable, then the scheme Π' is reduced, $card(\Pi') = 20d(d-2)^3$ and at each point of Π' the surface Π has an ordinary double point ([MSV], Prop. 1.11). Let C be the closure of the Thom-Boardman loci $\Sigma^{1,1}(\psi) := \{x \in \Pi \backslash \Pi': \dim(\ker(d\psi|\Pi)(x) = 1\}$. If $\Sigma^{1,1}(\psi)$ is smooth (which is the case if ψ is stable by [MSV], 6.1), set $\Sigma^{1,1,1}(\psi) := \{x \in \Sigma^{1,1}(\psi): \dim(\ker(\psi|\Sigma^{1,1}(\psi)) = 1\}$; let Λ be the closure of $\Sigma^{1,1,1}(\psi)$. C is called the cuspidal curve and Λ the second-order cuspidal locus. By [MSV], §2 and §6, if ψ is stable, then $\Sigma^{1,1}(\psi)$ and $\Sigma^{1,1,1}(\psi)$ are smooth, C is a curve and Λ is a reduced set of 5d(d-2)(3d-7)(17d-36) points. A nice feauture of the paper [MSV] is that the methods used are powerful enough for the explicit geometric analysis of beautyful examples (for instance the cubic 3-folds studied in [MSV], §4) and for further generalizations (e.g. singularities of the intersection of a hypersurface of $\mathbf{CP}^n$ with linear spaces of fixed dimension).

§2. In [MK] it was studied from the point of view of versal deformations the contact behaviour of the family of all hyperplanes to a fixed variety $X \subset \mathbf{P}^n$. If X is a smooth curve embedded by $|K_X|$, we get also the classical case of Weierstrass points. Indeed a very particular case of [MK], Thm. 1.1, gives that in characteristic 0 or bigger than g a general curve of genus g has only ordinary Weierstrass points.

Let $X \subset \mathbf{P}^n$ be a smooth subvariety such that for every $H \in \mathbf{P}^{n*}$, the scheme $X \cap H$ has only isolated singularities (e.g. take as X a smooth curve). We will say that X is *hyperplane versal* (or *tangentially versal*) if for every $H \in \mathbf{P}^{n*}$ the family $\{X \cap M\}_{M \in \mathbf{P}^{n*}}$ is a versal deformation of the multigerm of singularities of $X \cap H$. The main point of [MK] is a very interesting and very powerful static characterization ([MK], Lemma 2.1) of hyperplane versality. In the case of the canonical embedding in $\mathbf{P}^{g-1}$ of a genus g smooth curve X, using duality he translates his criterion just in the check (easy for most even very particular curves) that for every hyperplane H, the set $(X \cap H)_{red}$ spans H.

Since for dimensional reasons X cannot be hyperplane versal for large dim(X), in [MK] the author introduced a very technical definition for an important generalization of versality and call it *hyperplane semi versality.* A main result of [MK] is that a general complete intersection of hypersurfaces of large degree is hyperplane semi versal.

In [Ba] reducible curves (used also independently in [MK] in the proof of [MK], Thm. 1.1) and monodromy arguments were used to obtaine results for general embedding of degree d curves of genus g in $\mathbf{P}^m$ for suitable (d,g,m). In [Ba] it was considered only the case of a curve. Fix a curve $C \subset \mathbf{P}^m$; a hyperplane $H \in \mathbf{P}^{m*}$ is said to have type $(m_1,...,m_k)$ with respect to C if the scheme $C \cap H$ has k connected components of length $m_1,...,m_k$. Since the set $\mathbf{P}^{m*}$ of hyperplanes of $\mathbf{P}^m$ is smooth, to check that a smooth complex curve $C \subset \mathbf{P}^m$ is tangentially versal we need to check the following property ($):

($) For every $(P,H) \in I(C)$ (say of type $(m_1,...,m_k)$) the differential at (P,H) of the morphism from $\mathbf{P}^{m*}$ to the versal deformation space for the type $(m_1,...,m_k)$ is surjective.

In particular only the integers m_i with $m_i \geq 2$ are important. Assume that $C \subset \mathbf{P}^m$ is tangentially versal; then, by the structure of the smooth miniversal spaces involved, the following condition ($$) (which will be also called "the right codimension condition") is satisfied:

($$) The tangential behaviour occurs in the right codimension i.e. if there is $(P,H) \in I(C)$ with type $(m_1,...,m_k)$, then near H the set of hyperplanes of type $(m_1,...,m_k)$ for C has the expected codimension $\Sigma_i (m_i-1)$ (i.e. the minimal possible one); furthermore, near H there are hyperplanes with all possible types $(m_1',...,m_k')$ with $m_i' \leq m_i$ for every i and strict inequality for at least one index i; the last part of this condition is the completeness of the family near (P,H).

The paper [Ba] contains the proof of the following results 2.1, 2.2, 2.3, 2.4 and 2.5.

Theorem 2.1. *Fix integers d, g, m. There is an irreducible generically smooth component M(d,g,m) of the Hilbert scheme of degree d and genus g smooth curves in* $\mathbf{P}^m$ *such that a general* $C \in M(d,g,m)$ *satisfies the right codimensional condition ($$) (in any characteristic) and is tangentially versal (in characteristic 0) if one of the following numerical conditions on (d,g,m) is satisfied:*

(a) Assume m = 3; let $\sigma(d)$ *be the unique integer such that* $\sigma(d)(\sigma(d)+1)/2 \leq d < (\sigma(d)+1)(\sigma(d)+2)/2$; *set* $F_d(t) = dt+1 - (t+3)(t+2)(t+1)/6$; *assume* $0 \leq g \leq F_d(\sigma(d)-1)$; *there exists such a component M(d,g,3) with the right number of moduli (i.e. such that the morphism* $M(d,g,3) \rightarrow M_g$ *has image of codimension* $max(0,-\rho(g,d,3)) := max(0,3g+12-4d)$.

(b) Assume $m \geq 4$; *there is a function w(g) with* $lim_{g\rightarrow\infty} w(g) = (m-2)/2$ *such that M(d,g,m) exists if* $d \geq g \cdot w(g)$.

Theorem 2.2. *Assume characteristic 0. Fix integers m, g, d with* $m \geq 3$ *and* $g \geq 0$ *and* $d \geq g+m$. *Then a general curve* $C \subset \mathbf{P}^m$ *with degree d, genus g and* $h^1(C,\mathcal{O}_C(1)) = 0$ *is tangentially versal.*

For all positive integers d, g, r, set $\rho(g,d,r)$: g - (r+1)(g+r-d) (the Brill - Noether number). A curve of genus g has a g^r_d if and only if $\rho(g,d,r) \geq 0$; if this condition is satisfied for some $r \geq 3$, it is known (see e.g. [BE]) that such a curve has a non degenerate degree d embedding into $\mathbf{P}^r$; if $\rho(g,d,r) > 0$ the set of all g^r_d on C is irreducible; if $\rho(g,d,r) = 0$, this set is finite, but, moving C, the corresponding finite cover of a Zariski open part of M_g is integral

([EH2]) and hence, strectching the terminology, we will speak of "generic" g_d^r even in this case: there is even in this case a unique irreducible component (call it W(d,g;r)) of the Hilbert scheme Hilb($\mathbf{P}^r$) of $\mathbf{P}^r$ containing curves with general moduli.

For all integers d, g, m with m≥3, d≥m, d-m<g≤d-m+[(d-m-2)/(m-2)], in [BE] it was proved the existence of an irreducible component W(d,g;m) of the Hilbert scheme Hilb($\mathbf{P}^m$) of curves in $\mathbf{P}^m$ with degree d and genus g with very nice properties; for instance, this component is generically smooth and contains curves with the right number of moduli (hence, if ρ(g,d,r)≥0, it is the component containing curves with general moduli). By the results in [BE], §2, we have W(d,g;m) = M(d,g,m) (the component claimed to exists in Theorem 0.1) for all integers d, g and m for which both components are defined.

Theorem 2.3. *Assume characteristic 0. For every integer m≥3 there is an integer a(m) and a function* ε_m*:* $N \to R_+$ *with* $\lim_{t\to\infty} \varepsilon_m(t) = 0$*, such that for all integers d, g with g≥0, d≥a(m) and* $md \geq (m+2)g(1+\varepsilon_m(g))$*, a general C∈W(d,g;m) has no point of type* $(m_1,...,m_k)$ *with* $\Sigma_i(m_i-1)>m$.

For the case x = 1, i.e. for the higher order ramification points ("Weierstrass points") of a general embedding it is given the following more precise result ; if ρ(g,d,m)≥0 the corresponding result was previously proved by D. Eisenbud and J. Harris ([EH1], Th. 2).

Theorem 2.4. *For all integers d, g, m with m≥3, d≥m, d-m<g≤d-m+[(d-m-2)/(m-2)] (i.e. for all integers d, g and m for which W(d,g;m) is defined), a general C∈W(d,g;m) has the property that for every P∈C the osculating flag to C at P has either contact order (2,3,...,m-1,m) or contact order (2,3,...,m-1,m+1).*

It is very cheap to find non linearly normal embeddings for which ($$) fails. Here is an interesting case with linearly normal (but not general) embeddings for which ($$) fails.

Theorem 2. 5. *Fix positive integers d, g, r with r≥3 and g≥2. Assume ρ(g,d,r+1)≥0 and let C be a smooth curve of genus g with general moduli. Then there are linearly normal embeddings of C into* $\mathbf{P}^r$ *which are not tangentially versal; more precisely, there are linearly normal embeddings with a point, P, which is hyperosculating with* $m_1>r+1$ *and hence which does not satisfy condition ($$).*

In the last section of [Ba] there are a few monodromy problems for the ramification points. In suitable ranges (either curves of high genus g with degree d≥g+m in $\mathbf{P}^m$ or curves with degree much greater than g+m) the methods of [Ba], §3, shows that the monodromy is huge for many particular subfamilies of such embeddings of all genus g curves. As an example it is given ([Ba], Thm. 3.1) the case of the family of all hyperelliptic curves. It was explicitly claimed in [Ba] that changing the numerical bounds the same proof works for many other families. In [Ba], Thm. 3.2, it was considered the existence and monodromy problem for ramification points for which the condition ($$) is not satisfied.

In [Ba] it was considered also the following notion of tangential versality with respect to r-dimensional subspaces.

Definition 2.6. Fix integers m, and r with $m \geq 3$ and $0<r<m$. Let $C \subset \mathbf{P}^m$ be a smooth curve; let $G(r) := G(r+1,m+1)$ be the Grassmannian of r-dimensional linear subspaces of $\mathbf{P}^m$; let $I(C)(r) := \{(P,V) \in C \times G: P \in V\}$ be the incidence correspondence and $\pi(C)(r): I(C)(r) \to G(r)$ be the projection; write I(r) and $\pi(r)$ if C is clear. We will say that C is *tangentially r-versal* if I(r) is versal. We define in the obvious way the equivalent condition ($)(r) and "the right codimension condition" ($$)(r) (with ($$)(m-1):= ($$)).

To extend in another direction the range in which the notions related to hyperplane versality may give very useful informations we introduce the following definition of *Hilbert r-versality* (by far the more intersting case being the case r = m-1 corresponding to the hyperplanes).

Definition 2.7. Let Γ be an irreducible component of the reduction of the Hilbert scheme $Hilb(\mathbf{P}^m)_{red}$. We will say that Γ is *Hilbert r-versal* if for a general $X \in \Gamma$ and every $A \in G(r+1,m+1)$ the family of all (Y,B) with $Y \in \Gamma$ and $B \in G(r+1,m+1)$ is versal at (X,A).

§3. Fix a smooth projective curve X, a rank r vector bundle E on X and a vector space $V \subseteq H^0(X,E)$ spanning E (hence inducing a morphism from X to the Grassmannian G(V,r), of r-dimensional subspaces of V). Canuto in [Ca], §2, introduced 3 measures of inflectional behaviour of a curve X mapped into a Grassmannian G. The first one comes from the Plücker embedding of G into a projective space **P** and taking the ramification loci for the composition $X \to G \to \mathbf{P}$. The second set of invariants are the order of vanishing of the sections of E coming from V. The third set of invariants comes from the order of contact of the curve with the Schubert cycles. He shows in [Ca] that these 3 notions are independent on a general Grassmannian (while they are equivalent in a projective space). The approach of [Pe] gives another sequence of ramification loci associated to the morphism $X \to G$.

Now in our opinion there is in characteristic 0 "the" theory of "Weierstrass poins" for spanned vector bundles on a smooth curve. It was introduced and studied in the thesis [Pe]. This theory is inspired from Piene's work ([Pi]) on the rank 1 case. In positive characteristic it seems better to modify slightly the main definitions. Here, following [Pe], we will describe this theory in characteristic 0.

We assume that the algebraically closed based field has characteristic 0. Let X be a smooth complete curve, E a rank r vector bundle on X and $V \subseteq H^0(X,E)$ a vector space of sections spanning E. Let $f: X \to G := G(V,r)$ be the associated morphism. On the Grassmannian G there is a "tautological" exact sequence:

$$0 \to S \to V_G \to Q \to 0 \qquad (1)$$

with V_G trivial vector bundle of rank dim(V), Q universal rank r quotient bundle and S universal rank dim(V)-r subbundle. By definition of f we have $f^*(Q) = E$. Let

$$0 \to S_E \to V_X \to E \to 0 \qquad (2)$$

be the pull-back of (1) by f. Since $TG \cong Hom(S,Q)$, the differential $TX \to f^*(TG)$ induces a morphism $\partial': S_E \to \Omega_X \otimes E$ or equivalently a morphism $\partial: S_E \otimes TX \to E$. Consider Coker($\partial$); it is the direct sum of a vector bundle (perhaps of rank 0) and a torsion sheaf. Set $E_1 :=$ Coker(∂)/Tors(Coker(∂)) and $r_1 :=$ rank(E_1). The torsion part of Coker(∂) is the first ramification locus we are looking for. The surjection $V_X \to E$ factors through the natural projection from the bundle of first order principal part $P^1(E)$ of E to E. Let

$$0 \to \Omega_X \otimes E \to P^1(E) \to E \to 0 \qquad (3)$$

be the exact sequence associated to $P^1(E)$ and u: $V_X \to P^1(E)$ "the first order Taylor map". Since $u(S_E) = 0$, u induces $\partial'': S_E \to \Omega_X \otimes E$. It was checked in [Pe], Prop. II.1.2, that $\partial' = \partial''$, giving a geometric reason for the definition of E_1. Note that E_1, as quotient of E, is a quotient of V. Hence, if $r_1 \neq 0$, we may iterate the construction obtaining another vector bundle E_2 quotient of E_1 and another ramification locus corresponding to the torsion part. Hence we get a chain of surjections

$$V \to E \to E_1 \to E_2 \to E_3 \to \cdots \qquad (4)$$

Set $E_0 :=$ E. Set $r_i :=$ rank(E_i). The bundle E_i is called the *i-th derived bundle* of (X,E,V). The integer $r_i - r_{i+1}$ is called the *i-th differential rank* of (X,E,V). The non increasing sequence of integers $\{r_i\}$ stabilizes to a certain value $\alpha \geq 0$. By [Pe], Remark II.5.1.2, the integer α is exactly the rank of a trivial factor, W, of E (hence the geometric situation is determined by (V/W,E/W)). From now on we assume $\alpha = 0$ (without loss of essential informations). If $E_{i+1} \neq 0$ the i-th differential rank is at least the (i+1)-th one ([Pe], Cor. II.3.6). These definitions capture the geometry of f since at a general point of X we have the following normal form for the triple (X,E,V).

Theorem 3.1. ([Pe], Th. II.4.2) *Assume $E_t = 0$ for large t; set $e := r - r_1$. Then near a general point of X we may take local coordinates $u_1, \ldots u_e$, such that the map f_0: $V_X \to E$ is a matrix with rows*

$$(u_1, u'_1, \ldots, u_1^{(i(1))}, \ldots, u_e, \ldots u_e^{(i(e))}) \qquad (5)$$

with $u_i^{(j)}$ j-th derivative of u_i; the integer e is the minimal one with this property. The m-th differential rank $r_{m-1} - r_m$ is card$\{j: i(j) \geq m-1\}$. The map f_j: $V_X \to E_j$ can be given in local coordinates as a matrix with rows

$$(u_1, u'_1, \ldots, u_1^{(i(1)-j)}, \ldots, u_e, \ldots u_e^{(i(e)-j)}) \qquad (6)$$

(in which the terms $u_s^{(a)}$ with $a<0$ are omitted).

Let $P^t(E)$ be the bundle of t-order principal parts of E. We have a chain of surjections

$$V_X \to \cdots \to P^t(E) \to P^{t-1}(E) \to \cdots \to E$$

The image $G^t(f)$ of V_X in $P^t(E)$ is called ([Pi]) the *osculating bundle of order t.* In [Pe], Prop. II.3.4, it was proved that for every i>0 the surjection $G^i(E) \to G^{i-1}(E)$ factors through the surjection $G^i(E) \to G_i(E)_1$. A major result contained in [Pe] is the following duality theorem.

Theorem 3.2. (Duality theorem [Pe], II.6.1) *The kernel of the map* f_1: $V_X \to E_1$ *is the dual of the first osculating bundle of* $V_X{}^* \to S_E{}^*$. *The kernel of the map* $V_X \to G^1(E)$ *is the dual of the first derived bundle of* $V_X{}^* \to S_E{}^*$.

§4. The notion of Weierstrass point for a smooth projective variety X with dim(X)>1 was introduced in [Og]. This notion was used as an essential tool to obtain bounds on card(Aut(X)) in [Co] (when X is a smooth surface). In [Co] it is also shown the link of the set of Weierstrass points of X with other ramification loci (for instance the parabolic curve of [MSV] discussed in §1) and how to use for his problem these ramification loci.

In [X1] and [X2] it was discussed the theory of Weierstrass points with respect to the line bundle $\mathcal{O}_X(1)$ for a smooth hypersurface $X \subset \mathbf{P}^n$ (with a very strong attention to the positive characteristic case with its "pathologies"). In the case of curves there are two equivalent ways for defining Weierstrass loci. One of them is the study of inflections, i.e. the points at which the tangent line has contact of order at least 3. The other one is the study of the ramification locus of the dual map (hence this is the approach linked to the theory of singularities of maps). In higher dimension these two approaches are not equivalent. In [X1] and [X2] it was used the first static approach. Contrary to the case of curves the locus of inflection points is empty for general X (see [X2], Remark 2 at page 589) as shown by an easy dimensional count.

The non enumerative part of [X1] and [X2] was generalized to the case in which X is singular (but with the codimension 1 part of Sing(X) of low degree) in [BGR], §6.

References

[ACGH] E. Arbarello, M. Cornalba, P. Griffiths, J. Harris, *Geometry of Algebraic Curves, Vol. I,* Springer-Verlag, 1985.

[Ba] E. Ballico, *On tangent hyperplanes to complex projective curves,* preprint.

[BE] E. Ballico and Ph. Ellia, *On the existence of curves with maximal rank in* $\mathbf{P}^n$, J. reine angew. Math. **397** (1989), 1-22.

[BGR] E. Ballico, F. Giovanetti, B. Russo, *On the projective geometry of curves in positive characteristic,* preprint.

[Ca] G. Canuto, *Associated curves and Plücker formulas in Grassmannians,* Invent. math. **53** (1979), 77-90.

[Co] A. Corti, *Polynomial bounds for the number of automorphisms of a surface of general type,* Ann. scient. Ec. Norm. Sup. (4) **24** (1991), 113-137.

[MSV] C. McCrory, T. Shifrin, R. Varley, *The Gauss map of a generic hypersurface in* $\mathbf{P}^4$, J. Diff. Geom. **30** (1989), 689-759.

[EH1] D. Eisenbud and J. Harris, *Divisors on general curves and cuspidal rational curves,* Invent. math. **74** (1983), 371-418.

[EH2] D. Eisenbud and J. Harris, *Irreducibility and monodromy of some families of linear series,* Ann. scient. Ec. Norm. Sup. (4) **20** (1987), 65-87.

[Ga] L. Gatto, *Weierstrass loci and generalizations, I,* in this volume.

[MK] J. McKernan, *Versality for canonical curves and complete intersections,* preprint.

[Og] R. Ogawa, *On the points of Weierstrass in dimension greater than one,* Trans. Amer. Math. Soc. **184** (1973), 401-417.

[Pi] R. Piene, *Numerical characters of a curve in projective n-space,* in: Real and Complex Singularities - Oslo 1976, pp. 475-495, Sijthoff and Noordhoff, 1977.

[Pe] D. Perkinson, *Jet bundles and curves in Grassmannians,* Ph. D. thesis, Chicago 1990.

[X1] M. Xu, *The hyperosculating points of surfaces in* $\boldsymbol{P}^3$, Compos. Math. **70** (1989), 27-49.

[X2] M. Xu, *The hyperosculating spaces of hypersurfaces,* Math. Z. **211** (1992), 575-591.

A Problem of Trigonality for Some Fano Varieties

E. AMBROGIO - L. PICCO BOTTA
Dipartimento di Matematica. Università di Torino. Italia.

Let $W \subseteq \mathbb{P}^N$ be a 3-fold satisfying the following conditions :

(i) W is projectively normal,

(ii) for a sufficiently general hyperplane H the section $W \cdot H = F$ is a smooth Enriques surface,

(iii) the general curve section $W \cdot H \cdot H'$ has genus $p > 5$,

(iv) W is not a cone .

Such a 3-fold is extensively studied in [CM2] , where, under suitable assumptions of generality, it is proved that:

- W is a variety of degree 2p-2 in $\mathbb{P}^p$ with only 8 quadruple points $P_1, \ldots, P_8$ as singular points ;
- W contains a linear system $|\varphi|$ of dimension p-1 with $P_1, \ldots, P_8$ as base points, whose generic member is a K3 surface;
- the associated map $\lambda = \lambda_{|\varphi|}$ is a birational morphism and the image $V = \lambda(W) \subseteq \mathbb{P}^{p-1}$ is a 3-fold of degree 2p-6 whose general hyperplane sections are smooth K3 surfaces;
- V contains 8 planes $\pi_1, \ldots, \pi_8$ (the "images" of the points P_i);

- V is smooth except (possibly) for finitely many singular points coming from the contraction of the lines P_iP_j when these lines are contained in W; in this case the points P_i e P_j are called associated and the planes π_i and π_j intersect exactly in the corresponding singular point.

In particular, in [CM2] the singular points P_i are assumed to be non collinear and "similar", i.e. they "behave" in the same way and they have the same numerical characters (cfr [CM2] 4.1) . It follows that every point is associated to the same number of other points , so V cannot have a unique singular point .

We recall that a (classical) Fano Variety is a irreducible complete variety of dimension 3, of degree 2g-2 in $\mathbb{P}^{g+1}$ having a K3 surface as general hyperplane section and having the canonical curve of genus g in $\mathbb{P}^{g-1}$ as general curve section .
It follows that $V = \lambda(W)$ is a Fano variety with $g = p-2 \geqslant 4$.
The Fano varieties are extensively studied in [I] . In particular the trigonal ones are examinated , i.e. the Fano varieties whose general curve section is a trigonal curve .
Then the question arises whether the variety $V = \lambda(W)$ is trigonal .
If p = 6 , $W \subseteq \mathbb{P}^6$ has degree 10 and is studied in [CM1] and in [CM2] , where the corresponding $V_6 \subseteq \mathbb{P}^5$ is proved to be the intersection of a quadric and of a cubic , i.e. , by identifying the quadric with the Grassmanian G(1,3) , V_6 is a cubic complex.
V_6 is trigonal since the curve section is the well known trigonal curve obtained as intersection of a quadric and of a cubic in $\mathbb{P}^3$.

We want to prove the following

THEOREM : Under the above assumptions the general variety V is not trigonal for p>6 (i.e. g>4) .

We need the following

LEMMA : If V is trigonal, then V contains a pencil of cubic surfaces whose general member does not intersect the planes $\pi_1,..,\pi_8$.

Proof of the Lemma : The section curve Γ of V is a trigonal canonical curve, therefore it is not an intersection of quadrics (Petri's theorem [ACGH], pag 131) . It follows that V itself is not an intersection of quadrics . Let X be the intersection of the quadrics containing V : X is a 4-dimensional variety in $\mathbb{P}^{g+1}$ and the same arguments used in the proof of the proposition 3.3, cap.2 of [I] show that $\deg X = \operatorname{codim} X + 1 = g-2$.

(Indeed the prop. 3.3 proves the same fact for a non singular trigonal Fano variety using the existence of a non singular curve section, but such a section does exist also in our case)

The projective varieties Y satisfying the condition $\deg Y = \operatorname{codim} Y+1$ are completely classified (th. 1.1 of [SD] , th. 3.11, cap 1 of [I]).
According to this classification, the 4-dimensional variety $X \subseteq \mathbb{P}^{g+1}$ $(g+1 \geq 6)$ is a rational ruled variety, i.e. there exists a rank 4 bundle $E = \mathcal{O}(d_1) \oplus \mathcal{O}(d_2) \oplus \mathcal{O}(d_3) \oplus \mathcal{O}(d_4)$, $d_1 \geq d_2 \geq d_3 \geq d_4 \geq 0$ on $\mathbb{P}^1$ such that $X = \varphi_M(\mathbb{P}(E))$ where $M = \mathcal{O}_{\mathbb{P}(E)|\mathbb{P}^1}(1)$ is the tautological bundle of $\mathbb{P}(E)$.
Moreover the variety X either is smooth if $d_i > 0$ $\forall i$ or is a cone on a non singular rational ruled variety of dimension m if $d_1 = \ldots = d_m \neq 0$ and $d_{m+1} = 0$.
Let $L = f^*\mathcal{O}(1)$, where $f: X \to \mathbb{P}^1$ is the natural projection and let L and M be the divisors associated to the bundles L and M . It is known that :

$$\operatorname{Pic}(X) = \mathbb{Z}M \oplus \mathbb{Z}L \quad , \quad L^2 = 0 \quad , \quad M^3 \cdot L = 1 \quad , \quad M^4 = \Sigma d_i \ .$$

(For the properties of the rational ruled variety, see [SD] or [I], cap.1 §3).

By similar arguments as in the prop. 3.3, cap 2 of [I] we can prove that X is non singular except (possibly) at the singular points of V.
(In fact if x is a non singular point of V (or if $x \notin V$) it is possible to find a $\mathbb{P}^{g-1}$ through x such that the curve $\Gamma = V \cap \mathbb{P}^{g-1}$ is non singular. It follows that the surface $Y = X \cap \mathbb{P}^{g-1}$ is also non singular ([I] ,th. 3.7 cap.1) . Therefore x cannot be a singular point of X , otherwise x would be also a singular point of Y).

Case I : V is non singular.
Then X is also non singular and in Pic(X) $V = aM+bL$.

Since the linear system g^1_3 of the general curve section Γ is cut out by the fibres of the rational normal scroll containing Γ ([ACGH] , pag. 124), it follows that :

$$3 = (aM+bL)\cdot M^2\cdot L = a \quad \text{and therefore} \quad V = 3M+bL .$$

Then , the trace on V of the pencil |L| is a pencil |S| of pairwise disjoint cubic surfaces.
We note that the 8 planes π_i have to be contained in 8 different fibres otherwise the restriction of f to π_i would be a morphism of $\mathbb{P}^2$ on $\mathbb{P}^1$.

Case II : V is singular.
In this case we know that V has at least 2 singular points and X either is smooth or is a cone on a 3-dimensional non singular rational ruled variety $X' = \mathbb{P}(E') \subseteq \mathbb{P}^g$ with vertex at a singular point p of V.
We want to prove that the second possibility cannot occur.
Suppose that X is a cone with vertex p and let $g : V \to \mathbb{P}^g$ be the projection from p .
Since V is not a cone , g turns out to be a generically finite map $V \to X'$ contracting only the lines through p contained in V.
Moreover V has at least another singular point q which is the intersection of two planes π_1 , π_2 not passing through p , so the image of π_i , i = 1,2 is a plane contained in X'.
But X' is a non singular rational ruled 3-fold and $Pic(X') = \mathbb{Z}M' \oplus \mathbb{Z}L'$, where L' is the fibre of the natural map $f' : X' \to \mathbb{P}^1$, so the image of π_i , i = 1,2 is contained in a fibre L' .
Since π_1 and π_2 have a common point while the fibres L' are disjoint , their images have to be the same plane π' and they are contained in $\mathbb{P}^3 = \langle p,\pi' \rangle$. This is impossible because their intersection is a point instead of a line .
We can conclude that also in this case X is smooth and the fibres of the natural map $f : X \to \mathbb{P}^1$ cut out on V a pencil of cubic surfaces. The planes π_i are still contained in different fibres for the same reasons as in the case I.
This ends the proof of the lemma.

Proof of the theorem:
Let F be the Enriques surface hyperplane section of W , let φ be the K3 surface of the linear system $|\varphi|$ and let's denote by $\overline{F}$ and $\overline{\varphi}$ respectively their images under the map $\lambda : W \to V \subseteq \mathbb{P}^{p-1} = \mathbb{P}^{g+1}$.
Then $\overline{\varphi}$ is the hyperplane section of V and under our assumptions the general $\overline{F}$ and the general $\overline{\varphi}$ are smooth.
From the results of [CM2] ,it follows that in Pic(V) :
$2\overline{F} = 2\overline{\varphi}+\pi_1+ \dots +\pi_8$. (lemma 3.12 and 5.20 di [CM2])
Let's suppose that V is trigonal.

Then , the lemma states the existence of a base points free pencil $|S|$ whose general member S is a cubic surface non intersecting the planes $\pi_1, \ldots, \pi_8$.
In Pic(S) :
$2\bar{F}\cdot S = 2\bar{\varphi}\cdot S$ and , since Pic(S) is torsion free , $\bar{F}\cdot S = \bar{\varphi}\cdot S$.

On the other hand , if H is the hyperplane containing $\bar{\varphi}$,
$E = \bar{\varphi}\cdot S = \bar{F}\cdot S = H\cdot S$ is a plane elliptic cubic lying on $\bar{F}$.
Therefore the Enriques surface $\bar{F}$ contains a pencil $|E|$ of elliptic curves generically of degree 3 , i.e. $E\cdot H' = 3$, where H' is a hyperplane section of $\bar{F}$.
We get a contradiction because it is known that on a Enriques surface $E\cdot D$ is even for any divisor D . ([SH] coroll. to the prop.2, cap 10) We can conclude that V is not trigonal , as required.

References

[ACGH] ARBARELLO—CORNALBA-GRIFFITHS-HARRIS: *Algebraic Curves*, Springer (1985),

[CM1] A. CONTE- J.P. MURRE: *Three- dimensional algebraic varieties whose hyperplane sections are Enriques surfaces*, Institut Mittag-Leffler, Report n° 10, 1981,

[CM2] A. CONTE- J.P. MURRE: *Algebraic varieties of dimension three whose hyperplane sections are Enriques surfaces*, Annali Scuola Norm. Sup. Pisa, Serie IV, Vol 12, n.1 (1985),

[I] V.A. ISKOVSKIH: *Anticanonical models of three dimensional algebraic varieties*, Itogi Nauki i Tekniki, Sov. Pr. Math., 12(1979), pp 59-157, engl. trans. in J. Soviet Math., 13(1980), pp 745-814,

[SD] B. SAINT-DONAT: *Projective models of K3 surfaces*, Amer. J. Math., 96(1972), pp 602-639,

[SH] I. R. SHAFAREVIC: *Algebraic Surfaces* , Proc. of Steklov Inst. of Math., n. 75 (1965)

Families of Singular Meromorphic Foliations

E. Ballico
Dept. of Mathematics, University of Trento, 38050 Povo (TN), Italy
e-mail: (bitnet) ballico@itncisca or ballico@itnvax.science.unitn.it
fax: italy + 461881624

Let X be a connected complex manifold of dimension n. Recall (see e.g. [2] or [5] or [7]) that a codimension q singular meromorphic foliation on X is given by the inclusion of a coherent sheaf, A, with rank(A) = n-q into Ω^1_X; such a singular meromorphic foliation is called *saturated* if Ω^1_X/A has no torsion. In this note we show the existence of a priori bounds for the number of the irreducible components (and their dimension and degree) of the "variety" of foliations "with give invariants". We first need to clarify the terms between quotations marks. In [2], §1, it was shown how to give the structure of algebraic variety to the set of codimension 1 meromorphic singular foliations of type L (for a fixed line bundle L) on any compact complex manifold X. Their construction is essential for the proof of codimension 1 case (theorem 1.2). In 1.3 their construction is extended to the case of codimension q foliations, allowing us to give a meaning (and then to prove) theorem 1.4 on codimension q foliations with no restriction on q. For the exact statements of the results, see theorem 1.2 for codimension 1 foliations and theorem 1.4 for the case of foliations with arbitrary codimension q (but 1.4 does not cover 1.2, while 1.2 does not cover completely the case q = 1 of 1.4). Suffice to say here, that the bounds are not explicit; the main content of the section is the existence of bounds which hold uniformly in a precise sense for all projective manifolds X with "bounded invariants". We believe that it would be very interesting to have more explicit bounds, and to have, even only in particular cases, computable and not too bad bounds). In general, we are more interested in saturated foliations. In this paper we will work mainly with saturated foliations and we always state explicitely if a foliation is not assumed a priori to be saturated. See theorem 1.5 for upper bounds on the data needed to describe the complements of the saturated foliations in the "variety" of all foliations.

The author was partially supported by MURST and GNSAGA of CNR (Italy). This paper is in final form and no part of it will be published elsewhere.

§1. Bounds on the number of components

Fix a compact complex manifold X and a line bundle L on X. In [2], §1, it was shown that, for fixed line bundle L, the set of meromorphic foliations with singularities on X corresponding to inclusion (as sheaves) of L^* into Ω^1_X is a closed algebraic subset, **t**, of a projective space $\mathbf{P}^N$ with $N := h^0(X, \Omega^1_X(L)) - 1$, and that **t** is defined by quadratic equations. Of course, to obtain the true object with a geometric meaning one should only consider the open subset **T** of **t** parametrizing the saturated foliations. The structure of **T** and **t** seems to be very important (in particular when X is a projective space (see 1.7 for a very partial motivation of the relevance of this case)). We will not review the construction in [2], §1, since in 1.3 we will give the general codimension q case (using only easy modifications of the construction in [2], §1).

To obtain the bound we will use the following very well known and elementary lemma (we heard it is attributed to A. Andreotti (early fifties)) (see [1], Lemma 1.28).

Lemma 1.1. *Fix positive integers N and d. Let $\{f_i\}$ be a family of homogeneous polynomials in N+1 variables with $deg(f_i) \le k$ for every i. Then the algebraic subset of $\mathbf{P}^N$ defined by all equations $\{f_i\}$ has at most k^N irreducible components.*

The quadratic equations defining **t** give a natural scheme structure on **t**; however we will use only its reduced structure.

Theorem 1.2. *Fix an integer n>0, a degree n polynomial $p(t) \in \mathbf{C}[t]$, and an integer a>0. Then there is a constant $k = k(n,p,a)$, k depending only on n, p(t) and a, such for every triple (X,H,M) with X compact connected complex manifold of dimension n, H ample line bundle on X such that $\chi(tH) = p(t)$ (i.e. H has p(t) as Hilbert polynomial), M line bundle on X with $h^0(X, H^{\otimes a} \otimes M) \neq 0$, the set of codimension 1 meromorphic singular foliations of type M on X has at most k irreducible components.*

Proof. By general principles (i. e. the existence of the Hilbert scheme (proven in [3], exp. Bourbaki 221; for a few words on that see [4], ch. III, Remark 9.8.1)) it is known that the set of such pairs (X,H) can be parametrized by an algebraic scheme. The condition "$h^0(X, H^{\otimes a} \otimes M) \neq 0$" for a fixed integer a means exactly that M cannot be "arbitrarly negative". By [6], p. 84 and p. 113, indeed the set of such triples (X,H,M) can be parametrized by an algebraic scheme, D. By the semicontinuity theorem and the fact that D has only finitely many irreducible component, we know that varying M in D, the set of integers $\{h^0(X, Hom(A, \Omega^1_X))\}$ is bounded. Hence the thesis follows from 1.1. ♦

(1.3) (The construction): Consider a codimension q holomorphic foliation $\boldsymbol{F}$ with singularities on a compact connected complex manifold with $\boldsymbol{F}$ induced by an inclusion of sheaves $A \to \Omega^1_X$. Fix A and consider the set $\mathbf{T} = \mathbf{T}(A)$ of all such saturated foliations corresponding to inclusions of the fixed coherent rank q sheaf A (call them "foliations of type A "). Set $N := h^0(X, Hom(A, \Omega^1_X)) - 1$. We will show how to give to $\mathbf{T}$ a structure of quasi-projective scheme. More precisely, we will describe $\mathbf{T}$ as an open part in an algebraic subscheme $\mathbf{t}$ of $\mathbf{P}^N$ defined by equations of degree q+1, with $\mathbf{t}$ parametrizing all foliations of type A. Fix a small ball B (or a polydisc) in X such that A is locally free of rank q on B. Let $z_1, \ldots, z_n$ be holomorphic coordinates for X on B. Fix a basis τ_i, $0 \le i \le N$, of $H^0(X, Hom(A, \Omega^1_X))$. See $\tau_i|B$ as a q-ple of holomorphic 1 - forms, i.e. write $\tau_i|B := (\Sigma_j a_{ijt}(z)dz_j)_{1 \le t \le q}$ and $d(\tau_i|B) := (\Sigma_{j<k} a_{ijkt}(z)dz_j \wedge dz_k)_{1 \le t \le q}$. Consider any foliation $\boldsymbol{F} \in \mathbf{t}$. $\boldsymbol{F}$ is induced (up to a multiplicative constant) by $\tau \in H^0(X, Hom(A, \Omega^1_X))$, say $\omega = c_0\tau_0 + \cdots + c_N\tau_N$. On B ω can be written as a q-ple of holomorphic 1 - forms, say $\omega|B = (\omega^{(1)}, \ldots, \omega^{(q)})$. Since all the objects involved are holomorphic , by the " principle of prolongations of identities " and the connectedness of X, the integrability condition for τ means exactly that for every t with $1 \le t \le q$ we have $d\omega^{(t)} \wedge \omega^{(1)} \wedge \cdots \wedge \omega^{(q)} = 0$. Evaluating this condition at each point of B we find that this condition is equivalent to a set of homogeneous equations degree q+1 for the variables $c_0, \ldots, c_N$; these equations describe $\mathbf{t}$. Of course, since every ring of polynomials is Noetherian only finitely many of these equations are necessary to describe $\mathbf{t}$, but what equations and their number is not determined (or bounded, a priori) (see remark 1.5 for the case $X \cong \mathbf{P}^n$). Furthermore, $\mathbf{t} \backslash \mathbf{T}$ may be expressed as a closed subset of $\mathbf{t}$ in the following way. The condition that the map $A \to \Omega^1_X$ corresponding to $\omega \in H^0(X, Hom(A, \Omega^1_X))$ is an inclusion of sheaves means exactly that on B a certain q×n matrix of holomorphic functions have generically rank q. For fixed $z \in B$, evaluating this matrix at z, we find that a necessary condition for being ω in $\mathbf{t} \backslash \mathbf{T}$ is given by the vanishing of all q×q minors of the evaluated matrix; this gives (a very much studied) system of degree q homogeneous equations on the variables $c_0, \ldots, c_N$. A subtle point is that there are two different definitions of singular foliation, according if one require the integrability condition at the singular points of the foliation or not; the integrability condition at the singularities is automatically satisfied (see e.g. [7], sentence just after eq. (1.4)) for saturated foliations.

Note that even in the case with q = 1 the assumptions in the next theorem are slightly different from the assumptions in 1.2.

Theorem 1.4. *For all integer n, q, with* $1 \leq q < n$, *all polynomials p(t) and* $\chi(t)$ *with* $deg(p(t)) = n$ *and* $deg(\chi(t)) = n-q$, *there is a constant* $d = d(n,q,p,\chi)$ *(d depending only on n, p(t) and* $\chi(t)$*) such that for every triple* (X,H,Δ) *with X compact connected complex manifold of dimension n, H ample line bundle on X with p(t) as Hilbert polynomial,* Δ *rank q coherent subsheaf of* Ω^1_X *with quotient sheaf having* $\chi(t)$ *as Hilbert polynomial, the set of singular foliations of type* Δ *on X has at most d irreducible components.*

Proof. The boundness of the set of pairs (X,H) was stated at the begining of the proof of 1.2. Since the set of all such (X,H) is parametrized by an algebraic scheme, the set of possible Hilbert polynomials of such subsheaves of Ω^1_X is finite and their number is bounded in terms of the data n and p(t). By Grothendieck's general theory of the Quot-scheme ([3]), the set of all such (X,H,Δ) can be parametrized by an algebraic scheme. By the semicontinuity theorem we see that, varying Δ, the integers $\{h^0(X,\mathcal{H}om(\Delta,\Omega^1_X))\}$ are bounded. Hence the thesis follows from 1.1. ♦

But in general we are more interested in saturated foliations; hence under the assumptions of 1.2 or of 1.4 we would like to know how "difficult" can be a priori the description of the set **t\T** of non saturated foliations (the garbage to dispose of).

Theorem 1.5. *(a) With the notations of 1.2 fix n, p(t) and the integer a. Let (X,H,M) be as in the statement of 1.2. As in the proof of 1.4, let* **t** *be the set of not necessarily saturated foliations of type M, with* **t** *seen as an algebraic subset of a big projective space,* Π*, as in the construction 1.3. Then* ***t\T*** *can be described as subset of* Π *by a bounded number of equations of bounded degree, with the bounds depending only on n, p(t), and a.*

(b) With the notations of 1.4 , fix n, q, p(t) and $\chi(t)$*. Let* (X,H,Δ) *be as in the statement of 1.2. As in the proof of 1.4, let* **t** *be the set of not necessarily saturated foliations of type* Δ*, with* **t** *seen as an algebraic subset of a big projective space,* Π*, as in the construction 1.3. Then* ***t\T*** *can be described as subset of* Π *by a bounded number of equations of bounded degree, with the bounds depending only on n, p(t), and* $\chi(t)$*.*

Proof. Both parts have the same proof. We fix (X,H,Δ) (or (X,H,M) if you prefer). Fix an effective divisor D and $\boldsymbol{F} \in \mathbf{t}$ with $\boldsymbol{F}$ having exactly D as 1 codimensional part of the singular locus. Hence $\boldsymbol{F}$ is induced by a saturated foliation $\boldsymbol{G}$ with $\boldsymbol{G}$ induced by an element of $\mathcal{H}om(\Delta(D),\Omega^1_X)$. By the boundness of **t** we get an upper bound on the possible D. For each fixed D, the corresponding set of $\boldsymbol{F}$ is given by the construction 1.3 as a subset of $H^0(X,\Delta(D),\Omega^1_X)$ to whom we may apply 1.4 (or 1.2 for part (a)). We conclude again by "general principles". An alternative proof can be easily given using the last part of the construction 1.3. ♦

Remark 1.6 There is a beautiful book ([5]) with a strong algebraic and algebro-geometric flavor considering codimension 1 holomorphic foliations on projective spaces. Here a miracle occurs. Since on $\mathbf{P}^n$ we may work with homogeneous forms, there is a "global" differential for holomorphic forms on $\mathbf{P}^n$ with value in a line bundle, i.e. for every p, t we have $\mathbf{d}_{p,t}$: $H^0(\mathbf{P}^n,\Omega^p(t))\rightarrow H^0(\mathbf{P}^n,\Omega^{p+1}(2t-1))$. Hence one can get more explicit bounds than in 1.2 and 1.4; the main obstruction to obtain good bounds in this way is the weakness of 1.1.

Remark 1.7 The case of foliations on $\mathbf{P}^n$ is interesting also because, in the following very weak sense, it is a universal case. Fix a codimension q meromorphic (saturated) foliation with singularties $\boldsymbol{F}$ on $\mathbf{P}^n$ and a projective manifold X with dim(X) = n. There is a holomorphic map with finite fibers π: $X\rightarrow\mathbf{P}^n$ (which makes X a branched covering of $\mathbf{P}^n$). One can produced a meromorphic codimension q saturated foliation with singularities on X taking the saturated foliation associated to the foliation obtained pulling - back $\boldsymbol{F}$ by π (i. e. if $\boldsymbol{F}$ corresponds to an inclusion i: $G\rightarrow\Omega^1_X$, $\boldsymbol{G}$ corresponds to the inclusion of the quotient of $\pi^*(G)$ by its torsion into Ω^1_X obtained from $\pi^*(i)$). Note that there are many such π; if we fix $x\in\mathbf{P}^n$ we may take π unramified over x (hence the singularity of $\boldsymbol{G}$ at each point of $\pi^{-1}(x)$ is analytically the same as the singularity of $\boldsymbol{F}$ at x. Similarly, if instead of X we take a m-dimensional subvariety Y of $\mathbf{P}^n$, the restriction of $\boldsymbol{F}$ gives a codimension q foliation on Y (if we move Y (if necessary) by the action of the projective linear group Aut($\mathbf{P}^n$) to be sure that Y meets at most in codimension 2 the singular set of $\boldsymbol{F}$).

REFERENCES

1. F. Catanese: *Chow varieties, Hilbert schemes, and moduli of surfaces of general type*, J. Algebraic Geom. **1** (1992), 561-596.
2. X. Gomez-Mont and J. Mucino: *Persistent cycles for holomorphic foliations having a meromorphic first integral,* Lect. Notes in Math. **1345**, (1988), 129-162.
3. A. Grothendieck: *Fondaments de la Géometrie Algébrique* , Seminaire Bourbaki 1957-1962, Secrétariat Math., Paris, 1962.
4. R. Hartshorne: *Algebraic Geometry* , Graduate Text in Math. **52**, Springer-Verlag, 1977.
5. J. P. Jouanolou: *Equations de Pfaff algébriques,* Lect. Notes in Math. **708**, Springer-Verlag, 1979.
6. D. Mumford: *Lectures on curves on an algebraic surface,* Annals of Math. Studies, Princeton, N.J. (1966).
7. T. Suwa: *Unfoldings of complex analytic foliations with singularities,* Japan J. Math. **9** (1983), 181-206.

On Stable Sheaves on Algebraic Surfaces

E. Ballico
Dept. of Mathematics, Università di Trento, 38050 Povo (TN), Italy
e-mail:ballico@itncisca.bitnet or ballico@itnvax.science.unitn.it
fax: italy + 461881624

The topic of this paper is: stable sheaves on singular algebraic surfaces.

We think that for one of the research topics (the study of vector bundles on *all* threefolds) on which we are working a strong necessity would be a better understanding of "good" vector bundles and reflexive sheaves on a singular surface Y; here "good" means "stable" and "understanding" may mean "existence, structure of moduli spaces, cohomological properties of a sufficiently general bundle, and so on". Many results can be easily translated from the case of smooth surfaces, if one has the "right" set up. This set up is the main content of this paper.

For the study of projective varieties a very important tool is Mori theory. In turn, Mori theory justifies to study certain kind of varieties and morphisms rather than certain other ones. We think this would be a necessary step for the understanding of many properties of moduli space of vector bundles and reflexive sheaves. In the case of surfaces from Mori theory two classes of singular surfaces appear in a natural way: surfaces with only rational double points and semi-smooth surfaces (see §1 for their definition).

Among the normal singular surfaces, the most important ones are the "canonical" ones, i.e. the ones which have rational double points (RDP) as only singular points; of course both their name and their very old importance came from the fact that RDP are exactly the singularities which appears in the (pluri)canonical models of surfaces of general type. The basic results on reflexive sheaves (and their moduli) on surfaces with RDP are given in [I]. Reflexive sheaves on such surfaces are considered in sections 2 and 3 (using only the local classification given in [AV] and [Ka]). In section 2 (following often quite closely at certain steps [Q1]) we study rank 2 reflexive sheaves which are stable for a polarization and unstable for another polarization; here the main point is theorem 2.7. In section 3, we connect (following [Br]) the stability properties of rank 2 reflexive sheaves on a surface, X', with only A_1 and A_2 singularities and the stability properties of vector bundles on the minimal desingularization of X'; here the main point is theorem 3.4.

In [KSB], §4, it was introduced an important type of singular non normal surfaces with very mild singularities ("the semi-smooth surfaces" (see here §1, definition 1.2)) which should be the natural " up to codimension 2 desingularization" for non normal surfaces (or varieties); see [KSB], §4. Section 2 is devoted to the introduction and study of "vector bundle like" sheaves on semi-smooth surfaces; main points: theorem 1.4, definition 1.5 and proposition 1.6.

The author was partially supported by MURST and GNSAGA of CNR (Italy).

§1. In the classification of threefolds and in the deformation theory of surface singularities certain non normal surfaces with very mild singularities appeared in a very natural way (see definition 1.2). These surfaces were introduced in [KSB], §4, and called there semi-smooth. Recall the following terminology (in this section we will always assume for simplicity char(**K**) $\neq$ 2).

Definition 1.1. A surface singularity (x,X) is called a normal crossing point (resp. a pinch point) if it is formally (or analytically, if you prefer) isomorphic to the singularity at $0 \in \mathbf{K}^3$ with equation $xy = 0$ (resp. $x^2 = zy^2$).

Definition 1.2. A surface X is called semi-smooth if every closed point of X is either smooth or normal crossing or a pinch point.

We think that the content of this section (i.e. definition 1.5, theorem 1.4 and proposition 1.6) will be very useful to attack any moduli problem on "nice" sheaves on semi-smooth surfaces will appear in the near future in the study of sheaves on threefolds fibered over a curve.

In this section we will fix the following notations. Let X be a semi-smooth surface, D_X its singular locus and π: $S \to X$ its normalization. D_X is a smooth curve; D_X is called the double curve of X. S is a smooth surface and $C_X := \pi^{-1}(D_X)$ is a smooth curve. Set $\pi' := \pi|(C_X)$; hence π': $C_X \to D_X$ is generically 2:1 and ramifies exactly at the pinch points. Let $R \subset D_X$ be the set of pinch points; we will see R also as embedded in C_X, hence as a finite subset of S (or denote it R').

In the first part of this section we will handle the following descent problem (which is the first necessary step for a detailed study of moduli spaces of stable vector bundles on semi-smooth surfaces). Let E be a vector bundle on S. Find when there is a vector bundle F on X such that $E \cong \pi^*(F)$. First we need to descend $E|C_X$ to a vector bundle on D_X. Note that C_X and/or D_X may be disconnected.

Lemma 1.3. *Let f: C→D be a 2:1 ramified covering of smooth curves (hence a* **Z/2Z** *Galois covering); we do not assume that C is connected. Let* σ *be the generator of the Galois group of f. Let* $R' := f^{-1}(R) \cong R$ *be the ramification divisor of f. Let E' be a vector bundle on C. There is a vector bundle F' on D with* $E' \cong f^*(F')$ *if and only if the following two conditions are satisfied:*

(i) $E' \cong \sigma^*(E')$*;*

(ii) making the identification of E' and $\sigma^*(E')$ *whose existence is given by (i),* σ *acts as the identity on E|R.*

Proof. It is obvious that the two conditions are necessary. By standard Galois cohomology (or just local computations) condition (i) is equivalent to the existence of a vector bundle F" on D\R with E|(C\R') ≅ $f^*(F")$. Condition (ii) is exactly the condition that F" is trivial in a punctured neighborhood of every point of R according to the data given by E' on C near R', i.e. that we may fill in the fibers of F" over R (there is no patching problem for different charts since R is discrete). ♦

Theorem 1.4. *Let E be a vector bundle on S. There is a vector bundle F on X with E ≅ $\pi^*(F)$ if and only if there is a vector bundle F' on D_X such that E':= E|C_X ≅ $\pi'^*(F')$, i.e. if and only if the two conditions of lemma 1.3 are satisfied.*

Proof. The "only if" part is trivial. Note that X\R is seminormal in the sense of [T]. Note that F exists over X\R as a set, because it is obvious that E|(S\R') has a family of local trivialization on open sets, U, with U = $\pi^{-1}(\pi(U))$. By definition of seminormality this implies the existence of F over X\R. Fix P∈R (seen as a point of D_X⊂X and of R'⊂C_X⊂S); let r be the rank of E. By the assumption on E', there is an affine neighborhood V of P in C_X with V saturated for π' (i.e. saturated by π) such that E'|V is trivialized by r sections coming from D_X. These sections extend to sections on an open affine U⊂S, with U∩C_X = V; by the definitions of C_X and π, both U and these sections are π-invariant. By Nakayama lemma, restricting if necessary V and U, we may assume that these sections induce a trivialization of E|V. By the seminormality of X\R these sections come from sections of F' on π(V)\R. Thus F'|(p(V)\R) is trivial. Since R is discrete, this means that F' extends to a vector bundle F on X. Since S is normal and E|(S\R') ≅ $\pi^*(F)$|(S\R'), we have E ≅ $\pi^*(F)$. ♦

Definition 1.5. Let U be a sheaf on X. We will say that U is good if it is locally free over X\R and the natural map U→U^{**} is an isomorphism.

It is important for our aims that any notion of goodness for a sheaf, U, on X implies that U is locally free outside finitely many points. Another possibility instead of definition 1.5 would be to allow that U is not locally free at finitely many points of D_X\R; we choose 1.5 mainly because D_X may be disconnected and if it is disconnected one should fix not the total number of non locally free points on D_X\R, but the one of each component of D_X\R (hence notations and language is worst without any advantage in sight to the author).

The following proposition seems to be a key technical tool for everything in the non locally free case.

Proposition 1.6. *Let U be a good sheaf on X. Then $\pi^*(U)/Tors(\pi^*(U))$ is locally free on S.*

Proof. Set r:= rank(U). Set U':= U|D_X, T:= Tors(U') and U":= U'/T. Since D_X is a smooth curve, U" is locally free. By the degeneration on D_X of the spectral sequence connecting local *Ext* and global Ext, we have U' ≅ U"⊕T. Set V:= $\pi'^*(U)$, V':= $\pi'^*(U')$ and V":= $\pi'^*(U")$ (hence V' ≅ Tors(V')⊕V" and Tors(V') ≅ T); we will identify Tors(V') and T. Fix P∈R ≅ R'. Lift a basis of sections of U" near P (using π') to a basis τ of V" in an affine neighborhood of P in C_X. Then lift τ to a family τ' of r sections of $\pi^*(U)$ in an affine neighborhood of P in S. By the choice of τ and Nakayama lemma, τ' defines a rank r free summand of $\pi^*(U)$ in a neighborhood of P. Hence the thesis. ♦

§2. We assume as known the more elementary parts of the theory of stable vector bundles on a smooth algebraic surface. F. Takemoto ([Ta1], [Ta2]) and Z. Qin (see [Q1], [Q2], [Q3]) studied rank 2 vector bundles on a smooth surface which are stable with respect to a polarization but unstable with respect to another polarization. In this section we will extend their results to the case of rank 2 reflexive sheaves on a surface X with RDP as only singularities. First we need a definition (due to Tyurin for smooth surfaces) and two remarks on the Cayley - Bacharach property.

Fix a Weil divisor D of X and a quotient Z of $\mathcal{O}(D)$ with zero dimensional support; note that if D is not a Cartier divisor, Z may not be the structural sheaf of a zero dimensional subscheme of X; let $\mathbf{I}\{Z,D\}$ be the kernel of the quotient map $\mathcal{O}(D)\to Z$. Recall the following definitions due to Tyurin for smooth surfaces (hence D Cartier divisor) ([Ty], Def. 1.1 and Def. 1.2); however we change the name in the second one from D-stable to D-minimal, since stability here has a different meaning.

Definition 2.1. **(a)** Set $\delta(D,Z) := h^1(\mathbf{I}\{Z,D\}) - h^1(\mathcal{O}_X(D))$; $\delta(D,Z)$ is called the *index of D-speciality of Z.*

(b) Z is called *D-minimal* if $\delta(D,Z')<\delta(D,Z)$ for every proper quotient sheaf Z' of Z.

Note that $\mathrm{Ext}^1(\mathbf{I}\{Z,D\},K_X) \cong H^1(\mathbf{I}\{Z,D\})^*$. Hence, fixing a basis of $H^1(\mathbf{I}\{Z,D\})^*$ we obtain a unique extension

$$0\to H^1(\mathbf{I}\{Z,D\})\otimes K_X\to E(Z,D)\to \mathbf{I}\{Z,D\}\to 0 \qquad (1)$$

which defines (up to isomorphisms) the sheaf E(Z,D).

Remark 2.2. If D is a Cartier divisor the proof of [Ty], lemma 1.2, shows that E(Z,D) is reflexive if and only if Z is D-minimal. If D is not a Cartier divisor, using [AV], lemma 1.9, and the local to global spectral sequence for the Ext-functor and the duality in the Cohen - Macaulay case it is very easy to obtain a Cayley - Bacharach type result.

Remark 2.3. We will apply the Cayley - Bacharach property to construct sheaves (see theorem 2.7). In the statement of 2.7 we may take as, say, $\mathbf{I}(Z,c_1)$, the sheaf $\mathcal{I}_Z\otimes\mathcal{O}_X(c_1)$ with Z general set of points of X with large cardinality, a cardinality choosen so large to satisfy the following condition (which kills every Cayley - Bacharach type obstruction: "we have $h^0(\omega_X\otimes\mathcal{O}(c_1)\otimes\mathcal{I}_W) = 0$ for every $W\subset Z$ with card(W) = card(Z)-1 (or equivalently if Z is general: " $h^0(\omega_X\otimes\mathcal{O}(c_1)<\mathrm{card}(Z)$ "). In the (omitted) proof of 2.7 we will take as Z the disjoint union $Z'\cup Z''$ with Z' reduced, Z' general and with card(Z') so large to kill the Cayley - Bacharach obstruction just given and such that the proofs in [Z], [OG] and [GL] work for card(Z'), while Z" is chosen to satisfy the singularity type conditions as in [AV], lemma 1.9 (ii)).

Definition 2.4. Let B be the completion of the local ring at a singular point P of the normal surface Y; an isomorphism class \$ of reflexive rank r modules over B is called *admissible* (or *admissible for the germ (Y,P)*) if there is a reflexive sheaf F on Y with germ at P whose completion is in the class \$ (or equivalently if there is a reflexive module over $\mathcal{O}_{Y,P}$ whose completion is in the class \$).

This is certainly the right definition if the surface Y has only rational double points as singularities (hence if the corresponding ring B has only finitely many isomorphism classes of irreducible reflexive modules). However if for the corresponding complete ring B there are families of non isomorphic reflexive sheaves with the fixed rank, r, another possible definition would be to take as \$ a stratum of a "useful and meaningful" stratification of the set of all isomorphism classes of rank r reflexive sheaves over B.

Fix two polarizations R and H on X. Fix a rank 2 reflexive sheaf E on X; set $M := \det(E)$ (as reflexive rank 1 sheaf); M will denote also the associated Weil divisor; assume that E is R-stable but not H-stable (resp. properly H-unstable), i.e. assume the existence of an exact sequence

$$0 \to D \to E \to E/D \to 0 \qquad (2)$$

with D rank 1 and reflexive, E/D torsion free and, seeing D as Weil divisor, $2D \cdot H \geq M \cdot H$ (resp. $2D \cdot H > M \cdot H$); denote. The proof of [Q1], th. 1.2.5, works with no change in our situation, because it is based only on Hodge index theorem; hence we have the following key lemma:

Lemma 2.5. *Assume that E is R-stable but not H-stable with (2) as H-destabilizing sequence; let F be the rank 1 reflexive sheaf associated to the Weil divisor M-2D (and the corresponding Weil divisor); then we have* $R \cdot F < 0 < H \cdot F$ *and* $c_1(E)^2 - 4c_2(E) < 0$.

Note that for the last inequality in 2.5 it is not necessary to use Bogomolov theorem (hence there is no restriction on char(**K**) here).

We believe that lemma 2.5 shows that we have given the right set up. Now it is easy to translate the relevant definitions from the case of a smooth surface, and obtain with no effort the needed "working properties" of the definitions; with that "working properties" and with the right interpretation, it is very easy to check that almost all the proofs extend to the singular case. Then, the main general theorems on this topic (e.g. [Q1], ch. II, th. 2.2.5) are asymptotic theorems or "birational geometry theorems" and it is even easier to check their proofs using Cayley - Bacharach (i.e. Remark 2.2) (keeping in mind Remark 2.4 for the moduli spaces with fixed singularity type). Hence we will be very, very brief. Everything will be collected here in (2.6).

(2.6) Fix $c_1 \in \mathrm{Pic}(X) \otimes \mathbf{R}$, $c_2 \in \mathbf{Q}$, polarizations R, H and a numerical class $\zeta \in \mathrm{Num}(X) \otimes \mathbf{R}$ with $c_1^2 - 4c_2 \leq \zeta^2 < 0$; as in [Q2] and [Q1] let $\mathbf{C}_X \subseteq \mathrm{Num}(X) \otimes \mathbf{R}$ be the ample cone; W^ζ is the intersection of $\mathbf{C}_X$ with the orthogonal of ζ and $W(c_1, c_2)$ ("a wall") is the union of all W^ζ for every ζ with $c_1^2 - 4c_2 \leq \zeta^2 < 0$. A chamber of type (c_1, c_2) is a connected component of $\mathbf{C}_X \backslash W(c_1, c_2)$. Two polarizations are called equivalent for (c_1, c_2) if they have the same rank 2 stable reflexive sheaves with Chern classes c_i. One can also fix an admissible singularity type \$ for rank 2 reflexive sheaves on X and speak of equivalence of polarizations with respect to $(c_1, c_2, \$)$. Now the relevant parts of [Q1] (i.e. [Q1], chapter II, lemma 1.2.2, th. 1.2.3, prop. 1.2.5 (and this even if we do not fix the singularity type of sheaves)) work and gives [Q2],

Remark 3, i.e. that a chamber is formed by equivalent polarizations and that every equivalence class of polarizations is a union of chambers and possibly some ample Weil divisor lying on walls. The magic formula is the substitution of "invertible subsheaf" with "rank 1 reflexive subsheaf". At this point one has (for every fixed admissible singularity type) the following interesting results of [Q1], chapter II: prop. 1.3.1, Cor. 1.3.2 and Th. 1.3.3). Everything is ready for the following interesting "existence result" ([Q1], ch. II, th. 2.2.5) which we want to state explicitly.

Theorem 2.7. *Fix c_1, a polarization H and a singularity type \$ admissible for X. There is an integer $c = c(X,c_1,H)$ (c depending on X, c_1 and H) such that for all rational numbers $c_2 \geq c$ and for every numerical equivalence class ζ which defines a non empty wall, either $E_{\zeta}(c_1,c_2)$ or $E_{(-\zeta)}(c_1,c_2)$ is not empty.*

Indeed, to follow the proof of this theorem in [Q1] in our setting, use the Cayley - Bacharach type results 2.2 and 2.3.

§3. In this section we will consider in detail the case of rank 2 reflexive sheaves on a surface with only A_1 and A_2 singularities; furthermore we need to consider only the case in which $c_1(E)$ (with $E:= \pi^*(F)/\mathrm{Tors}(\pi^*(F))$ contains all the exceptional divisors exactly with multiplicity +1.
In the case of a singularity A_2 we will impose another restriction; remark 3.2 will show why it is relevant and (in the spirit of this reduction to smaller singularities) absolutely not restrictive for the applications here. Let {P(i)} (resp. {Q(j)}) be the set of singular points of type A_1 (resp. A_2) of X; usually an index i (resp. an index j) will run over the set of all singular points of type A_1 (resp. A_2) of X. We denote by Σ' (resp. Σ'') a sum which runs over the set of all singular points of type A_1 (resp. A_2) of X. Let E_i be the exceptional divisor over the singular point P(i) and $A_j \cup B_j$ the exceptional divisor over Q(j); usually n_i in a summation Σ' will denote something related to E_i, while n_{j1} (resp. n_{j2}) in a summation Σ'' will denote something related to A_j (resp. B_j).

Convention and Remark 3.1. Let D an effective divisor (with integer coefficients) on S; set $C':= \pi(D)$ (as Weil divisor with integer coefficient on X); let $C:= \pi^*(C')$ be the total transform of C' (see e.g. [S]); by definition $C \cdot E_i = C \cdot A_j = C \cdot B_j = 0$ for every i, j. We have $D = C + \Sigma' b_i E_i + \Sigma''(b_{j1}A_j + b_{j2}B_j)$ for some b_i, b_{j1}, b_{j2}. However, in general (i.e. for any normal surface) the coefficients b_i, b_{j1}, b_{j2} are only rational numbers; they are integers if (and only if in the case of rational singularities) C' is a Cartier divisor; in particular if the only singularities are RDP, one has a good bound for the denominators of these coefficients; if there are only A_1 and A_2 singularities, these coefficients have denominators which divide 6. We note explicitly that for every Weil divisor T of X, (j+1)T is a principal divisor in a neighborhood of every singularity of type A_j. We will call *standard pull-back family* the family of rank 2 vector bundles, E, on S obtained as $\pi^*(F)/\mathrm{Tors}(\pi^*(F))$ when F varies among a "prescribed" family of reflexive sheaves of fixed singularity type on X; here "prescribed" means "universal" if the corresponding moduli functor has a universal family, or may be "the family of all families" in the functorial sense for the moduli functor or a versal local family or ...

Remark 3.2. Let $C = T' \cup T''$ be a reducible, reduced curve with $p_a(C) = 0$ (i.e. a reducible conic). In [Da] it is proven that there are exactly two isomorphism classes, say E' and E", of rank 2 vector bundles on C whose restriction to each irreducible component is the direct sum of a line bundle of degree 0 and a line bundle of degree -1; they are distinguished by the fact that , say, $h^0(C,E') = h^1(C,E') = 0$, while $h^0(C,E'') = h^1(C,E'') = 1$. The second one may be deformed to the first one.

With the notations of the remark just given, we will assume that for every singularity of type A_2 of X, our vector bundle E on S restricts to a bundle like E' (i.e. with no section) to the exceptional reducible conic coming in the resolution of that singularity. Note that only this type of bundles arises as pull-backs modulo torsion of reflexive sheaves on X.

A rank 2 vector bundle E on S will be called *normalized* at a point $x \in \mathrm{Sing}(X)$ with respect to a suitable fixed class of formal isomorphism classes of rank 2 reflexive modules at x if there is a reflexive sheaf F on the germ (X,P) with $E \cong (\pi^*(F)/\mathrm{Tors}(\pi^*(F)))$ near $\pi^{-1}(P)$; under the assumptions of this section this means that $c_1(E)$ contains with multiplicity 1 every exceptional divisor over x and E is "balanced on D" (i.e. the most general one with that c_1 (see Remark 4.2 for the A_2 singularities).

We stress that in this section we will try to follow as much as possible [Br], §3. Set $H\hat{}:= nH-\Sigma' n_iE_i - \Sigma''(n_{j1}A_j+n_{j2}B_{j2})$. Note that obvious necessary conditions for the ampleness of $H\hat{}$ are:

$$n_i>0,\ n_{j1}>0,\ n_{j2}>0,\ n^2H^2>2(\Sigma' n_i^2)+\Sigma''(2n_{j1}^2+2n_{j2}^2-n_{j1}n_{j2}).$$

From Nakai's numerical ampleness criterion ([Ha], ch. V, th. 1.10) it is easy to find sufficient conditions for the ampleness of $H\hat{}$. Here is one of them (very rough, because we do not need a good one, we need only to fix one).

Lemma 3.3. If H is very ample, then $H\hat{}$ is ample if

$$n^2H^2>2(\Sigma' n_i^2)+\Sigma''(2n_{j1}^2+2n_{j2}^2-n_{j1}n_{j2}) \text{ and } n\geq 4\Sigma' n_i+6\Sigma''(n_{j1}+n_{j2}) \qquad (3)$$

Proof. We apply Nakai's criterion ([Ha], ch. V, th. 1.10). The first condition of (3) means that $H\hat{}^2>0$. Fix a curve D on S with D numerically equivalent to $C - \Sigma' a_iE_i - \Sigma''(a_{j1}A_j+a_{j2}B_j)$ and C total transform of an effective Weil divisor on X. Since H is a very ample Cartier divisor we have $2H\cdot C\geq 2\max_i(\mathrm{mult}_{P(i)}(C))\geq \max(a_i)$ and $3H\cdot C\geq 3\max_j(\mathrm{mult}_{Q(j)}(C))\geq \max_j(a_{j1}+a_{j2})$. ♦

Fix $c_1\hat{} \in \mathrm{Pic}(S)$ and the integer $c_2\hat{}$. Set $\Delta\hat{}:= 4c_2\hat{}-c_1\hat{}^2$ and set:

$$n\hat{} = \max((4\Sigma' n_i+6\Sigma''(n_{j1}+n_{j2}),9((\Sigma' n_i^2+\Sigma''(n_{j1}^2+ n_{j2}^2-n_{j1}n_{j2})(|\Delta\hat{}|+1/H^2)^{1/2}.$$

Theorem 3.4. Assume H very ample and $n>n\hat{}$. Let E be a rank 2 normalized vector bundle on S. If $\pi_*(E)^{**}$ is H-stable, then E is $H\hat{}$-stable.

Proof. Set $c_1\hat{} = \pi^*L+\Sigma' d_iE_i+\Sigma''(d_{j1}A_j+d_{j2}B_j)$. Take any saturated filtration

$$0\to F\hat{}\to E\to \mathcal{I}_Z(c_1\hat{}-F\hat{})\to 0 \qquad (4)$$

and set $F^\wedge := \pi^*F+\Sigma' b_iE_i+\Sigma''(b_{j1}A_j+b_{j2}B_j)$. Set $\delta := H(L-2F)$. By Hodge index theorem we have

$$\delta^2 \geq H^2(L-2F)^2 \qquad (5)$$

The relation given by (5) on Chern classes and the definition of $\Delta^\wedge$ gives:

$$(L-2F)^2 = 4\text{length}(Z) + 2\Sigma'(2b_i-d_i)^2 +$$
$$+\Sigma''(2(2b_{j1}-d_{j1})^2+2(2b_{j2}-d_{j2})^2-(2b_{j1}-d_{j1})(2b_{j2}-d_{j2})) - \Delta^\wedge \qquad (6)$$

Pushing down (4) we obtain a morphism from $\boldsymbol{O}(F)$ to $\pi_*(E)^{**}$; hence by the condition of H-stability we have $\delta>0$, i.e. by Remark 3.1 $\delta\geq 1/9$. By the choice of $n^\wedge$ and Schwarz inequality we get

$$n\delta>9(2(\Sigma'(2b_i-d_i))+(\Sigma''(2n_{j1}(2b_{j1}-d_{j1}-b_{j2})+2n_{j1}(2b_{j2}-d_{j2}-b_{j1})))$$

which implies that (4) does not $H^\wedge$-destabilize E. ♦

Corollary 3.5. *For every $n>n^\wedge$ the standard pull-back family is an open non empty subfunctor of the corresponding moduli functor.*

Proof. This follows from theorem 3.4, the description of the Mc Kay correspondence and the definitions given. ♦

References

[AV] M. Artin, J.-L. Verdier, *Reflexive modules over rational double points,* Math. Ann. **270** (1985), 79-82.

[Br] R. Brussee, *Stable bundles on blown up surfaces,* Math. Z. **205** (1990), 551-565.

[Da] M. Daoudi, *Fibrés bi-uniforms sur* $\boldsymbol{P}_2(\boldsymbol{C})$, J. Reine Angew. Math. **321** (1981), 25-35.

[GL] D. Gieseker, J. Li, *Irreducibility of moduli of rank two vector bundles on algebraic surfaces*, preprint.

[Ha] R. Hartshorne, *Algebraic Geometry,* Graduate Text in Math. **52**, Springer-Verlag, 1977.

[I] A. Ishii, *On the moduli of reflexive sheaves on a surface with rational double points,* Math. Ann. **294** (1992), 125-150.

[Ka] C. Kahn, *Reflexive moduln auf einfach-elliptischen flächensingularitäten,* Max-Planck-Institut preprint MPI 87-25.

[KSB] J. Kollar, N. I. Shepherd-Barrow, *Threefolds and deformations of surface singularities,* Invent. Math.**91** (1988), 299-338.

[Q1] Z. Qin, *Equivalence classes of polarizations and moduli spaces of stable locally free rank two sheaves,* J. Diff. Geometry **37** (1993), 397-415.

[Q2] Z. Qin, *Chamber structures of algebraic surfaces with Kodaira dimension zero and moduli spaces of stable rank two bundles,* Math. Z. **207** (1991), 121-136.

[Q3] Z. Qin, *Birational properties of moduli spaces of stable locally free rank-2 sheaves on algebraic surfaces,* Manuscripta Math. **72** (1991), 163-180.

[OG] K. O'Grady, *The irreducible components of moduli spaces of vector bundles on surfaces,* Invent. Math. **112** (1993), 585-617.

[S] F. Sakai, *Weil divisors on normal surfaces,* Duke Math. J. **51** (1984), 877-887.

[Ta1] F. Takemoto, *Stable vector bundles on algebraic surfaces,* Nagoya Math. J. **47** (1972), 29-48.

[Ta1] F. Takemoto, *Stable vector bundles on algebraic surfaces II,* Nagoya Math. J. **52** (1973), 173-195.

[T] C. Traverso, *Seminormality and Picard groups,* Ann. Scu. Norm. Pisa **75** (1970), 585-595.

[Ty] A. N. Tyurin, *Cycles, curves and vector bundles on an algebraic surface,* Duke Math. J. **54** (1987), 1-26.

[Z] K. Zuo, *Smoothness of the moduli of rank two stable bundles over an algebraic surface,* Math. Z. **207** (1991), 629-643.

Cohomology of Generic Rank 2 Stable Bundles on Fano 3-folds of Index 2

Edoardo Ballico - Rosa M. Miró -Roig
Dept. of Mathematics, University of Trento, 38050 Povo (TN), Italy
e-mail: ballico@itncisca.bitnet or ballico@itnvax.science.unitn.it
fax: Italy + 461881624
Facultat de Matemàtiques, Universitat de Barcelona, Gran Via 585, 08007 Barcelona, Spain
e-mail: miro@cerber.ub.es fax: Spain + 343 4021601

Here we are concerned with the cohomological properties of "generic" rank 2 stable bundles on a complex 3-fold, V. Here V will be usually a Fano 3-fold of index 2. We want to prove the existence (for almost all Chern classes c_i) of a rank 2 stable vector bundle E with $c_i(E) = c_i$ and natural cohomology in the following sense. Let H be the ample generator of Pic(V) (or a "natural" ample line bundle on V); set $E(t) := E \otimes H^{\otimes t}$; we will say that E has natural cohomology if for every integer t we have $h^i(V,E(t)) \neq 0$ for at most one integer i with $0 \leq i \leq 3$. Of course, by Serre duality, such a notion in this form has sense only because the canonical bundle K_V is rather negative. For 3-folds X with K_X "non-negative", very often $h^0(X,E(t)) \neq 0$ implies $h^3(X,E(t)) \neq 0$ and this definition must be modified (and even with the obvious modifications only weaker results seems to be true in general). Furthermore, if we take as E a rank 2 reflexive sheaf which is not locally free the "right" notion is not the notion of natural cohomology, but the following one of seminatural cohomology (see [HH2], §1, or [Hi]); let E be a rank 2 reflexive sheaf on a Fano 3-fold of index r; E is said to have seminatural cohomology if for every integer $t \geq -(r+1)/2$ we have $h^i(V,E(t)) \neq 0$ for at most one integer i with $0 \leq i \leq 3$.

Here are our main results. Let V_d (or V(d)) be a Fano 3-fold of index 2 and degree d (over the complex number field) (hence $1 \leq d \leq 7$). Let H be the ample line bundle such that $H^{\otimes(-2)} \cong K_{V(d)}$ (hence $d = H^3$); for any second Chern class, say τ, we will consider τ as the integer $\tau \cdot H$; this is a loss of information if $B_2(V_d) > 1$ (e.g. for d = 6 if $V_d = \mathbf{P}^1 \times \mathbf{P}^1 \times \mathbf{P}^1$). A rank 2 vector bundle E on V_d is called H-stable (in the sense of Mumford and Takemoto) if for every $L \in Pic(V_d)$ with $h^0(V_d, E \otimes L) \neq 0$ we have $c_1(E) \cdot H \cdot H + 2(L \cdot H \cdot H) > 0$. Let $M(V_d; c_1, c_2)$ be the moduli scheme of rank 2 H-stable vector bundles on V_d with Chern classes c_1 and c_2. We will consider only the case in which c_1 is a multiple, m, of H and take this integer m as c_1; up to a twist we may assume m = 0 or -1.

Theorem 0.1. *Assume $3 \le d \le 7$. Then for every integer $t \ge 67d+2$ there is $E \in M(V_d;0,t)$ with natural cohomology.*

Theorem 0.2. *Assume $3 \le d \le 6$. Then for every even integer $t \ge 532d+1$ there is $E \in M(V_d;-1,t)$ with natural cohomology.*

The restriction to even integers t in Theorem 0.2 is due to a well-known topological restriction (see 1.1) for rank 2 vector bundles on any 3-fold.

Theorems 0.1 and 0.2 will be proved respectively in §3 and §4 using the method introduced in [HH2] to solve the corresponding problem for $\mathbf{P}^3$. For a few words on this method (needed to fix the notations) see the beginning of §1. Our debt to the authors of [HH2] will be obvious to any reader. In our situation we need to use Iskovskih classification of Fano 3-folds of index 2 (see [I1], [I3], or [MM2], p. 106, or [Mu], §2). Indeed the geometry of V_d and of its smooth surface sections (Del Pezzo surfaces) and curve sections (with arithmetic genus 1) will play a big role. Among the curve sections, the more important ones for Theorem 0.2 are suitable reducible ones whose existence is proved in Remarks 2.2 and 2.3. This is the main reason why for odd c_1 for d = 7 we are able to prove only a far weaker result (see Proposition 7.7). For the case d = 1 and 2, see Remark 7.6. In § 5, just for completeness, we will consider the case of a smooth quadric 3-fold $Q \subset \mathbf{P}^4$ (i.e. the unique Fano 3-fold of index 3). In § 6 we will discuss briefly the case of rank 3 vector bundles. In § 7 we will consider (see Theorem 7.8) the existence on any Fano 3-fold of generically smooth components of the moduli scheme of stable bundles with the "expected" dimension. Such a problem (for very special Chern classes) on any 3-fold will be discussed in a different way in § 8.

A strong motivation for our search of rank 2 vector bundles on V_d with natural cohomology came from [BMR], Thm. 2.2, which gives (for any rank 2 stable bundle E on V_d) an integer m such that $h^0(V_d,E(m)) \neq 0$; the existence (for given V_d and Chern classes) of a rank 2 stable bundle with natural cohomology is equivalent to the sharpness of that non-vanishing theorem [BMR], Thm. 2.2 (for the same V_d and Chern classes).

The first named author was partially supported by MURST and GNSAGA of CNR (Italy); he would like to thank the Department of Mathematics of the University of Zaragoza for making our collaboration possible. The second named author was partially supported by DGICYT PB91-0231-C02-02.

This paper is in final form and no part of it will be published elsewhere.

1. Preliminaries.

We work over an algebraically closed field with characteristic zero. Let X be a projective variety on which one may want to prove the existence of rank 2 locally free sheaves with "good" cohomology. A rank 2 bundle may be associated to a suitable codimension 2 object, Z ("Serre construction"). A key idea contained in [HH2] is to use not only all possible Z's in X but similar objects contained in the total space of a suitable vector bundle on X. Here are more details (just enough to fix some notations). Let $\mathcal{O}_X(t)$, t integer, be the family of line bundles such that the cohomology of all $E(t):= E\otimes\mathcal{O}_X(t)$ measures the goodness of the cohomology of X. There is a small integer $b\geq 0$ with $h^0(X,\mathcal{O}_X(b)) \neq 0$ and with the following property; every $s\in H^0(X,\mathcal{O}_X(b))$, $s \neq 0$, induces a linear map $H^0(X,\mathcal{O}_X(t-b))\rightarrow H^0(X,\mathcal{O}_X(t))$ for every integer t; hence, composing with the addition s induces a linear map α_s: $H^0(X,\mathcal{O}_X(t))\oplus H^0(X,\mathcal{O}_X(t-b))\rightarrow H^0(X,\mathcal{O}_X(t))$; the existence of a rank 2 locally free sheaf with natural cohomology is reduced to the proof of the existence of a suitable codimension 2 object $A\subset X$ such that for every integer t the general linear map $\rho_A(t)$: $H^0(X,\mathcal{O}_X(t))\oplus H^0(X,\mathcal{O}_X(t-b))\rightarrow H^0(A,\mathcal{O}_A(t))$ obtained composing α_s for general s with the restiction map has maximal rank. In 2.8 we will state the integer b and the curve A needed in this paper to apply this key idea of [HH2]. Usually, it will be sufficient to check that $\rho_A(t)$ has maximal rank just for one "critical" integer t. In the rest of the paper the scheme $\mathbf{V}(\mathcal{O}_X(-b))$ is the total space (in the sense of Grothendieck) of the locally free sheaf $\mathcal{O}_X(-b)$. The condition on $\rho_A(t)$ may be translated into a condition on an object, A', inside $\mathbf{V}(\mathcal{O}_X(-b))$ which projects isomorphically onto A; even more (and this is a key point): by semicontinuity it is often sufficient to prove the existence of a suitable $A'\subset\mathbf{V}(\mathcal{O}_X(-b))$ to whom no A is associated. In the following sections to point out explicitely the integer b, the map $\rho_A(t)$ will be denoted with $\rho_A(t+(t-b))$. Of course, here there is X in the background; in the following 4 sections we will take as X a linear section of a Fano 3-fold, say V_d; if the linear section has dimension j (hence j = 1 or 2 or 3) we will write sometimes $\rho_A(j;t+(t-b))$ instead of just $\rho_A(t+(t-b))$ to avoid misunderstanding.

Since we want to prove the existence of suitable rank 2 vector bundles, it is worthwhile to point out the following topological restriction.

(1.1) Parity condition: Recall (see e.g. [BP]) that every rank 2 topological vector bundle E on a complex compact 3-fold, X, has Chern classes $c_i(E)$ satisfying the parity condition:

$$(c_1(E)+c_1(X))\cdot c_2(E) \equiv 0 \bmod(2) \qquad (1)$$

On a projective 3-fold (in any characteristic) the congruence relation (1) can be proved very easily in the following way; twist E by a large power of an ample line bundle; take (Bertini) a section of the twisted bundle whose zero locus, Z, is a smooth curve; by the adjunction formula the congruence (1) is equivalent to the fact that ω_Z has even degree. Recall also ([BP], p. 428) that when $c_1(E) \cong K_X$ there are two different topological rank 2 vector bundles with the same Chern classes (distinguished by the "Atiyah - Rees invariant") and which are K_X-symplectic.

2. The method.

Let $V_d, 1\leq d\leq 7$, (or V(d)) be a Fano 3-fold of index 2 and let H be the ample generator of $\mathrm{Pic}(V_d)$ (hence $2H \cong K_{V(d)}^{-1}$) with $d:= H^3$; see e.g. [Mu], §2.3, th. 2, for their classification; note that if d = 6 there are two different manifolds (a smooth divisor of type (1,1) of $\mathbf{P}^2\times\mathbf{P}^2$ and

$\mathbf{P}^1\times\mathbf{P}^1\times\mathbf{P}^1$) denoted with the same notation, V_6. Let S_d (or S(d)) be a smooth hyperplane section of V_d. Hence S_d is a Del Pezzo surface of degree d isomorphic to the blowing-up π: $S_d\to\mathbf{P}^2$ of $\mathbf{P}^2$ at 9-d sufficiently general points P_i, $1\le i\le 9$-d.

Remark 2.1. Note that by the adjunction formula we have $D^2+K_{S(d)}\cdot D = -2$, i.e. $D^2 = \deg(D)-2$. Hence every smooth rational curve D of S_d with $\deg(D)\ge 2$ moves in a positive dimensional family inside the fixed S_d.

Remark 2.2. We will check here that for every d with $3\le d\le 7$ every smooth S_d contains a reducible, reduced curve section $C_d = A\cup B$ with A and B smooth and rational and $\deg(A)\le\deg(B)\le\deg(A)+1$. If d = 7, take as A the strict transform of a line of $\mathbf{P}^2$ not passing through P_1 or P_2 and as B the strict transform of a smooth conic of $\mathbf{P}^2$ containing P_1 and P_2. If d = 6 take as A the strict transform of a line not passing through P_1 or P_2 and as B the strict transform of a smooth conic containing P_1, P_2 and P_3. If d = 5 take as B the strict transform of a line not passing through any P_i and as A the strict transform of a smooth conic containing every point P_i. If d = 4 take as A the strict transform of a line containing P_1 and as B the strict transform of a conic containing the other P_i's. If d = 3 take as A the strict transform of a line through P_1 and P_2 and as B the strict transform of a conic containing the other 4 points P_i's. Of course, there are many other possible decompositions $A\cup B$.

Remark 2.3. Here we will check that if $3\le d\le 6$ every smooth S_d contains a reduced curve section C_d which is union of conics and, for odd d, a line A (with respect to H). If $3\le d\le 4$, this was checked in 2.2. If d = 5 take as A the strict transform of a line of $\mathbf{P}^2$ containing P_3 and P_4 and as conics the strict transforms of lines D_i of $\mathbf{P}^2$, i = 1 and 2 with $P_i\in D_i$. If d = 6 take as conics the strict transforms of 3 lines of $\mathbf{P}^2$ each of them containing a different point P_i.

(2.4) Using Riemann - Roch and standard vanishing we obtain (in any characteristic) the following table of cohomology groups:

$h^0(C_d,\mathcal{O}(n)) = dn$ for every $n\ge 1$;

$h^0(S_d,\mathcal{O}(n)) = dn(n+1)/2 + 1$ for every $n\ge 0$;

$h^0(V_d,\mathcal{O}(n)) = dn(n+1)(n+2)/6 + (n+1)$ for every $n\ge -1$.

Remark 2.5. In S_d there are different families of smooth rational curves with the same degree (think about the 27 lines on a cubic surface). Hence often the parameter spaces appearing in the two-dimensional and one-dimensional assertions (as in M(1,n), M(2,n), A(2,n)) are not integral; however the proof of each assertion allows us to find one particular solution and this particular solution varies in an irreducible family; this is sufficient to obtain the corresponding assertion on V_d and here the parameter space is integral and the general method of [HH2] works (see the very beginning of the proof of [HH2], 7.1, for a place where this is stressed). However, if the goal is the construction of objects on S_d, it can be obtained in this way with only formal changes; varying a little bit the construction gives solutions which vary in different integral parameter spaces; usually, this is not dangerous.

(2.6) Construction EATP: ("Eating the points"). Here we will explain a well-known trick in the "Horace Market" and stress another reason to use it: often using it we will be able to bypass in the inductive procedure several unknown initial cases. This is extremely important because working with many varieties and many numerical data (not only in $\mathbf{P}^3$) the amount of work/ink to check these initial cases is huge; furthermore sometimes these initial cases are false (see e.g. on $\mathbf{P}^3$ the cases $c_1 = -1$, $c_2 = 2$ or 4) and this construction seems to be a simple, general recipe to bypass them. With EATP the input to start the induction may be a far weaker statement (sometimes even a trivially true one). Fix, say, $V := V_d$ and a hyperplane section S. Suppose you have (or you may construct) a reduced curve T in V with no component in S and a finite set, B, in S with $B \cap T = \emptyset$ and such that $A := T \cup B$ has a good cohomological property with respect to the integer n-1, say $\rho_A(3;n-1+(n-3))$ bijective. The goal is to find a nodal curve $U \subset S$ with $B \subseteq \mathrm{Sing}(U)$ and with good $\rho_U(2;n+(n-2))$; then for every $P \in B$ take a two dimensional smooth surface germ, Y_P, at P, not contained in S, and let $\chi(P)$ be the first infinitesimal neighborhood of P in Y_P; the union, Z, of T, U and these card(B) nilpotents has A as residual scheme with respect to S. We may use Z to apply "Horace". Usually it is very easy to smooth Z. In every case needed in this paper $T \cup U$ will be a nodal curve and the union,W, of U and the card(B) nilpotents will be the flat limit inside V of a flat family of curves $\{U_t\}$ in S with $p_a(U_t)+\mathrm{card}(B) = p_a(U)$; the smoothing of the nodal points of $T \cup U$ not in B is always proved very easily as in [HH2] or [Se], because the normal bundles in V of all the curves involved in the construction will have (for obvious reasons) strong properties of vanishing for H^1.

We stress again the use of this known construction to avoid to check initial cases before starting the inductive "Horace" procedure.

(2.7) (Intersection with a hyperplane section S_d). Several approaches are possible and past experience in $\mathbf{P}^N$ shows strongly that it is extremely useful to have many approaches, each one being tuned for a specific situation. In our situation an easy way is the following one. We are working in a range in which H is very ample and each curve T appearing in the construction is a deformation (with very, very low arithmetic genus) of union of parts of linear curve sections. Each curve section is embedded by H as a non degenerate curve, C, of degree d in $\mathbf{P}^{d-1}$; hence the intersection of C with a hyperplane of $\mathbf{P}^{d-1}$ is formed by d points in the open orbit of the corresponding symmetric product of $\mathbf{P}^{d-2}$; hence it is easy to control the postulation of $T \cap S_d$ (for sufficiently general T). For the case of lines and conics, we may be more specific. By deformation theory (and the adjunction formula) there is a 2-dimensional family of lines in V_d; hence for each point P of S_d with P not contained in one of the finitely many lines of S_d there is a line $D \subset V_d$ with $P \in D$ (hence intersecting S_d transversally and exactly at P). Again, by deformation theory, we see easily the following fact; fix $P \in S_d$; then there is an irreducible component, U, of the set of all conics of V_d not contained in S_d and such that the closure of the set $\{U \cap (S_d \backslash \{P\}))$ is a curve (indeed with high degree) of S_d. In section 7 we will need (implicitely) a similar control for any Fano 3-fold, hence with as general hyperplane section, S, a K3 surface; however we will need to control the postulation of the intesection with S of a union A of quasi-lines or quasi-conics or elliptic curves only with respect to the linear system associated to a very high multiple of S, say $|tS|$ with $t>\deg(A)$: a very trivial case.

Recall ([Hi], Def. 1.1) that a rank 2 reflexive sheaf F on a 3-fold V is called *convenable* if locally around each point, P, at which it is not locally free it is the cokernel of a morphism $\mathcal{O}_{V,P} \to \mathcal{O}_{V,P}^{\oplus 3}$ induced by 3 elements of $\mathcal{O}_{V,P}$ spanning the maximal ideal of $\mathcal{O}_{V,P}$.

A key input for the method of [HH2] is the following kind of results.

Proposition 2.8. *Fix integers d, b, g, s with $1 \le d \le 7$ and b = 0, 1 or 2. Assume the existence of a nodal curve $Y \subset V_d$ with $p_a(Y) = g$, $\deg(Y) = s$ and satisfying the following conditions:*

i) The normal bundle N_Y of Y in V_d has $H^1(N_Y) = 0$;

ii) $H^1(\mathcal{O}_Y(b)) = 0$;

iii) $H^1(\mathcal{I}_Y(b-2)) = 0$;

iv) $\omega_Y(2-b)$ has a section with at most simple zeroes and non vanishing on the nodes of Y;

v) if either b = 2 or g>0, then Y is connected; if b = 1 and g = 0, then the irreducible components of Y are smooth and rational;

vi) for a general $\beta: \mathcal{O}_{V(d)}(-b) \to \mathcal{O}_Y$ and for every integer n, the linear map

$$H^0(\alpha(n)): H^0(V_d, \mathcal{O}(n-b)) \oplus H^0(V_d, \mathcal{O}(n)) \to H^0(Y, \mathcal{O}_Y(n))$$

induced by $\alpha := (\beta, 1): \mathcal{O}(-b) \oplus \mathcal{O} \to \mathcal{O}_Y$, is of maximal rank.

Then $\mathcal{O} \oplus \mathcal{I}_Y(b)$ is a flat limit of a family of rank 2 convenable reflexive sheaves with Chern classes $c_1 = b$, $c_2 = s$, $c_3 = (2-b)s+2g-2$ and whose general member, E, has seminatural cohomology. In particular if either b = 2 and Y is elliptic or b = 1 and the connected components of Y are conics or b = 0 and the connected components of Y are lines, then E is a rank 2 vector bundle on V_d with natural cohomology.

Proof. The proof is exactly the same as the one in [Hi], prop. 2.1, (which in turn is based essentially only on [HH2], §2 and §4). However there is a very important difference between the statement of 2.8 here and the statement of [Hi], 2.1 (i.e. of $\mathbf{P}^3$); here we allowed the case b = 2 for connected Y, while the corresponding case in $\mathbf{P}^3$ (which would corresponds, with the notations of [Hi], 2.1, to the integer b = 4) was not allowed in [Hi], 2.1. However, the proof (i.e. the smoothing part [HH2], 4.1 and 4.2) works also in this case since for Y connected we have $H^1(V_d, \mathcal{I}_Y) = 0$ (or equivalently, the condition iii) may have a solution, Y). ♦

3. Proof of Theorem 0.1.

In this section we will always assume $c_1 = 2$ (hence b = 2). Furthermore we will assume $d \ge 3$.

By 2.4 we have the following values for $h^0(C_d,\mathcal{O}(n)\oplus\mathcal{O}(n-2))$: 1 if n = 0, d if n = 1, 2d+1 if n = 2, d(2n-2) for every n≥3. Note that for the corresponding h^1, we have $h^1 \neq 0$ if and only if n = 0, 1 or 2.

By 2.4 we have the following values for $h^0(S_d,\mathcal{O}(n)\oplus\mathcal{O}(n-2))$: 1 if n = 0, d+1 if n = 1, $d(n^2-n+1)+2$ for every n≥2.

By 2.4 we have $h^0(V_d,\mathcal{O}(n)\oplus\mathcal{O}(n-2)) = 2n + dn(n^2+2)/3$ for every n≥0.

(3.1) (The inductive assumptions). Motivated by the values just given we consider the following assertions; in general in an assertion of the type ***(j,n) (with some fixed j with 1≤j≤3) the index j is equal to the dimension of the linear section of V_d involved in the assertion, i.e. C_d if j = 1, S_d if j = 2 and V_d if j = 3; usually the subscript "d" is omitted.

A(1,n) = $A_d(1,n)$, 3≤d≤7, n≥3: There exists in $\mathbf{V}(\mathcal{O}_{C(d)}(-2))$ the union A of a reduced connected curve T with projection C_d, $p_a(T) = 0$, deg(T) = d, and d(n-2)-1 points with $\rho_A(1;n+n-2)$ bijective. Furthermore, we may find as solution such curves T which are integral (i.e. smooth) and as other solutions curves which have 2 irreducible components.

B(1,n) = $B_d(1,n)$, 3≤d≤7, n≥2: There exists in $\mathbf{V}(\mathcal{O}_{C(d)}(-2))$ the union A of a reduced connected curve T with projection C_d, $p_a(T) = 1$, deg(T) = d, and d(n-2) points with $\rho_A(1;n+n-2)$ bijective. Furthermore, we may find as solution such curves T which are integral (i.e. smooth) and as other solutions curves with only a node as singularity and exactly 2 irreducible components, each of them smooth.

A(1,2)", 3≤d≤7: There exists in $\mathbf{V}(\mathcal{O}_{C(d)}(-2))$ the union A of a reduced connected curve T with projection C_d, $p_a(T) = 0$, deg(T) = d-1, and d(n-2)-1 points with $\rho_A(1;n+n-2)$ bijective. Furthermore, we may find as solutions such curves T which are integral (i.e. smooth) and other solutions which have 2 irreducible components; note that if T is not irreducible, we have to take C_d reducible; this may be done by Remark 2.2.

A(2,n) = $A_d(2,n)$, 3≤d≤7, n≥2: There exists in S_d the union A of a reduced connected curve T with $p_a(T) = 0$, deg(T) = (d-1)+d(n-2) and dn-1 points with $\rho_A(2;n+(n-2))$ bijective. Furthermore, for every integer x with 1≤x≤max(1,n-3) we may find such a curve with exactly x irreducible components.

A(2,n)' = $A_d(2,n)'$, 3≤d≤7, n≥2: There exists in S_d the union A of a reduced connected curve T with $p_a(T) = 0$, deg(T) = (d-1)+d(n-2) and dn-1 points with $\rho_A(2;n+(n-2))$ bijective. Furthermore, for every integer x with 1≤x≤max(1,2n-3) if n ≡ 0, 2 mod(3), 1≤x≤2n-3+2kd if n = 3k. with k≥(d/2) we may find such a curve with exactly x irreducible components.

B(2,n) = $B_d(2,n)$, 3≤d≤7, n≥2: There exists in S_d the union A of a reduced connected curve T with $p_a(T) = 1$, deg(T) = d(n-1) and dn-1 points with $\rho_A(2;n+(n-2))$ bijective. Furthermore, for every integer x with 1≤x≤max(1,n-3) we may find such a curve with exactly x irreducible components.

$B(3,n) = B_d(3,n)$, $3 \le d \le 7$, $n \ge 2$. If $n \equiv 1$ or 2 mod(3), there exists in V_d the union A of a smooth connected curve T with $\deg(T) = d[(n^2+2)/3]$, $p_a(T) = 1$, and 2n points with $\rho_A(3,n+(n-2))$ bijective. If $n = 3k$ for some integer k, there exists in V_d the union A of a smooth connected curve T with $\deg(T) = d[(n^2+2)/3]$, $p_a(T) = 1$, and $2n+((2d/3)-[(2d/3)]n$ points with $\rho_A(3,n+(n-2))$ bijective.

Although we will not use it in this paper, it seems worthwhile to state the following assertion B(3,n)' which follows from the proof of B(3,n) (Lemma 3.9).

$B(3,n)' = B_d(3,n)$, $3 \le d \le 7$, $n \ge 7$. If $n \equiv 1$ or 2 mod(3), there exists in V_d the union A of a reduced, connected curve T with $\deg(T) = d[(n^2+2)/3]$, $p_a(T) = 1$, and 2n points with $\rho_A(3,n+(n-2))$ bijective; furthermore, for every integer x with $1 \le x \le 3n-3$ we may find such a curve T with only nodes as singularities with exactly x irreducible components, each of them smooth. If $n = 3k$ for some integer k, there exists in V_d the union A of an reduced, connected curve T with $\deg(T) = d[(n^2+2)/3]$, $p_a(T) = 1$, and $2n+((2d/3)-[(2d/3)]n$ points with $\rho_A(3,n+(n-2))$ bijective; furthermore, for every integer x with $1 \le x \le 3n-3$ we may find such a curve T with only nodes as singularities and with exactly x irreducible components, each of them smooth.

$H(3,n) = H_d(3,n)$, $3 \le d \le 7$, $n \ge 2$. If $n \equiv 1,2$ mod(3), there exists in V_d a smooth connected curve T with $p_a(T) = 1$, $\deg(T) = d[(n^2+2)/3]+2$, $p_a(T) = 1$, and with $\rho_T(3,n+(n-2))$ bijective. If $n = 3k$ for some integer k, there exists in V_d the union A of a smooth connected curve T with $p_a(T) = 1$, $\deg(T) = d[(n^2+2)/3]+2+[(2d/3]$ and $((2d/3)-[(2d/3)]n$ points with $\rho_A(3,n+(n-2))$ bijective.

Theorem 0.1 follows in a standard (and here omitted) way if we have proved H(3,n) for every $n \ge 14$.

The proof of the next 3 lemmas is easy and omitted; for a detailed explanation of the proof, see the proof of [HH2], prop. 7.1.

Lemma 3.2. *A(1,2)" is true.*

Lemma 3.3. *A(1,n) is true for every $n \ge 3$.*

Lemma 3.4. *B(1,n) is true for every $n \ge 2$.*

The next lemma is, up to inessential change of notations, just [HH2], lemma 6.2; anyway, its proof (just the 5-lemma) is a standard part in any weak form of Horace method and will be recalled in later parts of this paper just as "by Horace".

Lemma 3.5. *Fix $n \ge 3$; if A(2,n-1) and A(1,n) are true, then A(2,n) is true.*

Lemma 3.6. *A(2,3) is true.*

Proof. Use A(1,2)". ♦

Lemma 3.7. *B(2,n) is true for every $n \ge 3$.*

Proof. Just note how weak is B(2,3) and then apply Horace as in the previous lemma. ♦

Lemma 3.8. *A(2,n)' is true for every $n \ge 14$.*

Proof. In the inductive proof at each step increase the number of irreducible components. ♦

Lemma 3.9. *Assume $3\le d\le 7$ and $n\ge 5$. If A(2,n) and B(3,n-1) are true, then B(3,n) is true.*

Proof. If n is congruent to 1 or 2 modulo 3, this is plain Horace. If $n = 3k+3$ for some integer $k\ge 0$, we need to apply the EATP construction 2.6 to "eat" at most n-4 points. Hence we need to obtain from A(2,n-1) reducible curves with arithmetic genus 0 and such that the total number of their singular points is at least n-4; this is possible by definition of A(2,n-1) and the assumption $n\ge 5$; for the position of these singular points, see 2.7. ♦

Lemma 3.10. *H(3,n) is true for every $n\ge 14$.*

Proof. Use B(3,n-1), A(2,n)' and a very easy smoothing. To obtain H(3,n) from B(3,n-1) we need "to eat" 2(n-1) points if $n \equiv 0$ or 2 mod(3), while if $n = 3k+1$ we need "to eat" $2(n-1)+(n-1)((2d/3)-[(2d/3)])$ points. This is exactly the reason for the statement of A(2,n)'. ♦

4. Proof of Theorem 0.2.

In this section we will consider the case of odd c_1. Hence in the statement of 2.8 we will take $b = 1$ (i.e. $c_1 = 1$); unfortunately we are able (see 2.3) to find enough conics in a reducible curve section of V_d only if $3\le d\le 6$. Thus in theorem 0.2 we assumed $3\le d\le 6$. For the case $d = 7$, see Proposition 7.7 for a far weaker result; for V_d with $d = 1, 2$, see Remark 7.6.

We fix d with $3\le d\le 6$ and set $z:= [(d+1)/2]$. Let C+ (or C_d+) be a reduced but reducible curve section of V_d with z irreducible components; hence C+ consists of $[(d/2)]$ conics and, if d is odd, a line. The existence of C+ for $3\le d\le 6$ was proven in Remark 2.3. By 2.4 we have:

$h^0(C_d+,\mathcal{O}(n))\oplus h^0(C_d+,\mathcal{O}(n-1)) = d(2n-1)$ for every $n\ge 2$;

$h^0(S_d,\mathcal{O}(n))\oplus h^0(S_d,\mathcal{O}(n-1)) = dn^2 + 2$ for every $n\ge 1$;

$h^0(V_d,\mathcal{O}(n))\oplus h^0(V_d,\mathcal{O}(n-1)) = dn(n+1)(2n+1)/6 + (2n+2)$ for every $n\ge 0$.

Consider the following assertions M(j,n) (with j = 1,2 or 3):

M(1,n), $n\ge 2$. Fix an integer x with $0\le x<z$. Then there exists in $\mathbf{V}(\mathcal{O}_{C+}(-1))$ a reduced scheme A which is the disjoint union of a degree d curve, T, which projects surjectively onto C+, with z irreducible components (smooth conics and if d odd a line), with $p_a(T) = 0$, with exactly x nodes as singularities, and with d(n-1)-z+x points (with different projections contained in $C+_{reg}$), and such that $\rho_A(n+(n-1))$ is bijective.

Note that, since C+ is reducible, $\mathbf{V}(\mathcal{O}_{C(d)}(-2))$ is reducible and the points claimed to exist in M(1,n) do not form (a priori) an irreducible parameter space; however the bijectivity of ρ_A almost force the parameter space for their images in C+ (compare with Remark 2.5).

To state M(2,n) it is useful to introduce the following notations; set $a_d(2,n):= [(dn^2+2)/(2n+1)]$, $b_d(2,n):= dn^2+2-(2n+1)a_d(2,n)$, $a_d(3,n):= 2 + [dn(n+1)/6]$ and $b_d(3,n) = dn(n+1)(2n+1)/6 - (2n+1)a_d(3,n)$.

M(2,n), $n\ge 3$. There exists in S_d the disjoint union A of $a_d(2,n)$-4 conics and $8n+4+b_d(2,n)$ points with $\rho_A(n+(n-1))$ bijective.

The main difference with respect to the case of even c_1 arises at the level of the curve section, hence in M(1,n); now the contribution of the curve part with image C+ is higher than the contribution, $h^0(C_d,\mathcal{O}(n))+1$, arising in the even c_1 case; here if we have u connected components, the contribution is $h^0(C+,\mathcal{O}_{C+}(n))+u$; hence to have a chance, we need $h^0(C+,\mathcal{O}(n-b))\geq u$; fortunately, here we have b = 1. Also, in the inductive step from M(2,n-1) to M(2,n) (using a curve, T, for M(2,n-1) and a curve, U, for M(1,n)) several points of $T\cap C+$ are not on U; hence we need the presence of corresponding free conditions for $h^0(C+,\mathcal{O}(n)\oplus\mathcal{O}(n-1))$. We avoid these problems loosing 4 conics in the statement of M(2,n). Then the induction works, but of course the assertion corresponding to H(3,n) in the odd c_1 case is proven now only for integers n which are relatively large; we checked it for $n\geq 56$, the only problem being to have one initial case. In this "c_1 odd case" the numerology is more complicated; now it is not enough to consider nice cases for each congruence class modulo 3, but this is not a problem and a reason for loss of condition, since for the relatively high integers n involved here the number of conics or lines involved is much higher than 2n+1 (i.e. the contribution of a conic with respect to $\mathcal{O}(n)$).

5. Rank 2 vector bundles on the smooth quadric of $\mathbf{P}^4$.

In this section we will consider briefly (just to complete the picture) the far easier case of rank 2 bundles on a smooth quadric 3-fold $Q\subset\mathbf{P}^4$ (i.e. on the unique Fano 3-fold of index 3). We claim the following result.

Theorem 5.1. *Assume c_1+c_2 even. If either $c_1 = 1$ and $c_2\geq 26$ or $c_1 = 2$ and $c_2\geq 45$ there exists in the smooth quadric $Q\subset\mathbf{P}^4$ a rank 2 vector bundle E with $c_1(E) = c_1$, $c_2(E) = c_2$ and natural cohomology.*

The lower bounds for c_2 appearing in 5.1 are taken just to avoid some very easy check of initial cases; making these easy checks one can take c_2 very low, but the a priori optimal result depends on nasty initial cases (some of them may even be false, as in the case $c_1 = -1$ and $c_2 = 2$ or 4 on $\mathbf{P}^3$).

For a smooth quadric instead of 2.8 we use the following result.

Proposition 5.2. *Fix integers b, g, s with $0\leq b\leq 3$. Let Q be a smooth quadric hypersurface of $\mathbf{P}^4$. Assume the existence of a nodal curve $Y\subset Q$ with $p_a(Y) = g$, $deg(Y) = s$ and satisfying the following conditions :*

i) The normal bundle N_Y of Y in Q has $H^1(N_Y) = 0$;

ii) $H^1(\mathcal{O}_Y(b)) = 0$;

iii) $H^1(\mathcal{I}_Y(b-3)) = 0$;

iv) $\omega_Y(3-b)$ has a section with at most simple zeroes and non vanishing on the nodes of Y;

v) if either b = 3 or g>0, then Y is connected; if $b\leq 2$ and g = 0, then the irreducible components of Y are smooth and rational;

vi) for a general β: $\mathcal{O}_{V(d)}(-b)\to\mathcal{O}_Y$ and for every integer n, the linear map

$$H^0(\alpha(n)): H^0(Q,\mathcal{O}(n-b))\oplus H^0(Q,\mathcal{O}(n))\to H^0(Y,\mathcal{O}_Y(n))$$

induced by $\alpha := (\beta, 1)$: $\mathcal{O}(-b) \oplus \mathcal{O} \to \mathcal{O}_Y$, *is of maximal rank.*

Then $\mathcal{O} \oplus \mathcal{I}_Y(b)$ *is a flat limit of a family of rank 2 convenable reflexive sheaves with Chern classes* $c_1 = b$, $c_2 = s$, $c_3 = (3-b)s+2g-2$ *and whose general member, E, has seminatural cohomology. In particular if either* $b = 3$ *and Y is elliptic or* $b = 2$ *and the connected components of Y are conics or* $b = 1$ *and the connected components of Y are lines, then E is a rank 2 vector bundle on Q with natural cohomology.*

Let Q:= Q_3 be a smooth quadric hypersurface of $\mathbf{P}^4$, S:= Q_2 a smooth quadric surface, C a smooth conic and C+ a reduced but reducible conic. We have $h^0(C,\mathcal{O}(n)) = h^0(C+,\mathcal{O}(n)) = 2n+1$ for every $n \geq 0$, $h^0(S,\mathcal{O}(n)) = (n+1)^2$ for every $n \geq -1$ and $h^0(Q,\mathcal{O}(n)) = (n+2)(n+1)n/3 + (n+2)(n+1)/2$ for every $n \geq -2$.

First, we will consider the case b = 1 (hence we will use lines to construct bundles with natural cohomology). We have $h^0(C+,\mathcal{O}(n)) \oplus h^0(C+,\mathcal{O}(n-1)) = 4n$ for every $n \geq 1$, $h^0(S,\mathcal{O}(n)) \oplus h^0(S,\mathcal{O}(n-1)) = 2n^2+2$ for every $n \geq 0$ and $h^0(Q,\mathcal{O}(n)) \oplus h^0(Q,\mathcal{O}(n-1)) = (n+1)n(2n+1)/3$ for every $n \geq -1$. For any line D we have $h^0(D,\mathcal{O}(n)) = n+1$.

Consider the following assertions U(j,n), j = 1,2,3, and U(2,3)':

U(1,n), $n \geq 1$: There exists in $\mathbf{V}(\mathcal{O}_{C+}(-1))$ the union A of 2 disjoint lines (with projection on different lines of C+) and 2n-2 points (divided into two subsets of cardinality n-1, each of them with image contained in a different line of C+) with $\rho_A(n+(n-1))$ bijective.

U(2,n), $n \geq 1$: There exists in $\mathbf{V}(\mathcal{O}_S(-1))$ the disjoint union A of n-1 lines of type (1,0), n-1 lines of type (0,1) and 2n+3 points with $\rho_A(n+(n-1))$ bijective.

U(2,3)': There exists in $\mathbf{V}(\mathcal{O}_S(-1))$ the disjoint union A of a reducible conic, 2 lines of type (1,0), 2 lines of type (0,1) and a point with $\rho_A(3+2)$ bijective.

U(3,n), $n \geq 1$: There exists in Q the disjoint union A of $[n(2n+1)/3]$ lines and a set B with $B = \emptyset$ if $n \equiv 0$ or 1 mod(3), B formed by k+1 collinear points if n = 3k+2 for some integer k, and such that $\rho_A(n+(n-1))$ is bijective.

The next remark may be very useful (also for V_d and for similar results on any ambient variety); it can be used both as a tool to check initial nasty cases and as a tool to simplify a little bit the statements and/or the proofs of the general inductive assertions.

Remark 5.3. Using the irreducibility of the set of lines in Q, one can prove Theorem 5.1 for b = 1 using the following two assertions $U(3,n)_s$ and $U(3,n)_i$ instead of U(3,n). Look at the notations in U(3,n). Assertion $U(3,n)_s$ is exactly the same as U(3,n), except taking instead of A only the one dimensional part A' of A and asserting the surjectivity of the corresponding map for A'; in assertion $U(3,n)_i$ as scheme, A'', claimed to have $\rho_{A''}$ injective take A'' = A if $B = \emptyset$, A'' union of A and a general line if $B \neq \emptyset$.

Lemma 5.4. *U(1,n) is true for every* $n \geq 1$.

Lemma 5.5. *U(2,n) is true for every* $n \geq 1$.

Lemma 5.6. *U(2,3)' is true.*

Lemma 5.7. *U(3,n) is true for every $n \geq 1$.*

Proof. U(3,1) and U(3,2) are easy. As in §3 the inductive step is easy. The non trivial case is U(3,n) with $n \equiv 0$ mod(3) (and with low n). This is the only case with $n-1 = 3k+2$ for some integer k in which one have to check carefully (just because the numbers involved are very small) the existence of a "winning" configuration on S supporting k+1 nilpotents (as in the EATP construction 2.6) to apply Horace; indeed we stated explicitely U(2,3)'. ♦

Now assume b = 2. Here the numerology for the application of Horace in the 3-dimensional case is more complicated because the values $h^0(C,\mathcal{O}(n)) = 2n+1$ and $h^0(Q,\mathcal{O}(n))+h^0(Q,\mathcal{O}(n-2)) = 2n^2+2$ do not fit well. Anyway, the same outline works and we avoided to check a lot of initial cases claiming the statement of Theorem 5.1 only for $n \geq 7$; note that for c_1 odd we proved U(2,3)' (hence U(3,3)), but this was not claimed in the statement of 5.1, because if $n \geq 5$ the EATP construction 2.6 applies easily. The best strategy seems to use as much as possible reducible conics. To work on S one can use the statement of [HH2], 7.6, and in Q work always with reducible conics. Alternatively, one can use the proof of [HH2], 7.6, to obtain (with the notations of [HH2], 7.6, with $r_1 \leq r_2$) in $\mathbf{V}(\mathcal{O}_S(-1))$ a disjoint winning configuration of r_1 reducible conics, r_2-r_1 lines in the second family and the "right number" of points. When one applies Horace in Q some pair of lines, with one line outside S and one line inside S are linked to obtain reducible conics.

6. Existence of rank 3 vector bundles with seminatural cohomology.

In this section we will discuss briefly the case of rank 3 vector bundles (and reflexive sheaves). We can construct very easily a lot of rank 3 stable vector bundles on V_d with natural cohomology using the constructions of the previous sections and the following proposition.

Proposition 6.1. *Fix integers d, b, g, s with $1 \leq d \leq 7$ and b = 1 or 2. Assume the existence of a nodal curve $Y \subset V_d$ with $p_a(Y) = g$, $deg(Y) = s$ and satisfying the following conditions :*

i) The normal bundle N_Y of Y in V_d has $H^1(N_Y) = 0$;

ii) $H^1(\mathcal{O}_Y(b)) = 0$;

iii) $H^1(\mathcal{I}_Y(b-2)) = 0$;

iv) $\omega_Y(2-b)$ has two sections with at most simple zeroes and without common zeroes;

v) if either b = 2 or g>0, then Y is connected; if b = 1 and g = 0, then the irreducible components of Y are smooth and rational;

vi) for a general $\beta: \mathcal{O}_{V(d)}(-b) \to \mathcal{O}_Y$ and for every integer n, the linear map

$$H^0(\alpha(n)): H^0(V_d,\mathcal{O}(n-b)) \oplus H^0(V_d,\mathcal{O}(n))^{\oplus 2} \to H^0(Y,\mathcal{O}_Y(n))$$

induced by $\alpha := (\beta,1,1): \mathcal{O}(-b) \oplus \mathcal{O} \oplus \mathcal{O} \to \mathcal{O}_Y$, is of maximal rank.

Then $\mathcal{O} \oplus \mathcal{O} \oplus \mathcal{I}_Y(b)$ is a flat limit of a family of rank 3 stable vector bundles with Chern classes $c_1 = b$, $c_2 = s$, $c_3 = (2-b)s+2g-2$ and whose general member has seminatural cohomology.

Proof. In the rank 3 case the proof is exactly the same as in 2.8 (i.e. in [HH2], 4.1 and 4.2, and [Hi], 2.1) because in [HH2], 4.1 and 4.2, everything is reduced to a check on a sheaf $\mathcal{O}\oplus\mathcal{I}_Y(b)$ and the same proof applies with no change to the case $2\mathcal{O}\oplus\mathcal{I}_Y(b)$. Since $\mathrm{Pic}(V_d)$ is generated by H, the stability (in the sense of Mumford - Takemoto) of a rank 3 reflexive sheaf E can be checked just proving the vanishing of $h^0(E(t))$ and $h^0(E^*(m))$ for suitable t, m. ♦

The situation is easier than in the rank 2 case not only because we can use more curves, but also because at the curve section level, since we need a control for $\mathcal{O}(n-b)\oplus\mathcal{O}(n)^{\oplus 2}$ we can put on C two full sections (plus n-b+1 conditions), instead of only one. More precisely, under the assumptions on d of the previous sections for fixed c_1 and c_3 one can obtain bundles with good cohomology for every high c_2. Even easier is to prove the existence of a constant τ, a linear function u(t) and a function v(t) of order 1/2 such that for every $c_1 = 0, -1$ or -2, and every c_2 and c_3 with $c_2 \geq \tau$ and $v(c_2) \leq c_3 \leq u(c_2)$, there is a rank 3 vector bundle on V_d with Chern classes c_i. However, there are two compelling reasons for not doing in detail and full generality the rank 3 case:

(1) The absence of reasonable existence or non existence theorems for stable rank 3 bundles on V_d with given Chern classes;

(2) The fact that the components of the moduli scheme of stable bundle in which we can prove the existence of rank 3 bundles with good cohomology do not seem to be more important than any other component (if, as expected, there are several other components).

7. Existence of components in $M(V;2;c_1,c_2;K_V^{-1})$ with the expected dimension.

In this section we will consider the existence on a Fano manifold V of components of the moduli scheme $M(V;2;c_1,c_2;K_V^{-1})$ of rank 2 K_V^{-1}-stable vector bundles with Chern classes c_1, c_2 which are generically smooth and of the "expected" dimension. Here "expected" has the following meaning: "if E is a rank 2 vector bundle on V with Chern classes c_1 and c_2, by Riemann - Roch the expected dimension is $-\chi(\mathcal{E}nd(E))+1$"; note that if E is stable, then $h^0(\mathcal{E}nd(E)) = 1$, $h^3(\mathcal{E}nd(E)) = 0$; if $h^2(\mathcal{E}nd(E)) = 0$ then by standard deformation theory the moduli scheme is smooth and of dimension $h^1(\mathcal{E}nd(E)) = -\chi(\mathcal{E}nd(E))+1$ at E. Since $c_1(\mathcal{E}nd(E)) = c_3(\mathcal{E}nd(E)) = 0$ and $c_2(\mathcal{E}nd(E)) = 4c_2(E)-c_1(E)^2$, the "expected" dimension is easily computed using Riemann - Roch. The main result is stated as Theorem 7.8; here if $b_2(V)>1$ we will always take K_V^{-1} as polarization for the definition of stability and (unfortunately) we will allow as c_1 only multiples of K_V ; for a weaker statement for general c_1, see the "chat" in the next section. For the case of Fano manifolds of index two, see Proposition 7.7 for V_7 and Remark 7.6 for V_1 and V_2.

Remark 7.1. Fix $E\in M(X;2;c_1,c_2;H)$ (with no restriction on X and the polarization H) with $h^2(\mathcal{E}nd(E)) = 0$. Since $M(X;2;c_1,c_2;H)$ is smooth at E, E is contained in a unique component, Y, of $M(X;2;c_1,c_2;H)$. Again, by deformation theory and the fact that Y has the expected dimension, the scheme Y is everywhere locally a complete intersection (hence everywhere reduced).

Fix a rank 2 bundle E on a 3-fold V. To prove the vanishing of $h^2(End(E))$ one can do the following procedure. Fix $R\in Pic(V)$ such that $E\otimes R$ has a section, s, with zero locus $Z:= (s)_0$ of codimension 2. Hence s gives an exact sequence:

$$0\to E\otimes R^*\to E\otimes E\to \mathcal{I}_{Z,V}\otimes E\otimes(R\otimes c_1)\to 0 \qquad (2)$$

Since $E^* \cong E(-c_1)$, it is sufficient to check that $h^2(E\otimes(R\otimes c_1)^*) = h^2(E\otimes R\otimes \mathcal{I}_{Z,V}) = 0$; using Serre duality very often we have constructed E starting from Z (or as in [HH2] deforming a solution arising from a prescribed Z); for degree reasons very often $h^1(Z,E\otimes R|Z) = 0$. Hence by Serre duality we are reduced to prove vanishing of $h^1(V,E\otimes M)$ for suitable $M\in Pic(V)$. Sometimes, we may choose Z (e.g. several disjoint copies of elliptic curves) which gives this very weak vanishing. The general methods of [HH2] used in the previous sections greatly enlarge the range of possible solutions and gives much more precise vanishing results.

We fix a Fano manifold V; let $M(V;2;c_1,c_2;K_V^{-1})$ (or $M(V;c_1,c_2)$ or $M(c_1,c_2)$) be the moduli scheme of rank 2 K_V^{-1}-stable vector bundles on V with Chern classes c_1, c_2. Set $K:= K_V$.

In this section we will represent (except for Remark 7.6 and Proposition 7.7) any second Chern class c_2 as an integer $t:= -K\cdot c_2$; in particular if V has index $r>1$ and $B_2 = 1$, this integer is r times the degree of c_2 with respect to the ample generator of Pic(V); this convention conflicts with the one followed in the other sections of this paper if $r>1$, but this will not cause any confusion. It is standard to call quasi-line (resp. quasi-conic) an integral curve $C\subset V$ with $C\cdot K = -1$ (resp. -2); if -K is very ample these curves are just lines and conics with respect to the embedding induced by -K. By the adjunction formula and deformation theory if V contains a quasi-line (resp. a quasi-conic), then it contains at least a one dimensional (resp. a two-dimensional) family of quasi-lines (resp. quasi-conics).

Proposition 7.2. *Assume the existence on V of infinitely many pairwise disjoint quasi-lines with as normal bundle the direct sum of a line bundle of degree -1 and a line bundle of degree 0. Then for all odd integers $t\geq 10$ $M(K,t)$ contains a component which is generically smooth of the expected dimension.*

Proof. To apply as in the previous sections the methods of [HH2], use (in general singular) hyperplane sections of V containing a quasi-line and reducible curve sections containing a quasi-line. ♦

By the parity condition 1.1, the parity conditions in the statements of 7.1 and 7.2 are necessary condition for the existence of any topological bundle with that Chern classes.

Remark 7.3. It was proved in [Sh2] that (in characteristic 0) $\mathbf{P}^1\times\mathbf{P}^2$ is the only Fano manifold of index 1 without quasi-lines. The reader can easily check the existence on $\mathbf{P}^1\times\mathbf{P}^2$ of enough generically smooth components of the moduli schemes to prove Theorem 7.8 for this three-fold.

Fix an index 1 Fano 3-fold V with $B_2 = 1$ and K_V^{-1} very ample; the cases with K_V^{-1} not very ample are completely classified (see [I1], [I2] or [I3]): either K_V^{-1} has base points (and these V are described in [I1], 3.1 (a), (b)) or V is hyperellipitic (and classified in [I1], §7, (or see [MM2], p. 106)); by deformation theory V contains a one dimensional family of lines; assume also that V is not trigonal (the trigonal ones being described in [I2], §3, Thm. 3.4 at p. 779)); it was proved in [I2], §3 (or see [I3], ch. III, §2) that for the general line, D, the normal bundle has as factors a line bundle of degree -1 and a line bundle of degree 0 and that if V has genus $g \geq 4$ a general line intersects only finitely many lines ([I2], Thm. 3.4 (ii)). For the case g = 3 (i.e. any quartic hypersurface of $\mathbf{P}^4$) the existence of many conics, Proposition 7.5 and the sentence just after its statement may be very useful, since, if we fix a line D on such V, the general codimension 2 section of V will be the union of a line and a degree 3 smooth plane cubic. Also note that by [I2], Thm. 4.4, if V contains a line, then it contains a smooth conic.

Proposition 7.4. *Assume the existence on V of an arbitrarly large number of disjoint quasi-conics. Then for every even integer $t \geq 24$ M(0,t) contains a generically smooth component of the expected dimension.*

Proposition 7.5. *Fix an integer s. Assume that for every integer $a \geq s$ V contains an elliptic curve of degree a. Then there is an integer m(s) such that for every odd integer $t \geq m(s)$ M(K,t) contains a generically smooth component of the expected dimension.*

Usually, by the general smoothing techniques of [HH1] and/or [Se] if V contains an elliptic curve, I, of degree s and a line, then it is easy to show that for every integer $t \geq s$ it contains an elliptic curve of degree t; it is sufficient that the normal bundle of I in V is positive.

Remark 7.6. Note that by Proposition 7.2 and its proof (see in particular [HH2], statement of 4.2) the bundles with natural cohomology constructed in sections 3 and 4 belong to components of the corresponding moduli space which are generically smooth and of the expected dimension. Mimicking the proof of 7.2 or section 3 (see also the proof of Proposition 7.7 below) it is very easy to check for c_2 not very low the existence of stable bundles $E' \in M(V_d;2;-H,c_2)$ and $E'' \in M(V_d;2;0,c_2)$, d = 1 or 2, and such that the corresponding moduli spaces are smooth and of the expected dimension at E' and E".

Using the values of $h^0(V_7,\mathcal{O}(n))$ given in 2.4 we see the existence of a function $\phi(t)$ from the natural numbers into itself which grows at infinity as 7/3 and such every bundle $E \in M(V_7;2;-H,c_2)$ with natural cohomology has $h^1(V_7,E(n)) = 0$ for every $n \geq \phi(c_2)$. We are not able to prove the existence of such bundles with natural cohomology; here is what we can prove.

Proposition 7.7. *There is a function $\xi(t)$ with $\lim(\xi(t)/\phi(t)) = 7/6$ and such that for every even integer $c_2 > 0$ there is $E \in M(V_7;2;-H,c_2)$ with $h^1(V_7,E(n)) = 0$ for every integer $n \geq \xi(c_2)$.*

Proof. Fix a smooth hyperplane section $S = S_7$ of V_7 (hence S is isomorphic to $\mathbf{P}^2$ blown-up at two points P_1 and P_2). The strict transform in S of a line of $\mathbf{P}^2$ containing exactly one of the points P_i is a conic of V_7. Hence V_7 has a reducible curve section, C+, with 3 smooth rational curves as irreducible components, two of them being conics. We can mimick the construction of section 4 of bundles with odd c_1 and natural cohomology on V_d, $3 \leq d \leq 6$, (in particular assertion M(1,n)) but at each step adding as curves with support on C+ only two conics. At the end we obtain a "surjectivity" statement which by semicontinuity implies the vanishing of $h^1(V_7, E(m))$ for suitable bundles E and sufficiently large integers m. ♦

In summary, using also the last few lines of the next section for the case of a hyperelliptic V, we have the following result.

Theorem 7.8. *Let V be a Fano 3-fold. There is an integer t(V) such that for all integers t≥t(V) satisfying the parity condition 1.1 the moduli schemes M(0,t) and M(K,t) have a component which is generically smooth and of the expected dimension.*

8 Generalizations.

In this short section we will show how to find bundles with rather good cohomology (for certain Chern classes) on every 3-fold X, just using a fixed finite morphism f: $X \to \mathbf{P}^3$; the bundles are just pull-back of bundles on $\mathbf{P}^3$ (hence the restriction on the possible Chern classes). The stability is always with respect to the ample bundle L:= $f^*(\mathcal{O}(1))$. Set k:= deg(f). Since f is flat, $f_*(\mathcal{O}_X)$ is a rank k vector bundle which is an extension of a rank (k-1) vector bundle, E, on $\mathbf{P}^3$ and the trivial line bundle; at least in characteristic 0 (or if char(**K**)>k) this extension splits. Fix a stable rank s bundle F on $\mathbf{P}^3$ with s = 2 or 3 and with rather good cohomology (e.g. with natural cohomology, but much less is sufficient). Since s≤3, if Pic(X) is generated by L (e.g. for Fano 3-folds with B_2 = 1) to check the stability of G:= $f^*(F)$ it is sufficient to prove easy vanishing for $H^0(X, G \otimes L^t)$ and $H^0(X, G^* \otimes L^m)$ and suitable t, m. In general, vanishing of suitable $H^0(X, G \otimes L^t)$ prove smoothness and expected dimension for the local deformation space of G. By the projection formula, such a vanishing is equivalent to $H^0(\mathbf{P}^3, F(t))$ = $H^0(\mathbf{P}^3, F \otimes S^t(E))$ = 0. If E splits as direct sum of line bundles, these vanishing (and a rather good control of the cohomology of G) comes for free from the assumptions on the cohomology of F; this splitting arises in particular if f is an abelian covering (and char(**K**) = 0 or if the base field **K** has char(**K**)>k). Now assume that E is not a direct sum of line bundles; fix an integer y such that E(-y) is very negative; note that y depends only on f, not on any choice of a vector bundle, F, on $\mathbf{P}^3$. Hence if F has sufficiently high Chern classes, we still obtain $H^0(\mathbf{P}^3, F(t))$ = $H^0(\mathbf{P}^3, F \otimes S^t(E))$ = 0 for enough t to obtain smooth deformation spaces with the expected dimension. However, it seems hard to obtain in general extimates and statements which are both precise and non trivial.

Instead of $\mathbf{P}^3$ we may take another 3-fold on which we have proved the existence of bundles with good cohomology. Hence everything works for instance for hyperelliptic Fano 3-folds with B_2 = 1 (listed e.g. in [MM2], p. 106). If the group N(X) of numerical equivalence classes of line bundles is not **Z**, but the effective cone NE(X) is polyedral (e.g. for all Fano 3-folds), one can check again that a rank 2 bundle E is stable using suitable vanishing for

$H^0(X,E\otimes H^{\otimes t})$ using only a fixed ample line bundle H; of course one get conditions which are very far from being necessary condition for the stability and need much higher t than in the rank 1 case. For instance everything is fine (asymptotically) for hyperelliptic Fano 3-folds.

References

[BMR] E. Ballico, R. M. Miro'-Roig, Stable vector bundles on Fano 3-folds of index 2. preprint 1993.

[BP] C. Banica, M. Putinar, On complex vector bundles on projective threefolds. Invent. math. 88 (1987), 427-438.

[HH1] R. Hartshorne, A. Hirschowitz, Smoothing algebraic space curves. in: Algebraic Geometry - Sitgers 1983, pp. 98-131, Lect. Notes in Math. 1124, Springer-Verlag, 1984.

[HH2] R. Hartshorne, A. Hirschowitz, Cohomology of a general instanton bundle. Ann. Scient. Ec. Norm. Sup. 15 (1982), 365-390.

[Hi] A. Hirschowitz, Existence de faisceaux réflexifs de rang deux sur $\mathbf{P}^3$ à bonne cohomologie. Publ. Math. I.H.E.S. 66 (1987), 105-137.

[I1] V. A. Iskovskih, Fano 3-folds I. Math. USSR Izv. 11 (1977), 485-527.

[I2] V. A. Iskovskih, Fano 3-folds II. Math. USSR Izv. 12 (1978), 469-506.

[I3] V. A. Iskovskih, Anticanonical models of three-dimensional algebraic varieties. J. Soviet. Math. **13** (1980), 745-814.

[MM1] S. Mori, S. Mukai, Classification of Fano 3-folds with $B_2\geq 2$. Manuscripta Math. 36 (1981),147-162.

[MM2] S. Mori, S. Mukai, On Fano 3-folds with $B_2\geq 2$. in: Algebraic Varieties and Analytic Varieties, pp. 103-129, Advanced Studies in Pure Math. **1**, 1983.

[Mu] J. P. Murre, Classification of Fano threefolds according to Fano and Iskovskih. in: Algebraic Threefolds, Proceedings Varenna 1981, pp. 35-92, Lect. Notes in Math. 947, Springer-Verlag, 1982.

[Se] E. Sernesi, On the existence of certain families of curves. Invent. math. 75 (1984), 25-57.

[Sh1] V. V. Shokurov, Smoothness of the general anticanonical divisor a a Fano 3-fold. Math. USSR Izv. 14 (1980), 395-405.

[Sh2] V. V. Shokurov, The existence of lines on Fano threefolds. Math. USSR Izv. 15 (1980), 173-209.

Linear Series of Low Degree on Plane Curves and the Castelnuovo Function

Luisella CAIRE
Dipartimento di Matematica - Politecnico di Torino
Corso Duca degli Abruzzi, 24 - 10129 - Torino, Italy
e-mail:
LUISELLA@POLITO.IT. or
LUISELLA@ITOPOLI.BITNET

ABSTRACT - Looking at the behaviour of the Castelnuovo function of a particular 0-dimensional subscheme X of $\mathbb{P}^2$, we find conditions for the existence and the uniqueness of some linear series on a plane curve passing through X. In particular, we study plane curves of degree d and high geometric genus whose singularities are only nodes or ordinary cusps, or having only one multiple ordinary point, and we discuss on possible g^r_d's and g^r_{d-s}'s.

This paper is in its final form and no part of it will be submitted elsewhere.

1 - INTRODUCTION AND NOTATION

Let $C = C^g_d$ be an integral projective plane curve of degree d and geometric genus g; let $\tilde{C}$ be its normalization and $\nu: \tilde{C} \to C$ the normalization morphism. Let γ^2_d be the linear series on $\tilde{C}$ "cut out by the lines of $\mathbb{P}^2$".

It is a classical problem to find conditions in order that the γ^2_d be the unique g^2_d on $\tilde{C}$; this is interesting also because the uniqueness of the g^2_d implies the uniqueness of the plane model of degree d of $\tilde{C}$ (up to collineations).

If $d \geq 4$ and C is smooth, it is well known that the γ^2_d is the unique complete g^2_d, while it is very easy to find other g^2_d's if C has "many singular points".

One problem we examine here is to find a bound on the singularities of C in order to have the uniqueness of the γ^2_d.

Among the classical geometers who first met the problem (and gave a proof for smooth curves), we recall Küpper [11], Wiman [13], Snyder [12]; the first one used the theory of adjoints, Wiman treated "by hands" the cases of small degrees, the last one considered sections of a certain ruled surface.

More recently, Accola [1], Ciliberto [2], Coppens and Kato [3], have studied the problem (and put it in a wider framework).

The nature of our method is different from that used by the quoted Authors; it is based on some properties of the Castelnuovo-Hilbert function of certain zero-dimensional subschemes of $\mathbb{P}^2$, and developes in a natural way some techniques introduced in [8].

Some of the results we find are known (they can be found in particular in [3]); but the method we use is elementary and put the problem in the natural environment of the theory of adjoints; moreover, we think it could be useful for other applications.

Let us recall some definitions and fix some notation.

Let $X \subset \mathbb{P}^2$ be a 0-dimensional closed subscheme of degree deg(X), $\mathcal{I}_X$ the ideal sheaf associated to X.

We denote, as usual by H(X,t) the Hilbert function of X

$H(X,t) := h^o(\mathbb{P}^2, \mathcal{O}_{\mathbb{P}^2}(t)) - h^o(\mathbb{P}^2, \mathcal{I}_X(t))$

and by $\Delta H(X,t)$ the first difference function

$\Delta H(X,t) := H(X,t) - H(X,t-1)$.

This function is also called the Castelnuovo-Hilbert function, or simply the function of Castelnuovo, who used it to study the generic plane section of a curve of $\mathbb{P}^3$ and first recognized its relevancy for finding linear systems of curves passing through X.

For general facts on the properties of the Castelnuovo function, we refer to [4] and to [7], and we shall use the same notation.

We work in $\mathbb{P}^2 = \mathbb{P}_k$, where $k = \bar{k}$ and char(k) = 0.

As usual in the theory of adjoints, we indicate by Δ the "divisor of double points" on $\tilde{C}$. We have (by the Theorem of Apéry, Gorenstein, Samuel), $\deg(\Delta) = 2\delta$, where

$$\delta = \sum_{P \in C} \dim_k(\mathcal{O}_{P,C} / \mathcal{C}_P) = p - g$$

($\mathcal{C}_P$ is the conductor of $\mathcal{O}_{C,P}$ in $\bar{\mathcal{O}}_{C,P}$ and p is the arithmetic genus of C).

If B is the 0-dimensional subscheme of C corresponding to the conductor sheaf, we recall (see, e.g., [9]) that

$$\deg(B) = \delta = \sum_{P \in C} \frac{e_P(e_P - 1)}{2}$$

where e_P denotes the multiplicity of P and the sum is extended to all points of C, actual and infinitely near.

B is in a natural way also a subscheme of $\mathbb{P}^2$ (of the same degree).

2 - LINEAR SERIES OF DEGREE d ON A C_d

Let $C = C_d^g$ ($d \geq 4$) be a plane projective integral curve; let g_n^r ($r>0$) be a complete linear series without base points on $\tilde{C}$.

Let $D \in g_n^r$ be a divisor consisting of n distinct points (we can find such a D, recalling a Theorem of Bertini) and such that $\nu(\mathrm{supp}(D)) \subset \mathrm{reg}(C)$.

Let D' be the 0-dimensional subscheme of $\mathbb{P}^2$ "corresponding to D", that is: if

$$D = \sum_{n_P} n_P P$$

D' is the subscheme of C corresponding to the ideal sheaf $\mathcal{I}$ such that $\mathcal{I}_P = m_P^{n_P}$.

Since $D' \subset C \subset \mathbb{P}^2$, D' becomes in a standard way also a 0-dimensional subscheme of $\mathbb{P}^2$.

In what follows, we shall replace D' by D, just to make the notation lighter.

Let B be the subscheme of $\mathbb{P}^2$ corresponding to the conductor sheaf of $\mathcal{O}_C$.

Finally, let $X = D \cup B$: we shall study the behaviour of the Castelnuovo function $\Delta H(X,t)$.

We have:

$\deg(X) = n+\delta$.

We put, as in [5],

$$\sigma := \min \{t\colon \Delta H(X,t)=0\}.$$

We start by stating a lemma which links the dimension of our linear series g_n^r with the Castelnuovo function of the 0-dimensional subscheme $X = D \cup B$ of $\mathbb{P}^2$.

2.1 - LEMMA -
$$r = \sum_{t=d-2}^{\sigma-1} \Delta H(X,t) \qquad \text{(i)}$$

In particular,

$\forall n,\ 0<n<d$
$$r = \Delta H(X,d-2) \qquad \text{(ii)}$$

$\forall n,\ md \le n < (m+1)d\ (m \in \mathbb{N})$
$$r = \sum_{t=d-2}^{d+m-2} \Delta H(X,t) \qquad \text{(iii)}$$

PROOF - Apply the Riemann-Roch Theorem to the linear series $|D| = g_n^r$:

$$r = n-g+i(D) = n-\left(\binom{d-1}{2} - \delta\right) + \left(h^o(\mathbb{P}^2, \mathcal{O}_{\mathbb{P}^2}(d-3)) - H(X,d-3)\right) =$$
$$= n + \delta - H(X,d-3) = \deg(X) - H(X,d-3).$$ Furthermore:

$$H(X,d-3) = \sum_{t=0}^{d-3} \Delta H(X,t) = \deg(X) - \sum_{t=d-2}^{\sigma-1} \Delta H(X,t)$$

whence (i).

We now recall (see Cor. 2.5 in [6]) that:

$$\sigma-1 \le \frac{\deg(X)-\tau}{d} + d-2$$

where $\tau = \tau(\mathcal{J}) = \sum_{P \in X} \tau(\mathcal{J}_P)$ ($\mathcal{J}$ is the ideal sheaf of X in $\mathcal{O}_C$ and $\tau(\mathcal{J}_P)$ is defined in [6]). By Lemma 4.1 of [6], we have $\tau=\delta$ and $\sigma-1 \le (n/d)+d-2$. From this we have (ii) and (iii). ■

We now state a Lemma concerning a numerical property of the Castelnuovo function.

2.2 - LEMMA - Let $d \in \mathbb{N}$, $d \ge 6$, and let $a = E(\frac{d}{2})$.

Let X be a 0-dimensional subscheme of $\mathbb{P}^2$ such that

$$\deg(X) < (d-a)(a+1)-1$$

If $\Delta H(X,d-2) = h \ge 2$, then:

a) $\exists\ m,k \in \mathbb{N}$, $2 \le m \le k < \frac{d}{2}$, such that

$$\Delta H(X,d-m) = \Delta H(X,d-(m+1)) = k$$

b) If m is the minimum for which a) is verified, then there exists a curve γ of degree k such that

$$\deg(X \cap \gamma) \ge \frac{2kd+(m-2)(2h+m-3)-k(2m+k-3)}{2} := b.$$

PROOF - a) Take the proof of Lemma 1 of [8] and make in it the slight modification that $\Delta H(X,d\text{-}2)\geq 2$, in the place of $\Delta H(X,d\text{-}2)\geq 1$; this forces k to satisfy the inequality $2 \leq k < \frac{d}{2}$, instead of $2 \leq k < a$.

If we take as m the minimum value for which $\Delta H(X,d\text{-}m)= \Delta H(X,d\text{-}(m+1))=k$, we have that $m \leq k$.

b) By assumption, in the graph of $\Delta H(X,t)$ we have a "step of height k" for t=d-m, $2 \leq m < \frac{d}{2}$, and this is "the first step we meet from right to left", that is

$$\Delta H(X,d\text{-}m) = \Delta H(X,d\text{-}m\text{-}1) = k$$

and

$$\forall q, \quad 2<q\leq m, \qquad \Delta H(X,d\text{-}q) > \Delta H(X,d\text{-}q+1)$$

So, $\forall q$, $2\leq q\leq m$, $\Delta H(X,d\text{-}q) \geq h+q-2$.

Then $\Delta H(X,t)$ is greater than or equal to the function having the following graph:

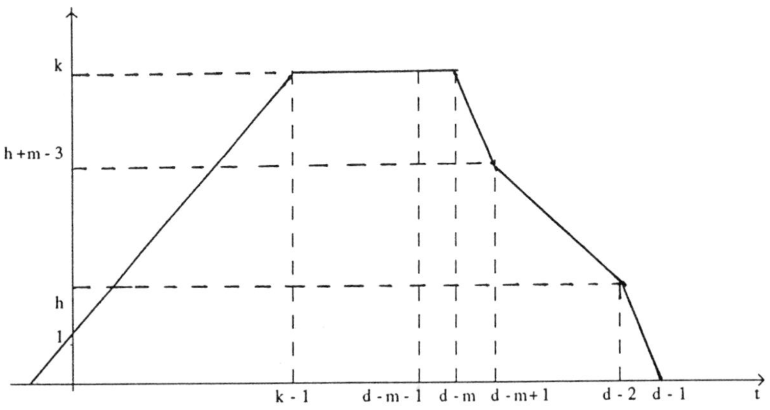

The step of height k means that (see [4] 2.2, 2.3, or [7] 3.4) there is a curve γ of degree k such that, if $X'=X\cap\gamma$, we have that deg(X') is greater or equal "to the area below the above graph", namely:

$$\deg(X') := b' \geq (1+2+...+(k-1))+k(d-m-k+2)+((h+m-3)+...+h) =$$

$$= \frac{2kd+(m-2)(2h+m-3)-k(2m+k-3)}{2} := b \qquad \blacksquare$$

Now we assume that C_d *(d ≥ 8) has only* δ *nodes or ordinary cusps as singularities,* and we use the previous notation.

2.3 - <u>THEOREM</u> - If $\delta < (d-a)(a+1)-d-1$, then:

a) $r \leq 2$

b) If $r=2$, our g^2_d coincides with the γ^2_d .

<u>PROOF</u> - a) In our hypothesis, we have $\deg(X) < (d-a)(a+1)-1$. By Lemma 2.2, if $\Delta H(X,d-2) = h \geq 2$, then there is a curve γ of degree k such that, if $X' = X \cap \gamma$, $\deg(X') := b' \geq b$.

Put $X'=Y' \cup Z'$, where $Y' \subseteq D$, $Z' \subseteq B$; let be $\deg(Y')=d' \leq d$ (and then $\deg(Z')=b'-d'$).

Since Y' consists of simple points and Z' consists of nodes (or ordinary cusps) of C, it must be

$$\sum_{P \in X'} i(P,\gamma \cap C) \geq d'+2(b'-d') = 2b'-d' \geq 2b-d =$$
$$= (2k-1)d+(m-2)(2h+m-3)-k(2m+k-3)$$

A long, but straightforward calculation shows that, since by 2.2, $2 \leq m \leq k < \frac{d}{2}$, then, $\forall d \geq 8$,

$$(2k-1)d+(m-2)(2h+m-3)-k(2m+k-3) > kd = \sum_{P \in C} i(P,\gamma \cap C)$$

and this contradicts Bézout theorem.

Then $\Delta H(X,d-2) \leq 1$, and also $\Delta H(X,d-1) \leq 1$; so, by Lemma 2.1, $r \leq 2$.

b) If $r=2$, we have $\Delta H(X,d-2)=\Delta H(X,d-1)=1$.

Then there is a straight line r containing a subscheme $X' \subseteq X$, and $\deg(X')= \sum_{t=0}^{d-1} 1 = d$ (see, e.g., [7]).

All the points of X' must belong to D, otherwise we should contradict Bézout theorem. So $X' \equiv D$ and $g^2_d \equiv \gamma^2_d$. ∎

2.4 - REMARK

Let us observe that, if $\Delta H(X,d-2)=2$, we cannot have $\Delta H(X,d-1)=2$, because, by reasoning as before on the step of height 2 and on the subscheme $X' \subseteq X$ belonging to the conic γ we should have, we again find $\deg(X') \geq 2d$, that is, a contradiction with Bézout theorem.

Then, if $\Delta H(X,d-2)=2$, we must have $\Delta H(X,d-1) \leq 1$ and $r \leq 3$.

2.5 - REMARK

Observe that the number $\delta_1 = (a+1)(d-a)-d-1$ is one less than the number δ_0 of nodes (or ordinary cusps) that a plane curve of degree d and genus π_o can have (we are using the notation of [2]: $\pi_o = g(d,3)$ is the genus of the Castelnuovo curve of degree d in $\mathbb{P}^3$). Then a plane curve C with $\delta < \delta_1 < \delta_0$ nodes cannot be the projection of a space curve of the same degree, and the γ_d^2 is a complete linear series on $\tilde{C}$.

Let us consider now a C_d $(d \geq 4)$ *having as singularities only one ordinary point of multiplicity* e.

As before, *we consider a complete linear series* g_d^r $(r>0)$ *without any base point on* $\tilde{C}$, *a divisor* $D \in g_d^r$, *and* $X = D \cup B$.

In this case the study of the Castelnuovo function $\Delta H(X,t)$ is easier than before, because the graph of $\Delta H(B,t)$ must be a part of it (see [4]), and it is easy to see that:

$\Delta H(B,t)=t+1$, if $0 \leq t \leq e-2$, and $\Delta H(\Delta,t)=0$, if $t \geq e-1$.

So, supposing $\Delta H(X,d-2)=h \geq 2$, the graph of $\Delta H(X,t)$ must be "higher" (in the sense stated before) than this one:

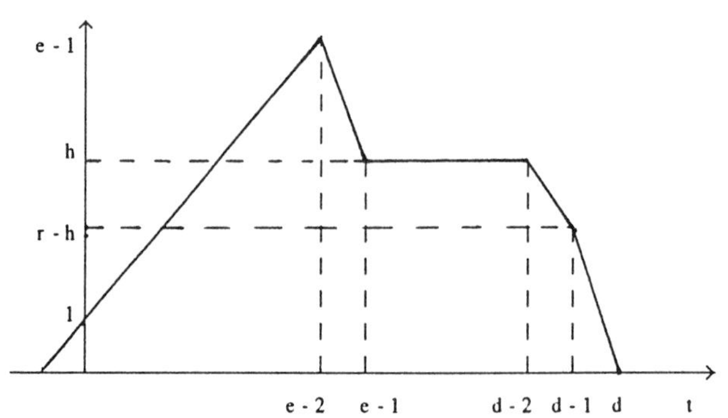

Then $\deg(X) \geq \frac{e(e-1)}{2} + h(d-e) + r - h$; since $\deg(X) = d + \frac{e(e-1)}{2}$, we shall have

$$d \geq h(d-e-1) + r \geq 2(d-e-1) + 2 = 2d - 2e$$

that is $e \geq d-a$.

So we get the following:

2.6 - PROPOSITION - If C_d has only one multiple ordinary point of multiplicity $e < d-a$, then:

a) $r \leq 2$

b) If $r=2$, our g^2_d coincides with the γ^2_d .

PROOF - a) We have $1 \geq \Delta H(X,d-2) \geq \Delta H(X,d-1)$; apply Lemma 2.1.

b) Here it is $\Delta H(X,d-2) = \Delta H(X,d-1) = 1$; so there is a line r containing a subscheme $X' \subseteq X$, with $\deg(X') = d$.

Suppose $X' = Y' \cup Z'$, where $Y' \subset D$, $Z' \subseteq B$, $Z' \neq \emptyset$; then we must have $\deg(Z') = e-1$ (and so $\deg(Y') = d-e+1$). So, we get a contradiction with Bézout theorem, since

$$\sum_{P \in X'} i(P, r \cap C) \geq (d-e+1) + e = d+1.$$

Then X' contains only points of D, and $|D| = \gamma^2_d$. ∎

2.7 - REMARK

If d is even and $e = \frac{d}{2}$, the conics, which are the union of couples of lines through the multiple point M, cut out on $\tilde{C}$ a g^2_d , composed with the γ^1_e of lines through M; (if d is odd and $e = \frac{d+1}{2}$, the same conics gives a g^2_{d-1}).

So, the number $e = \frac{d}{2}$ seems to be a good limit (at least in the even case) if we consider (as we always do in this paper) also composed g^r_d 's. If not, the limit can grow (see, e.g. the discussion in [12]).

2.8 - REMARK (historical)

In 1892 C. Küpper found the following result (see [11]):

" A plane C_d^g on which there is a g_d^2 (different from the γ_d^2 of lines) cannot have less than $\delta_1=(x_1-1)(d-x_1)-1$ double points, where $x_1 = E(\frac{2d}{3})$ ".

The proof is based on the fact that the g_d^2 must be cut out from a net of C_x's, where $\frac{d}{2} \leq x<d-3$, and that a net of C_x cannot have more than x^2-x+1 base points (as he showed in [10]).

Even if it contains some blanks or slight numerical imprecision, the proof is very interesting and significant (over all in the cases he examinates "by hands" making some wonderful geometrical projective construction).

In 1908 V. Snyder (in [12]) publishes the same result, with the same proof: he translates it in english, omitting some calculations, and using a lighter mathematical language; he takes some of the elegant geometrical examples and adds some very interesting other ones of his own. He knows Küpper papers, but he quotates only two intermediate results (one is the number x^2-x+1 seen above) and he annotates:

"This inequality is then proved, but a large number of errors have made the applications there made of it of small interest".

3 - LINEAR SERIES OF DEGREE SMALLER THAN d ON A C_d

We shall study now the complete, base point free, linear series g_{d-s}^r (0<s<d, r>0) on a C_d with only ordinary multiple points (or ordinary cusps).

We know by Lemma 2.1 that, if $D\in g_{d-s}^r$ and $X=D\cup B$,

- $\Delta H(X,d-1)=0$.
- $r=\Delta H(X,d-2)$

We have the following Lemma, which is analogous to Lemma 2.2.

3.1 - <u>LEMMA</u> - Let $d \in \mathbb{N}$, $d \geq 7$ and let $a = E(\frac{d}{2})$.

Let X be a 0-dimensional subscheme of $\mathbb{P}^2$ such that

$$\deg(X) < a(d-a)$$

If $\Delta H(X,d-3) = h \geq 2$, and $\Delta H(X,d-2) > 0$, then:

a) $\exists\ m \in \mathbb{N},\ 3 \leq m \leq a,\quad \exists\ k \in \mathbb{N},\ 2 \leq k < a$ such that

$$\Delta H(X,d-m-1) = \Delta H(X,d-m) = k$$

b) If m is the minimum for which a) is verified, then there exists a curve γ of degree k such that

$$\deg(X \cap \gamma) \geq \text{б} := \frac{2kd+(m-3)(2h+m-4)-k(2m+k-3)}{2} + 1$$

<u>PROOF</u> - a) As in Lemma 1 of [8], taking in account the fact that $\Delta H(X,d-2) > 0$. If we take as m the minimum value for which $\Delta H(X,d-m)=\Delta H(X,d-(m+1))=k$, we have that $m-1 \leq k$, and that $3 \leq m \leq a$.

b) Since a), in the graph of $\Delta H(X,t)$ we have a "step of height k" for $t = d-m$, and this is "the first step we meet from right to left", that is

$\Delta H(X,d-m-1) = \Delta H(X,d-m) = k$ and

$\forall q,\ 3<q\leq m,\qquad \Delta H(X,d-q) > \Delta H(X,d-q+1)$

So, $\forall q,\ 2<q\leq m,\ \Delta H(X,d-q) \geq h+q-3$.

Then the graph of $\Delta H(X,t)$ is higher than the graph below:

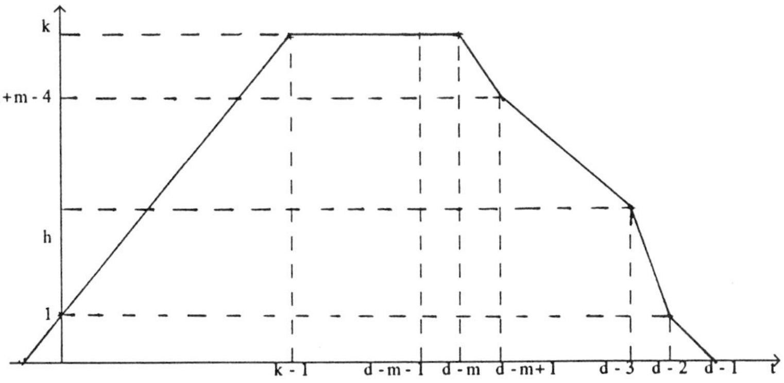

The step of height k again means that there is a curve γ of degree k such that, if $X' = X \cap \gamma$, we have:

$$\deg(X') := \text{б}' \geq (1+2+\ldots+(k-1))+k(d-m-k+2)+((h+m-4)+\ldots+h)+1 =$$

$$= \frac{2kd+(m-3)(2h+m-4)-k(2m+k-3)}{2} + 1 := \text{б} \ \blacksquare$$

Now we assume that C_d ($d \geq 7$) has only δ nodes or ordinary cusps as singularities, and we use the previous notation.

We get the following:

3.2 - <u>THEOREM</u> - If $\delta < a(d-a)-d+s$, then:

a) $r = 1$

b) If $s=1$ and $d \geq 12$, our g^1_{d-1} coincides with the one cut out from the lines through a simple point of C

c) If $s=2$ and d is odd, the only g^1_{d-2}'s are those cut out from the lines through a singular point of C.

d) If $s \geq 3$, there is no g^r_{d-s} on $\tilde{C}$.

Furthermore, if $s=2$ and d is even, or if $s=1$ and $8 \leq d \leq 11$, then, if $\delta<(a+1)(d-(a+1))-d+s$, the only g^r_{d-s}'s on $\tilde{C}$ are those cut out by a pencil of lines.

<u>PROOF</u> - In our hypothesis, we have $\deg(X) < a(d-a)$.
By Lemma 3.1, if $\Delta H(X,d-3) = h \geq 2$, then there is a curve γ of degree k such that $\deg(X \cap \gamma) \geq \bar{b}$.

As in Lemma 2.2, we see that here

$$\sum_{P \in X'} i(P,\gamma \cap C) \geq d-s+2(\bar{b}-d+s) = 2\bar{b}-d+s =$$

$$= (2k-1)d+(m-3)(2h+m-4)-k(2m+k-3)+s+2$$

-) If d is odd and $3 \leq m \leq a$, or if d is even and $3 \leq m < a$, then

$$(2k-1)d+(m-3)(2h+m-4)-k(2m+k-3)+s > kd = \sum_{P \in C} i(P,\gamma \cap C)$$

which contradicts Bézout Theorem.

Then $\Delta H(X,d-3) \leq 1$. Since $r>0$, it is then

$$\Delta H(X,d-3) = \Delta H(X,d-2) = r = 1.$$

Think now to the graph of $\Delta H(X,t)$: there is a straight line r containing a subscheme $X' \subseteq X$, with $\deg(X') = \sum_{t=0}^{d-2} 1 = d-1$.

Put $X'=Y' \cup Z'$, where $Y' \subseteq D$, $Z' \subseteq B$.

Assume $s=1$. If $Z' \neq \varnothing$, we must have $\deg(Z')=1$, and hence $\deg(Y')=d-1$, otherwise we get a contradiction with Bézout Theorem.

Then $Y'=D-P$, $P\in D$, and $|Y'|$ contains the linear series cut out by the lines through Z'. So $\dim|Y'| \geq 1$. But $\dim|D-P| = 0$, since $|D|$ has no base points, which is impossible. Then $X'=Y'=D$ and D is on a line through a further simple point of C.

If $s=2$, we must have $\deg(Z')=1$, hence $\deg(Y')=d-2$. So our g^1_{d-2} is cut out by the lines through a singular point of C.

If $s \geq 3$, we must have $\deg(Z') \geq 2$, and this contradicts the Bézout Theorem.

-) If d is even and $m = a = \frac{d}{2}$ (and then $k = \frac{d}{2} -1$), then

$$\Delta H(X,t) = \begin{cases} t+1 & \text{if } 0 \leq t < \frac{d}{2} -2 \\ \frac{d}{2} -1 & \text{if } \frac{d}{2} -2 \leq t \leq \frac{d}{2} \\ -t+d-1 & \text{if } \frac{d}{2} < t \leq d-1 \end{cases}$$

Then: $\Delta H(X,d-2)=1$; so $r=1$ and we get a). Furthermore, remark that $\deg(X)= \frac{d^2}{4} -1 = a(d-a)-1$.

Assume $s=1$: then $\deg(D)=d-1$, $\deg(B)= \frac{d^2}{4} - d = \delta$, and $|D|$ would be cut out by a curve γ of degree $\frac{d}{2} -1$ passing through the δ nodes and one simple point of C. But, if $d \geq 12$, there is no pencil (without fixed components) of curves of degree $\frac{d}{2} -1$ passing through the $\frac{d^2}{4} - d$ nodes and one simple point of a C_d (see Lemma 5.3 in [3]). So b) is proved.

If $s \geq 3$, then $\deg(D)=d-s$, $\deg(B)= \frac{d^2}{4} - d+s-1= \delta$, and, $\forall s>2$,

$$\deg(X\cap\gamma) = \frac{d^2}{2} -d+s-2 > d\left(\frac{d}{2} -1\right),$$ against the Theorem of Bézout, and we get c).

If we require in Lemma 3.1 that $\deg(X) < (a+1)(d-(a+1))$, we see that the case $m=a$ cannot occur.

This completes the proof. ■

3.3 - REMARK

a) The non existence of g^1_{d-3}'s and the uniqueness of the g^1_{d-2} and g^1_{d-1} on curves of high genus (with only nodes or ordinary cusps) have been studied in a complete way in [3].

Using different methods, the Authors find the same bound as ours for the number of nodes, and show its sharpness, by using a very interesting technique, but a little heavy for the calculations they need.

b) As an application of our method, we would like to describe here all the g^r_7's on $\tilde{C}$, for a C_8 with 8 nodes (as the only singularities); we shall find again a result of Theorem 3.4 in [3] (and, among other, we shall see that a bound of Theorem 3.2 is sharp).

Let $C_8 \subset \mathbb{P}^2$ be an integral curve with 8 nodes and no other singularity; let us suppose that the cubic curves through the nodes make a pencil whose ninth base point is a simple point $A \in C$ (it is easy to see that there are irreducible plane curves of degree 8 having 8 nodes in 8 of the intersection points of two cubics and containing the ninth as simple point).

We claim that the only complete g^r_7's without base points on $\tilde{C}$ are the g^1_7's cut out by the lines through a simple point of C or those cut out from the cubics through the nodes and A.

Let $D \in g^r_7$ and $X = D \cup B$. By Lemma 2.1, $\Delta H(X,7)=0$, $r=\Delta H(X,6)$. Since D is special, we have $r \leq 3$.

If r=3 or r=2, looking at the possible graphs of $\Delta H(X,t)$, we should have a contradiction with deg(X)=15, or with the Bézout Theorem.

If r=1, we have three graphs of $\Delta H(X,t)$ compatible with deg(X)=15; the first shows that there is a line containing 7 points of X (necessarily in D) - and this gives the g^1_7 cut out by the lines through a simple point of C - ; the second says that X is the intersection of a curve of degree 3 with another of degree 5 (and so we get the second case of g^1_7); the third one contradicts the Bézout Theorem.

We now study complete linear series g^r_{d-s} *(r>0) on a* C_d (d≥4) *having as singularities only one ordinary point* M *of multiplicity* e.

If ΔH(X,d-3)=h≥2, with the help of a graph analogous to the one used in the case of g^r_d, we can see that

$$\deg(X) \geq \frac{e(e-1)}{2} + h(d-e-1)+1 \geq \frac{e(e-1)}{2} + 2(d-e-1)+1$$

and, since $\deg(X)=\frac{e(e-1)}{2}+d-s$, we must have $e \geq \frac{d+s-1}{2}$.

So we get the following:

3.4 - PROPOSITION - If C_d has only one multiple ordinary point of multiplicity $e< \frac{d+s-1}{2}$, then

a) r = 1.

b) If s≠1 or s≠e, there is no g^1_{d-s}.

The only g^1_{d-s}'s are the g^1_{d-1} and the g^1_{d-e} cut out by the lines (through a simple point or through the multiple point of C).

PROOF - a) If $e < \frac{d+s-1}{2}$, then ΔH(X,d-3)≤1 and so

$$\Delta H(X,d-3)=\Delta H(X,d-2)=r=1.$$

So there exists a line r containing a subscheme X'⊂X with deg(X')=d-1.

b) If M∉r, on r we have d-1 points of D, whose degree is d-s. Then s=1 and $|D|=g^1_{d-1}$ cut from lines through a point A∈C, A≠M.

If M∈r, then X'=Y'∪Z', where Y'⊆D, Z'⊆B, deg(Z')=e-1 and deg(Y') = d-e ≤ d-s. So s ≤ e.

If s = e, the g^1_{d-s} is cut from the lines through M.

If s<e, an argument similar to the one used in the proof of 3.2 shows that a g^1_{d-s} is possible only if |D| has base points. ■

3.5 - REMARK - If s=e, Prop. 3.4 says that, if C is a non rational integral curve with only one multiple ordinary point M with multiplicity e, then the only g^1_{d-e} on $\tilde{C}$ is the one cut out from the lines through M.

REFERENCES

[1] R.D.M. Accola, *On Castelnuovo's inequality for algebraic curves I*, Trans. A.M.S. **251** (1979), 357-373

[2] C. Ciliberto, *Alcune applicazioni di un classico procedimento di Castelnuovo*, Seminari di Geometria dell'Università di Bologna (1982-1983), 17-43

[3] M. Coppens and T. Kato, *The gonality of smooth curves with plane models*, Manuscripta Math. **70** (1990), 5-25

[4] E. D. Davis, *0-dimensional subschemes of $\mathbb{P}^2$: new applications of Castelnuovo's function*, Ann. Univ. Ferrara **32** (1986), 93-107

[5] E. D. Davis and A. Geramita, *The Hilbert function of a special class of 1-dimensional Cohen-Macaulay gradeg algebras*, Queen's Pap. Pure Appl. Math. **67** (1984)

[6] S. Greco, *Remarks on the postulation of zero-dimensional subschemes of projective space*, Math. Ann. **284** (1989), 343-351

[7] S. Greco and G. Raciti, *The Lüroth semigroup of plane algebraic curves*, Pacific J. of Math. **151** (1991), 43-56

[8] S. Greco and G. Raciti, *Gap orders of rational functions on plane curves with few singular points*, Manuscripta Math. **70** (1990), 441-447

[9] S. Greco and P. Valabrega, *On the Theory of adjoints*, Springer Lecture Notes in Mathematics **732** (1979)

[10] C. Küpper, *Ueber die auf eine Curve m^{ter} Ordnung von Geschlecht p von dem ∞^2 Geraden G der Ebene ausgeschnittene lineare Schaar $g^{(2)}_{m}$*, Math. Ann. **31** (1887), 291-301

[11] C. Küpper, *Ueber das Vorkommen von linearen Schaaren $g^{(2)}_{n}$ auf Curven n^{ter} Ordnung deren Geschlecht p grösser als p_1, das Maximalgeschlecht einer Raumcurve $R^{n}_{p_1}$, ist*, Sitz. Kgl. Boem. Gesell. Wissen. (1892), 264-272

[12] V. Snyder, *On Birational Transformations of Curves of High Genus*, Am. J. of Math. **30** (1908), 10-18

[13] C. Wiman, *Ueber die algebraischen Kurven von den Geschlechten p=4,5,6, welche eindeutige Transformationen in sich besitzen*, Stockholm Ak. Handl., **21** (1895), 1-43

Rational Curves on Calabi-Yau Manifolds: Verifying Predictions of Mirror Symmetry

SHELDON KATZ Department of Mathematics, Oklahoma State University, Stillwater, OK 74078. Email: katz@math.okstate.edu

Recently, mirror symmetry, a phenomenon in superstring theory, has been used to give tentative calculations of several numbers in algebraic geometry[1]. This yields predictions for the number of rational curves of any degree d on general Calabi-Yau hypersurfaces in $\mathbf{P}^4$ [2], $\mathbf{P}(2,1^4)$, $\mathbf{P}(4,1^4)$, and $\mathbf{P}(5,2,1^3)$ [4, 9, 12]. The techniques used in the calculation rely on manipulations of path integrals which have not yet been put on a rigorous mathematical footing. On the other hand, there is currently no prospect of calculating most of these numbers by algebraic geometry.

Until this point, three of these numbers have been verified, all for the quintic hypersurface in $\mathbf{P}^4$: the number of lines (2875) was known classically, the number of conics (609250) was calculated in [7], and the number of twisted cubics (317206375) was found recently by Ellingsrud and Strømme [3].

Even more recently [6], higher dimensional mirror symmetry has been used to predict the number of rational curves on Calabi-Yau hypersurfaces in higher dimensional projective spaces which meet 3 linear subspaces of certain dimensions. Again, there is no known way to calculate these using algebraic geometry.

The purpose of this paper is to verify some of these numbers in low degree, giving more evidence for the validity of mirror symmetry. In §1, the number of weighted lines in a weighted sextic in $\mathbf{P}(2,1^4)$ is calculated, as well as the number of weighted lines in a weighted octic in $\mathbf{P}(4,1^4)$. In §2, the number of lines on Calabi-Yau hypersurfaces of dimension up to 10 which satisfy certain incidence properties is calculated. In §3, the number of conics on these same Calabi-Yau hypersurfaces satisfying the same incidence properties is calculated. These numbers are closely related to the Gromov-Witten invariants defined in [13, 11, 6]; and it is *these* numbers that are recorded here. In all instances, the calculations agree with

[1]See the papers in [14] for general background on mirror symmetry.

those predicted by mirror symmetry. Thus the number of verified predictions has increased from 3 to 65.

There are two parts to all of these calculations. The first part is to express the desired numbers in terms of the standard constructs of intersection theory. The second part is to evaluate the number using the Maple package SCHUBERT [8] (although the number of weighted lines in a weighted octic in $\mathbf{P}(4,1^4)$ was first found via classical enumerative geometry, using a classical enumerative formula). The short SCHUBERT code is not included here, but is available upon request.

While it is checked that the data being enumerated is finite, no attempt has been made here to check that the multiplicities are 1. All enumeration takes multiplicities into consideration. This suffices for comparison to the numbers arising in physics, since the Feynman path integrals would take account of any multiplicities greater than 1 as well.

Some of the Gromov-Witten invariants were computed in [6] using an intriguing relation between the various invariants. These relations arise in conformal field theory. A mathematical proof of the relations for the invariants corresponding to lines is sketched here.

It is appropriate to point out the recent work of Libgober and Teitelbaum [10], who have apparently correctly guessed the mirror manifold of complete intersection Calabi-Yau threefolds in an ordinary projective space. Their conjectured mirrors yield predictions for the numbers of rational curves. The predicted number of lines coincides with the results of a calculation done by Libgober 20 years ago, and the predicted number of conics coincides with the results of an unpublished calculation done by Strømme and van Straten in 1990.[2]

I'd like to thank D.R. Morrison for helpful suggestions and conversations, and for encouraging me to write this paper. I'd also like to thank S. Kleiman for his suggestions which have improved the manuscript.

1 WEIGHTED PROJECTIVE SPACES AND THEIR GRASSMANNIANS

Let $\mathbf{P}(k,1^n)$ denote an n-dimensional weighted projective space with first coordinate having weight $k > 1$, all other coordinates having weight 1. Thus $\mathbf{P}(k,1^n)$ consists of all nonzero $(n+1)$-tuples $(x_0,\ldots,x_n)$, with $(x_0,\ldots,x_n)$ identified with $(\lambda^k x_0, \lambda x_1, \ldots, \lambda x_n)$ for any $\lambda \neq 0$. Note that $\mathbf{P}(k,1^n)$ is smooth outside the singular point $p = (1,0,\ldots,0)$. There is a natural rational projection map $\pi : \mathbf{P}(k,1^n) - -- > \mathbf{P}^{n-1}$ defined outside p given by omitting the first coordinate. Let X be a weighted hypersurface of weight d. Assume $p \notin X$ (this implies that $k|d$, and that the monomial $x_0^{d/k}$ occurs in an equation for X). It is further assumed that X is smooth. The general weighted hypersurface whose weight is a multiple of k is an example of such an X.

DEFINITION A *weighted r-plane in* $\mathbf{P}(k,1^n)$ is the image of a section of π over an r-plane in $\mathbf{P}^{n-1}$.

---

[2] *Added in proof*: there have even more recently been more predictions, made in work by P. Candelas, X. de la Ossa, A. Font, S. Katz, and D.R. Morrison; by S. Hosono, A. Klemm, S. Theisen, and S.-T. Yau; by M. Bershadsky, S. Cecotti, H. Ooguri, and C. Vafa; and by V. Batyrev and D. van Straten. Many of these predictions have been verified as well.

Note that weighted r-planes do not contain p.

Let P be a weighted r-plane, with L its image in $\mathbf{P}^{n-1}$. Let $(q_0, \ldots, q_r)$ be any homogeneous coordinates on $L \simeq \mathbf{P}^r$. Then P may be thought of as the image of L via the mapping $x_0 = f_k(q_0, \ldots, q_r)$, $x_i = l_i(q_0, \ldots, q_r)$, where f_k is a form of degree k, and the l_i are all linear. Once L is fixed, we may fix in mind a choice of the q_i and l_i.

The moduli space of weighted r-planes can be represented (and compactified) as follows. Conventions have been chosen to be consistent with those in SCHUBERT [8]. Let $G = G(r+1, n)$ be the Grassmannian of r-dimensional linear subspaces L of $\mathbf{P}^{n-1}$. Let V be the n-dimensional vector space of linear forms on $\mathbf{P}^{n-1}$. This identifies $\mathbf{P}^{n-1}$ with $\mathbf{P}(V) = Proj(S^*V)$. G is then the space of $r+1$ dimensional quotients of V (since the space of linear forms restricted to L is an $r+1$-dimensional quotient of V). Let Q be the universal rank $r+1$ quotient bundle on G.

The equation $x_0 - f_k(q_0, \ldots, q_r) = 0$ which describes a section of π over an r-plane may be identified with a section s of the bundle $\mathbf{C} \oplus S^k(Q)$, where $\mathbf{C}$ denotes the trivial bundle. A scalar multiple of this section would correspond to the equation $ax_0 - af_k(q_0, \ldots, q_r) = 0$, which defines the same weighted r-plane. Note that (up to scalar) s does not depend on any of the choices which have been made. So $M = \mathbf{P}(\mathbf{C} \oplus S^k(Q)^*)$ gives a compactification of the space of weighted r-planes. Here, $\mathbf{P}(E)$ denotes the space of rank 1 quotients of the fibers of the bundle E; hence the need for dualizing in defining M.

In the sequel, we will also refer to $M^o \subset M$, the open subset which corresponds to the actual weighted $\mathbf{P}^1$'s, in other words, $M^o = M - \mathbf{P}(S^kQ^*)$, where $\mathbf{P}(S^kQ^*)$ is included in M via the map induced by the natural projection $\mathbf{C} \oplus S^kQ^* \to S^kQ^*$.

2 LINES ON WEIGHTED HYPERSURFACE CALABI-YAU THREEFOLDS

Now let $k > 1$, and let $X \subset \mathbf{P}(k, 1^4)$ be a smooth weighted hypersurface of weight $k+4$ with $(1,0,0,0,0) \notin X$. As has been noted in the previous section, this implies that $k | k+4$, which in turn implies that $k = 2$ or $k = 4$. The weight $k+4$ has been chosen to ensure that X is Calabi-Yau, i.e. that X has trivial canonical bundle.

The rational projection map π restricts to a morphism $\pi : X \to \mathbf{P}^3$. This is a 3-1 cover for $k = 2$, and a 2-1 cover for $k = 4$. The goal of this section is to enumerate the weighted $\mathbf{P}^1$'s contained in X.

Let us first consider the case $k = 4$. Then an equation for X has the form

$$F = ax_0^2 + g_4(x_1, \ldots, x_4)x_0 + g_8(x_1, \ldots, x_4) = 0, \tag{1}$$

where $a \in \mathbf{C}$ and g_i has degree i for $i = 4$ or 8. Such an equation naturally induces a section s of the bundle $\mathbf{C} \oplus S^4Q \oplus S^8Q$. Consider a point $C \in M^o$. We abuse notation by allowing C to also denote the corresponding curve. Let (q_0, q_1) be homogeneous coordinates on $\mathbf{P}^1$. Identifying C with $\mathbf{P}^1$, we may describe C by equations of the form

$$x_0 = f_4(q_0, q_1), \; x_i = l_i(q_0, q_1). \tag{2}$$

The equation $x_0 - f_4(q_0, q_1) = 0$ and its multiples for varying C form the tautological subbundle $\mathcal{O}_{\mathbf{P}}(-1) \subset \mathbf{C} \oplus S^4Q$ on $\mathbf{P} = \mathbf{P}(\mathbf{C} \oplus S^4Q^*)$. C is contained in X if and only if

Ambient space	Weighted degree	Number of lines
$\mathbf{P}(1^5)$	5	2875
$\mathbf{P}(2,1^4)$	6	7884
$\mathbf{P}(4,1^4)$	8	29504

Table 1: The number of lines.

an equation for X, when pulled back to $\mathbf{P}^1$ via a parametrization of C, vanishes. Substituting from the second of equations (2) into (1), it is seen that this happens if and only if $ax_0^2 + g_4(l_1(q_0,q_1),\dots,l_4(q_0,q_1))x_0 + g_8(l_1(q_0,q_1),\dots,l_4(q_0,q_1))$ is a multiple of $x_0 - f_4(q_0,q_1)$. Multiplication induces an inclusion of bundles

$$(\mathbf{C} \oplus S^4Q) \otimes \mathcal{O}_{\mathbf{P}}(-1) \hookrightarrow \mathbf{C} \oplus S^4Q \oplus S^8Q.$$

Putting all this together, we see that $C \subset X$ if and only the section $\bar{s}$ of

$$B = (\mathbf{C} \oplus S^4Q \oplus S^8Q)/((\mathbf{C} \oplus S^4Q) \otimes \mathcal{O}_{\mathbf{P}}(-1))$$

induced by s vanishes at C. Note that if $C \in M - M^o$, then C corresponds to a curve defined by equations of the form $f_4(q_0,q_1) = 0$, $x_i = l_i(q_0,q_1)$. Since such a curve would contain p, it follows that C is not in the zero locus of $\bar{s}$. Also note that $\dim(M) = rank(B) = 9$; so one expects finitely many zeros of such a section; hence finitely many weighted $\mathbf{P}^1$'s. It is easy to prove that this is indeed the case for general X. The actual number is the degree of $c_9(B)$. This may be calculated by standard techniques in intersection theory [5] and the calculation may be implemented via SCHUBERT [8].

The case $k = 2$ is similar. Changing the meaning of the notation in the obvious manner, one must consider $M = \mathbf{P}(\mathbf{C} \oplus S^2Q^*)$, and calculate the degree of $c_7(B')$, where

$$B' = (\mathbf{C} \oplus S^2Q \oplus S^4Q \oplus S^6Q)/((\mathbf{C} \oplus S^2Q \oplus S^4Q) \otimes \mathcal{O}_{\mathbf{P}}(-1)).$$

Combining these with the well-known number of lines on a quintic threefold, the calculation of some examples considered in [4, 9, 12] via mirror symmetry may be verified. The results are displayed in table 1.

PROBLEM: Verify the predictions of mirror symmetry for weighted $\mathbf{P}^1$'s in a weight 10 hypersurface in $\mathbf{P}(5,2,1^3)$. Also, verify the predictions of mirror symmetry for weighted conics on the weighted hypersurfaces considered in this section.[3]

Remark: The family of weighted conics on the general weighted octic in $\mathbf{P}(4,1^4)$ is positive dimensional (independently observed by Kollár); hence part of the problem in this case is to systematically assign numbers to positive dimensional families. This can be defined as the number of such curves that remain almost holomorphic under a general almost complex deformation; but it is desirable to give a purely algebraic description.

[3] *Added in proof*: the prediction for the number of weighted conics on the general weighted sextic in $\mathbf{P}(2,1^4)$ has recently been verified by P. Meurer.

3 LINES ON HIGHER DIMENSIONAL VARIETIES

In this section and the next, we consider rational curves on the generic Calabi-Yau hypersurface X in P^{k+1}. This is a hypersurface of dimension k and degree $k+2$. For $k > 3$, there will be infinitely many lines and conics contained in X. But there will only be finitely many lines or conics which satisfy certain incidence properties with fixed linear subspaces.

Since the normal bundle N of C in P^{k+1} has degree -2, one expects that for general X and any $C \subset X$, $N \simeq \mathcal{O} \oplus \ldots \oplus \mathcal{O} \oplus \mathcal{O}(-1) \oplus \mathcal{O}(-1)$ (with $k-3$ $\mathcal{O}$'s). Since $h^0(N) = k-3$ and $h^1(N) = 0$ in this case, the scheme of rational curves on X is expected to have dimension $k-3$.

For each i, let $L_i \subset \mathbf{P}^{k+1}$ denote a general linear subspace of codimension i. Pick positive integers a, b, c such that $a+b+c=k$. Following [13, 11, 6], define an invariant $n_b^a(d)$ of X as the number of holomorphic immersions $f : \mathbf{P}^1 \to X$ with $f(\mathbf{P}^1)$ of degree d such that $f(0) \in L_a$, $f(1) \in L_b, f(\infty) \in L_c$. These numbers, called "Gromov-Witten invariants" in [11], are expected to be finite. Note that the value of c is implicit in the notation $n_b^a(d)$ by virtue of the equation $a+b+c=k$.

These invariants are essentially the same as the number of reduced, irreducible rational curves of degree d in X which meet each of L_a, L_b, and L_c. The $n_b^a(d)$ differ from the corresponding numbers of curves by one factor of d for each of the indices a, b, or c equal to 1 (since C meets a general L_1 d times). There is no difference for lines; and for conics, we will see that in the calculation of the number of conics satisfying the required incidence properties, the Gromov-Witten invariants arise naturally. So the Gromov-Witten invariants will be calculated and tabulated, while the numbers of rational curves follow immediately by division by the appropriate power of d, if necessary.

In the remainder of this section, we specialize to $d=1$, i.e. lines. A theorem of Barth-van de Ven [1] states that the Fano variety of lines on a degree l hypersurface $X \subset \mathbf{P}^n$ is smooth of dimension $2n-l-3$ for generic X when $l+3 \leq 2n$. Applied in the present context of $X_{k+2} \subset \mathbf{P}^{k+1}$, we find that the variety of lines must be smooth of dimension $k-3$ whenever $k \geq 3$. From this, standard techniques show that a general X contains finitely many lines which meet each of L_a, L_b, and L_c.

So we can calculate the Gromov-Witten invariants by using the Schubert calculus. The lines are parametrized by the Grassmannian $G(2, k+2)$. The class of lines meeting L_a is the Schubert cycle σ_{a-1}; similarly for L_b and L_c. Let Q be the rank 2 universal quotient bundle on G. Since the class of the variety of lines on X is represented by $c_{k+3}(S^{k+2}Q)$ and dimensions work out correctly, the answer is the degree of $c_{k+3}S^{k+2}Q \cdot \sigma_{a-1} \cdot \sigma_{b-1} \cdot \sigma_{c-1}$. These may be easily worked out as integers using SCHUBERT. The answers obtained are displayed in table 2.

The original predictions for the numbers found in [6] resulted from a two-step process arising from mirror symmetry and conformal field theory. First, the $n_b^1(d)$ are found, followed by what amounts to an expression for any $n_b^a(d)$ in terms of the various $n_{b'}^1(d')$ for $d' \leq d$. Most of these expressions remain a mathematical mystery at present.[4] However, the case $d=1$ can be established mathematically as follows.

---

[4] *Added in proof*: a proof has been announced recently by Y. Ruan and G. Tian.

k	$n_b^a(1)$
3	$n_1^1(1) = 2875$
4	$n_1^1(1) = 60480$
5	$n_1^1(1) = 1009792,\ n_2^1(1) = 1707797$
6	$n_1^1(1) = 15984640,\ n_2^1(1) = 37502976,\ n_2^2(1) = 59021312$
7	$n_1^1(1) = 253490796,\ n_2^1(1) = 763954092,\ n_3^1(1) = 1069047153$
	$n_2^2(1) = 1579510449$
8	$n_1^1(1) = 4120776000,\ n_2^1(1) = 15274952000,\ n_3^1(1) = 27768048000$
	$n_2^2(1) = 38922224000,\ n_3^2(1) = 51415320000$
9	$n_1^1(1) = 69407571816,\ n_2^1(1) = 307393401172,\ n_3^1(1) = 695221679878$
	$n_4^1(1) = 905702054829,\ n_2^2(1) = 933207509234,\ n_3^2(1) = 1531516162891$
	$n_3^3(1) = 1919344441597$
10	$n_1^1(1) = 1217507106816,\ n_2^1(1) = 6306655500288$
	$n_3^1(1) = 17225362851840,\ n_4^1(1) = 28015971489792$
	$n_2^2(1) = 22314511245312,\ n_3^2(1) = 44023827234816$
	$n_4^2(1) = 54814435872768,\ n_3^3(1) = 65733143224320$

Table 2: Gromov-Witten invariants for lines.

THEOREM *Let X be any Calabi-Yau manifold of dimension k in any projective space. Define $n_b^a(1)$ as above. Assume that there are finitely many lines in X satisfying each of the respective incidence conditions needed to define the $n_b^a(1)$. Then*

$$n_j^i(1) = \sum_{l=0}^{j-1} n_{i+l}^1(1) - \sum_{l=1}^{j-1} n_l^1(1).$$

Proof (sketch). Follows immediately by intersecting the cycle class (in the appropriate Grassmannian) of the scheme of lines in X with the identity

$$\sigma_{i-1}\sigma_{j-1}\sigma_{k-i-j-1} = \sum_{l=0}^{j-1} \sigma_{i+l-1}\sigma_{k-i-l-2} - \sum_{l=1}^{j-1} \sigma_{l-1}\sigma_{k-l-2},$$

an identity which can be proven by a few applications of Pieri's formula.

k	$n_b^a(2)$
3	$n_1^1(2) = 4874000$
4	$n_1^1(2) = 1763536320$
5	$n_1^1(2) = 488959144352$, $n_2^1(2) = 1021575491286$
6	$n_1^1(2) = 133588638826496$, $n_2^1(2) = 448681408315392$ $n_2^2(2) = 821654025830400$
7	$n_1^1(2) = 39031273362637440$, $n_2^1(2) = 187554590257349088$
	$n_3^1(2) = 312074852318965368$, $n_2^2(2) = 506855012110118424$
8	$n_1^1(2) = 12607965435718224000$, $n_2^1(2) = 80684596772238448000$
	$n_3^1(2) = 200581960800610752000$, $n_2^2(2) = 295035175517918176000$
	$n_3^2(2) = 444475303469701680000$
9	$n_1^1(2) = 4565325719860021608624$, $n_2^1(2) = 37005001823802188657624$
	$n_3^1(2) = 127922335050535174614916$, $n_4^1(2) = 193693669320390878077186$
	$n_2^2(2) = 173901546566279203106468$, $n_3^2(2) = 364629304647788940660824$
	$n_3^3(2) = 498705676383823268404990$
10	$n_1^1(2) = 1861791822397620935737344$, $n_2^1(2) = 18415607624138339954786304$
	$n_3^1(2) = 83885220561474498867757056$, $n_4^1(2) = 179982840924749584358866944$
	$n_2^2(2) = 107227899142191919158312960$, $n_3^2(2) = 297755098999730079369412608$
	$n_4^2(2) = 417950364467570984815214592$, $n_3^3(2) = 527556832251612742800359424$

Table 3: Gromov-Witten invariants for conics.

4 CONICS ON HIGHER DIMENSIONAL VARIETIES

It can easily be shown that if X is a general hypersurface of degree $k+2$ in $\mathbf{P}^{k+1}$, then the variety of conics on X has the expected dimension $k-3$. Standard techniques show that given positive integers a, b, c with $a+b+c=k$, there will be a finite number of conics in X which meet each of L_a, L_b, and L_c. Thus the Gromov-Witten invariants are finite. They will be calculated here; the answers obtained are displayed in table 3.

We start with the well known description of the moduli space of conics in $\mathbf{P}^{k+1} = \mathbf{P}(V)$, where V is a $k+2$-dimensional vector space. To describe a conic, we first describe the 2 plane it spans, and then choose a quadric in that 2-plane (up to scalar). So let $G = G(3, V)$ be the Grassmannian of 2-planes in $\mathbf{P}(V)$ (that is, of rank 3 quotients of V), and let Q be the universal rank 3 quotient bundle of linear forms on the varying subspace. Then the moduli space of conics is $M = \mathbf{P}(S^2Q^*)$. Following the reasoning in section 2 (or [7]), the scheme of conics on X is given by the locus over which a certain section of $F = S^{k+2}Q/(S^kQ \otimes \mathcal{O}_{\mathbf{P}}(-1))$ vanishes. Here $\mathcal{O}_{\mathbf{P}}(1)$ is the tautological sheaf on $\mathbf{P}(S^2Q^*)$. Since F has rank $2k+5$, the conics on X are represented by $c_{2k+5}(F)$.

It remains to find the condition that a conic C meets L_a. One way to find this is to consider the moduli space $\mathcal{M}$ of pointed conics, i.e. pairs (p, C), with C a conic, $p \in C$. This may easily be constructed as a bundle over $\mathbf{P}^{k+1}$, with fiber over $p \in \mathbf{P}^{k+1}$ being the set of conics containing p. We start by constructing the moduli space of pointed 2-planes as follows. Consider the tautological exact sequence on $\mathbf{P}^{k+1}$:

$$0 \to K \to V_{\mathbf{P}^{k+1}} \to \mathcal{O}(1) \to 0,$$

where $V_{\mathbf{P}^{k+1}}$ is a trivial bundle of rank $k+2$ on $\mathbf{P}^{k+1}$ (more generally, E_Y will stand for the pullback of E to Y, the morphism used for the pullback assumed to be clear in context). Let $H = G(2, K)$ be the Grassmannian of rank 2 quotients of K, $\mathcal{Q}$ its universal rank 2 quotient, and $\mathcal{S} \subset K_H$ the universal subbundle. These fit into the exact sequence

$$0 \to \mathcal{S} \to K_H \to \mathcal{Q} \to 0$$

of sheaves on H. The natural quotient $V_H \to V_H/\mathcal{S}$ induces a map $H \to G$ by the universal property of the Grassmannian; since $V_H \to \mathcal{O}(1)_H$ clearly factors through $V_H/\mathcal{S}$, it is easy to see that H may be identified with the space of pointed 2-planes in $\mathbf{P}(V)$. Here $\mathcal{O}(1)$ denotes the tautological sheaf on $\mathbf{P}(V)$ as before.

The conics containing p globalize to a rank 5 bundle W on H. This bundle is in fact the the kernel of the natural map $S^2(V_H/\mathcal{S}) \to \mathcal{O}(2)_H$. Then the moduli space M' of pointed conics may be seen to be $\mathbf{P}(W^*)$. Let $\mathcal{O}_W(1)$ be its tautological bundle.

Let $h = c_1(\mathcal{O}(1)_{M'})$. Consider the natural morphism $f : M' \to M$. The variety of conics meeting L^a is represented by the class $f_*(h^a)$ for $a > 1$. Note that for $a = 1$, $f_*(h) = 2$. This factor exactly gives the factor needed to give the Gromov-Witten invariants rather than the number of conics meeting three linear subspaces. So the Gromov-Witten invariants are given by the formula $n_b^a(2) = \int_M c_{2k+5}(Q) f_*(h^a) f_*(h^b) f_*(h^c)$, which is valid since the dimensions work out correctly.

To compute these as numbers using SCHUBERT, everything is clear, except the description of the morphism f. But this may be described merely by knowing the pullbacks $f^*(Q)$ and $f^*(\mathcal{O}_{\mathbf{P}}(1))$. However, from the above description and the universal properties, this is just $V_H/\mathcal{S}$ and $\mathcal{O}_W(1)$. SCHUBERT takes care of the rest.

References

[1] W. Barth and A. van de Ven. Fano-varieties of lines on hypersurfaces. *Arch. Math.*, 31(1):96–104, 1978.

[2] P. Candelas, X.C. de la Ossa, P.S. Green, and L. Parkes. A pair of Calabi-Yau manifolds as an exactly soluble superconformal theory. *Nucl. Phys. B*, 359:21–74, 1991.

[3] G. Ellingsrud and S. A. Strømme. The number of twisted cubic curves on the general quintic threefold. Univ. of Bergen Preprint 63-7-2-1992.

[4] A. Font. Periods and duality symmetries in Calabi-Yau compactifications. *Nuc. Phys. B*, 391:358–388, 1993.

[5] W. Fulton. *Intersection Theory.* Springer-Verlag, Berlin Heidelberg New York Tokyo, 1984.

[6] B. R. Greene, D. R. Morrison, and M. R. Plesser. Mirror manifolds in higher dimension. Preprint, CLNS-93/1253.

[7] S. Katz. On the finiteness of rational curves on quintic threefolds. *Comp. Math.*, 60:151–162, 1986.

[8] S. Katz and S.A. Strømme. SCHUBERT: a MAPLE package for intersection theory. Available by anonymous ftp from ftp.math.okstate.edu or linus.mi.uib.no, cd pub/schubert.

[9] A. Klemm and S. Theisen. Considerations of one-modulus Calabi-Yau compactifications: Picard-Fuchs equations, Kähler potentials and mirror maps. *Nuc. Phys.*, B389:153–180, 1993.

[10] A. Libgober and J. Teitelbaum. Lines on Calabi Yau complete intersections, mirror symmetry, and Picard Fuchs equations. *Int. Math. Res. Not.*, 1:29–39, 1993.

[11] D. R. Morrison. Hodge-theoretic aspects of mirror symmetry. In preparation.

[12] D.R. Morrison. Picard-Fuchs equations and mirror maps for hypersurfaces. In S.-T. Yau, editor, *Essays on Mirror Manifolds*, pages 241–264. International Press, Hong Kong, 1992.

[13] E. Witten. Topological sigma models. *Commun. Math. Phys.*, 118:411–449, 1988.

[14] S.-T. Yau, editor. *Essays on Mirror Manifolds*, Hong Kong, 1992. International Press.